Great Yarmouth Libraries

REFERENCE CUPBOARD

THIS BOOK MUST NOT BE TAKEN AWAY

Scientific
Technical
Arts

and

ITEM 006 947

Reference Department

THIS BOOK MUST NOT BE TAKEN AWAY

GENERAL VIEW OF THE
AGRICULTURE OF THE
COUNTY OF NORFOLK

GENERAL VIEW of the AGRICULTURE of the COUNTY OF NORFOLK

A Reprint of the Work Drawn up for the Consideration of the Board of Agriculture and Internal Improvement

by

ARTHUR YOUNG

Secretary of the Board

DAVID & CHARLES REPRINTS

7153 4374 2

This book was first published in 1804

This edition published by David & Charles (Publishers) Limited 1969

Printed in Great Britain by
Clarke Doble & Brendon Limited Plymouth
for David & Charles (Publishers) Limited
South Devon House Railway Station
Newton Abbot Devon

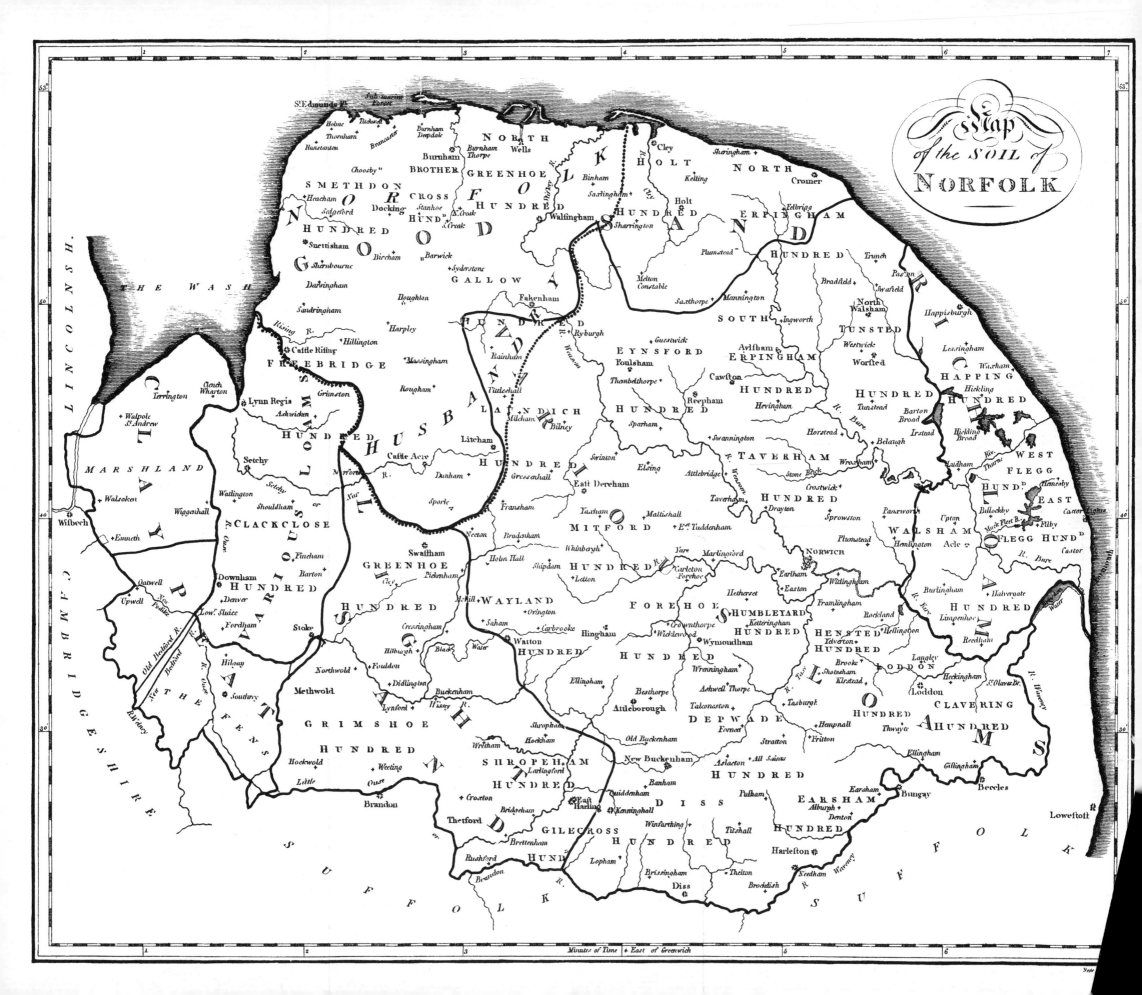

GENERAL VIEW

OF THE

AGRICULTURE

OF THE

COUNTY OF NORFOLK.

DRAWN UP FOR THE CONSIDERATION OF

THE BOARD OF AGRICULTURE

AND INTERNAL IMPROVEMENT.

BY

THE SECRETARY OF THE BOARD.

LONDON:

PRINTED BY B. M^cMILLAN, BOW-STREET, COVENT-GARDEN,
PRINTER TO HIS ROYAL HIGHNESS THE PRINCE OF WALES;
FOR G. AND W. NICOL, PALL-MALL, BOOKSELLERS TO HIS
MAJESTY, AND THE BOARD OF AGRICULTURE; AND SOLD
BY G. AND J. ROBINSON, PATERNOSTER-ROW;
J. ASPERNE, CORNHILL; CADELL AND DAVIES,
STRAND; W. CREECH, EDINBURGH; AND
JOHN ARCHER, DUBLIN.

1804.

ADVERTISEMENT.

THE great desire that has been very generally expressed, for having the AGRICULTURAL SURVEYS of the KINGDOM reprinted, with the additional Communications which have been received since the ORIGINAL REPORTS were circulated, has induced the BOARD OF AGRICULTURE to come to a resolution of reprinting such as may appear on the whole fit for publication. It is proper at the same time to add, that the Board does not consider itself responsible for any fact or observation contained in the Reports thus reprinted, as it is impossible to consider them yet in a perfect state; and that it will thankfully acknowledge any additional information which may still be communicated: an invitation, of which, it is hoped, many will avail themselves, as there is no circumstance from which any one can derive more real satisfaction, than that of contributing, by every possible means, to promote the improvement of his Country.

N. B. *Letters to the Board, may be addressed to Lord* SHEFFIELD, *the President, No. 32, Sackville-Street, Piccadilly, London.*

PLAN

FOR RE-PRINTING THE

AGRICULTURAL SURVEYS.

BY THE PRESIDENT OF THE BOARD OF AGRICULTURE.

A BOARD established for the purpose of making every essential inquiry into the Agricultural State, and the means of promoting the internal improvement of a powerful Empire, will necessarily have it in view to examine the sources of public prosperity, in regard to various important particulars. Perhaps the following is the most natural order for carrying on such important investigations; namely, to ascertain,

1. The riches to be obtained from the surface of the national territory.
2. The mineral or subterraneous treasures of which the country is possessed.
3. The wealth to be derived from its streams, rivers, canals, inland navigations, coasts, and fisheries ;— and
4. The means of promoting the improvement of the people, in regard to their health, industry, and morals, founded on a *statistical* survey, or a minute and careful inquiry into the actual state of every parochial district in the kingdom, and the circumstances of its inhabitants.

Under

Under one or other of these heads, every point of real importance that can tend to promote the general happiness of a great nation, seems to be included.

Investigations of so extensive and so complicated a nature, must require, it is evident, a considerable space of time before they can be completed. Differing indeed in many respects from each other, it is better perhaps that they should be undertaken at different periods, and separately considered. Under that impression, the Board of Agriculture has hitherto directed its attention to the first point only, namely, the cultivation of the surface, and the resources to be derived from it.

That the facts essential for such an investigation might be collected with more celerity and advantage, a number of intelligent and respectable individuals were appointed, to furnish the Board with accounts of the state of husbandry, and the means of improving the different districts of the kingdom. The returns they sent were printed, and circulated by every means the Board of Agriculture could devise, in the districts to which they respectively related; and in consequence of that circulation, a great mass of additional valuable information has been obtained. For the purpose of communicating that information to the Public in general, but more especially to those Counties the most interested therein, the Board has resolved to re-print the Survey of each County, as soon as it seemed to be fit for publication; and, among several equally advanced, the Counties of Norfolk and Lancaster were pitched upon for the commencement of the proposed publication; it being thought most advisable to begin with one County on the Eastern, and another on the Western Coast of the island. When all these Surveys shall have been thus reprinted, it will be attended with little difficulty to draw up

an

an abstract of the whole (which will not probably exceed two or three volumes quarto) to be laid before HIS MAJESTY, and both Houses of Parliament; and afterwards, a General Report on the present state of the country, and the means of its improvement, may be systematically arranged, according to the various subjects connected with Agriculture. Thus every individual in the kingdom may have,

1. An account of the husbandry of his own particular county; or,
2. A general view of the agricultural state of the kingdom at large, according to the counties, or districts, into which it is divided; or,
3. An arranged system of information on agricultural subjects, whether accumulated by the Board since its establishment, or previously known:

And thus information respecting the state of the kingdom, and agricultural knowledge in general, will be attainable with every possible advantage.

In re-printing these Reports, it was judged necessary, that they should be drawn up according to one uniform model; and after fully considering the subject, the following form was pitched upon, as one that would include in it all the particulars which it was necessary to notice in an Agricultural Survey. As the other Reports will be reprinted in the same manner, the reader will thus be enabled to find out at once where any point is treated of, to which he may wish to direct his attention.

PLAN OF THE RE-PRINTED REPORTS.

Preliminary Observations.

Chap. I. *Geographical State and Circumstances.*
  Sect. 1. Situation and Extent.
       2. Divisions.
       3. Climate.
       4. Soil and Surface.
       5. Minerals.
       6. Water.

Chap. II. *State of Property.*
  Sect. 1. Estates, and their Management.
       2. Tenures.

Chap. III. *Buildings.*
  Sect. 1. Houses of Proprietors.
       2. Farm Houses and Offices, and Repairs.
       3. Cottages.

Chap. IV. *Mode of Occupation.*
  Sect. 1. Size of Farms. Character of the Farmers.
       2. Rent—in Money—in Kind—in Personal Services.
       3. Tithes.
       4. Poor Rates.
       5. Leases.
       6. Expense and Profit.

Chap. V. *Implements.*

Chap. VI. *Enclosing—Fences—Gates.*

Chap. VII. *Arable Land.*

Sect. 1. Tillage.
2. Fallowing.
3. Rotation of Crops.
4. Crops commonly cultivated, such as Corn, Pulse, Artificial Grasses; their Seed, Culture, Produce, &c*.
5. Crops not commonly cultivated.

Chap. VIII. *Grass.*

Sect. 1. Natural Meadows and Pastures.
2. Hay Harvest.
3. Feeding.

Chap. IX. *Gardens and Orchards.*

Chap. X. *Woods and Plantations.*

Chap. XI. *Wastes.*

---

* Where the quantity is considerable, the information respecting the crops commonly cultivated may be arranged under the following heads—for example, Wheat:

1. Preparation { tillage, manure. }
2. Sort.
3. Steeping.
4. Seed (quantity sown).
5. Time of sowing.

6. Culture whilst growing { hoe, weeding, feeding. }
7. Harvest.
8. Threshing.
9. Produce.
10. Manufacture of bread.

In general, the same heads will suit the following grains: Barley. Oats. Beans. Rye. Pease. Buck-wheat.
Vetches — Application.
Cole-seed — { Feeding, Seed. }
Turnips — { Drawn, Fed, Fed on grass, —— in houses. }

Chap.

Chap. XII. *Improvements.*

Sect. 1. Draining.
2. Paring and Burning.
3. Manuring.
4. Weeding.
5. Watering.

Chap. XIII. *Live Stock.*

Sect. 1. Cattle.
2. Sheep.
3. Horses, and their Use in Husbandry, compared to Oxen.
4. Hogs.
5. Rabbits.
6. Poultry.
7. Pigeons.
8. Bees.

Chap. XIV. *Rural Economy.*

Sect. 1. Labour—Servants—Labourers—Hours of Labour.
2. Provisions.
3. Fuel.

Chap. XV. *Political Economy, as connected with, or affecting Agriculture.*

Sect. 1. Roads.
2. Canals.
3. Fairs.
4. Weekly Markets.
5. Commerce.
6. Manufactures.
7. Poor.
8. Population.

CHAP. XVI. *Obstacles to Improvement; including General Observations on Agricultural Legislation and Police.*

CHAP. XVII. *Miscellaneous Observations.*

SECT. 1. Agricultural Societies.
    2. Weights and Measures.
    3. Supply of London.
    4. Experimental Farm.

CONCLUSION. *Means of Improvement, and the Measures calculated for that Purpose.*

APPENDIX.

PERFECTION in such inquiries is not in the power of any body of men to obtain at once, whatever may be the extent of their views or the vigour of their exertions. If LOUIS XIV. eager to have his kingdom known, and possessed of boundless power to effect it, failed so much in the attempt, that of all the provinces in his kingdom, only one was so described as to secure the approbation of posterity[*], it will not be thought strange that a Board, possessed

---

[*] See VOLTAIRE's Age of LOUIS XIV. vol. ii. p. 127, 128, edit. 1752. The following extract from that work will explain the circumstance above alluded to:

"LOUIS had no COLBERT, nor LOUVOIS, when, about the year 1698, for the instruction of the Duke of BURGUNDY, he ordered each of the intendants to draw up a particular description of his province. By this means an exact account of the kingdom might have been obtained, and a just enumeration of the inhabitants. It was an useful work, though all the intendants had not the capacity and attention of Monsieur DE LAMOIGNON DE BAVILLE. Had what the King directed been as well executed, in regard to every

sessed of means so extremely limited, should find it difficult to reach even that degree of perfection which perhaps might have been attainable with more extensive powers. The candid reader cannot expect in these Reports more than a certain portion of useful information, so arranged as to render them a basis for further and more detailed inquiries. The attention of the intelligent cultivators of the kingdom, however, will doubtless be excited, and the minds of men in general gradually brought to consider favourably of an undertaking which will enable all to contribute to the national stores of knowledge, upon topics so truly interesting as those which concern the agricultural interests of their country; interests which, on just principles, never can be improved, until the present state of the kingdom is fully known, and the means of its future improvement ascertained with minuteness and accuracy.

every province, as it was by this magistrate in the account of Languedoc, the collection would have been one of the most valuable monuments of the age. Some of them are well done; but the plan was irregular and imperfect, because all the intendants were not restrained to one and the same. It were to be wished that each of them had given, in columns, the number of inhabitants in each election; the nobles, the citizens, the labourers, the artisans, the mechanics; the cattle of every kind; the good, the indifferent, and the bad lands; all the clergy, regular and secular; their revenues, those of the towns, and those of the communities.

"All these heads, in most of their accounts, are confused and imperfect; and it is frequently necessary to search with great care and pains, to find what is wanted. The design was excellent, and would have been of the greatest use, had it been executed with judgment and uniformity."

INTRODUCTION.

A second Report for the County of Norfolk, by a different Writer from the Gentleman who executed the first, demands a short explanation, to obviate any idea tending to lower the estimation in which the Original Report is justly held. There have been various instances of second, and even third Reports of the same County, all by different hands; but in the case of Norfolk a new Report was demanded, for a local reason. The introduction of a new breed of Sheep, and the rapidity with which the practice of Drilling spread in the County, had effected so great a change in the state of Norfolk Husbandry, that all former works on the Agriculture of that celebrated County must necessarily be deficient, however excellent in other respects. The present Report does not appear to the exclusion of the former, but merely in assistance of it; that such objects as were unnoticed, or but little

little attended to, from their being at that moment in their infancy, should now be registered, for the use of such other Counties as may not yet have made similar exertions.

July 14, 1804.

ERRATA.

In p. 527, third line from the bottom, after Rent 35, *read* per cent.

P. 529, in the col. of Uninhabited Houses in the Hundred of Eynesford, for 39 *read* 32.

CONTENTS.

CHAPTER I. GEOGRAPHICAL STATE.
                                                                PAGE
SECT. 1. Extent, ... 1
      2. Climate, ... 2
      3. Soil, ... ibid.
           District of various Loams, ... 5
           Light Sand, ... 12
           Light Loam, ... ibid.
           Marshland Clay, ... 14
      4. Water, ... ibid.

CHAP. II. PROPERTY, ... 17

CHAP. III. BUILDINGS, ... 19

Cottages, ... 24
Farm-yard, ... ibid.

CHAP. IV. OCCUPATION.
SECT. 1. Farms, ... 26
      2. Rent, ... 36
      3. Tithes, ... 40
      4. Poor-rates, ... 41
      5. Leases, ... 47

CHAP. V. IMPLEMENTS, ... 52
Ploughs, ... ibid.
Harrows, ... 55
Roller, ... 56
Waggons, ... ibid.
Drill-roller, ... 58
Drill Machine, ... ibid.
Fixed Harrow, ... 59
Horse-hoes, ... ibid.

Barrow

CONTENTS.

PAGE

Barrow to drop Wheat, ... 59
Double Barrow, ... 60
Scuffler, ... 61
Hoe-plough, ... 62
Rake, ... 63
Threshing-mill, ... ibid.
Chaff-cutter, ... 72
Steam Engine, ... 73

CHAP. VI. ENCLOSING, ... 75
Fences, ... 185

CHAP. VII. ARABLE LAND.

SECT. 1. Tillage, ... 187
2. Fallowing, ... 192
3. Course of Crops, ... 193
4. Turnips, ... 219
5. Barley, ... 238
6. Clover, ... 256
    Rye-grass, ... 267
    Burnet, ... 268
    Cock's-foot, ... 269
    Chicory, ... 270
7. Wheat, ... 271
8. Rye, ... 304
9. Oats, ... ibid.
10. Pease, ... 308
11. Beans, ... 311
12. Buck-wheat, ... 317
13. Tares, ... 318
14. Cabbages, ... 320
    Cole-seed, ... 323
15. Carrots, ... 324
16. Mustard, ... 325
17. Hemp and Flax, ... 326
18. Sainfoin, ... 338

SECT.

CONTENTS.

		PAGE
SECT. 19.	Lucerne,	342
20.	Mangel Wurzel,	347
21.	Potatoes,	ibid.
22.	Drill Husbandry,	348
23.	Norfolk Arable System,	362

CHAP. VIII. GRASS, ... 370

Rouen, ... 376

CHAP. IX. WOODS AND PLANTATIONS, 381

CHAP. X. WASTES, ... 385

CHAP. XI. IMPROVEMENTS.

SECT. 1.	Draining,	389
2.	Irrigation,	395
3.	Manuring,	402
	Marle,	ibid.
	Lime,	412
	Gypsum,	413
	Oyster-shells,	ibid.
	Sea-Ouze,	414
	Sea-weed,	ibid.
	Pond-weed,	416
	Burnt earth,	ibid.
	Sticklebacks,	417
	Oil-cake,	ibid.
	Ashes,	421
	Soot,	ibid.
	Malt-dust,	ibid.
	Buck-wheat,	ibid.
	Yard-dung,	422
	Littering,	432
	Leaves,	433
	Burning Stubbles,	ibid.
	River Mud,	ibid.
	Town Manure,	434

		PAGE
SECT. 4.	Paring and Burning,	434
5.	Embanking,	435

CHAP. XII. LIVE STOCK.

SECT. 1.	Cattle,	444
2.	Sheep,	447
	Breeds, Crosses, &c.	449
	Food, Wool,	470
	Fold,	473
	Distempers,	476
3.	Hogs,	478
4.	Horses and Oxen,	479

CHAP. XIII. RURAL ECONOMY.

SECT. 1.	Labour,	483

CHAP. XIV. POLITICAL ECONOMY.

SECT. 1.	Roads,	489
2.	Canals,	ibid.
3.	Fairs and Markets,	490
4.	The Poor,	492
	Houses of Industry,	494
	House of Correction,	503
5.	Comparison of Times,	504
	Day-work,	505
	Wages,	506
	Piece-work,	507
	Blacksmith,	508
	Artisans,	509
	Rent, Tithe, and Parish Taxes,	510
	Cultivation.	511
	Manure,	*511
6.	Population,	529

CHAP. XV. OBSTACLES TO IMPROVEMENT.

Rooks, ... 531

AGRICULTURAL SURVEY

OF

NORFOLK.

CHAPTER I.

GEOGRAPHICAL STATE.

SECT. I.—EXTENT.

ACCORDING to Templeman's Survey of the Globe, Norfolk is 57 miles in length and 35 in breadth, and contains 1426 square geographic miles. Mr. Kent makes the greatest length, from east to west, 59 miles, and its greatest breadth, from north to south, 38; containing 1710 square miles, and 1,094,400 statute acres. It contains 33 hundreds, one city, four sea-port towns, 25 other market-towns, and 756 parishes; a greater number than any other county in the kingdom. Mr. Howlet contends that Norfolk is larger than Essex, which is estimated at 1,240,000 acres. As this point was therefore doubtful, I had the area of the county on the new and very accurate map, measured carefully by the map-engraver to the Board, whose measurement gives 1830 square miles.

SECT. II.—CLIMATE.

THERE are several points of the compass from which he north and north-east winds blow more directly on this county than on any other in the kingdom: we feel these winds severely in Suffolk; but Norfolk is still more exposed to them, and the climate consequently colder, and more backward in the Spring. Another circumstance which must have some effect on the climate, is the whole western boundary being the fens and marshes of Lincolnshire and Cambridgeshire, to the amount of 5 or 600,000 acres; but this is more likely to affect the salubrity of the air relative to the human body, than to the products of the earth.

SECT. III.—SOIL.

THE annexed Map will best explain the great divisions of the soil of Norfolk. I travelled many miles, in order to give it as much accuracy as such a sketch admits, short of an attention that would demand years rather than months, perfectly to ascertain.

Entering the county from the south-west, the district of sand extends from Garboldsham to the fens of Hockwold and Methwold; contracting its breadth between Stoke and Pickenham, continues to Congham and South Rainham; and then diverging both to the east and west, fills up the whole north-eastern part of the county to Hindringham, and then taking a south-easterly direction, bends again to the north to Barningham, and strikes the sea at Overstrand. But this sandy district is again divisible into light and good sand; the former to the south of the line from Winch to Swaffham, and the latter to the north. The southern part comprehends by far the poorest part of the county, a considerable portion of which is occupied by rabbit-

rabbit-warrens and sheep-walk heaths, and has a most desolate and dreary aspect. It is, however, highly improveable by the marle and chalk, or cork, which is almost every where found under the surface. Much has been broken up in the last 20 years; but much remains to be done. The improvement of this district has been long impeded, from an idea that the white chalky marle would not answer carrying; and what they call good clay is scarce; but modern experiments have, or rather are, at present, working out these erroneous ideas; and under the article of *Monures*, will be found some satisfactory trials, which speak a different language.

The north-eastern angle, of better sand, contains large tracts of excellent land, intermixed with a good deal of an inferior quality. Here is found the agriculture to which the general epithet of Norfolk husbandry peculiarly belongs. The improvements wrought here from 60 to 70 years ago, first gave rise to, and afterwards established, the celebrity of the county: rents have risen from 1s. to to 15s. and from 1s. 6d. and 2s. to 20s. A country of rabbits and sheep-walk has been covered with some of the finest corn in the world; and, by dint of management, what was thus gained, has been preserved and improved, even to the present moment.

To the west of the light sand district is a small tract of various soils, between the Stoke and the Sechy rivers, and bounded on the west by the river Ouze. Here is much good sand, some strong wet clayey loams, and on the Ouze a line of rich marsh.

Still more to the west, and cut off by the Ouze from the rest of the county, lies the rich district of marshland; the larger part of which is a marine silt on a clay basis, at various depths. In some tracts the clay mixes with the silt to the surface, and forms the richer grazing lands.

The district of various loams comprehends the larger

part

SOIL.

part of the county, and includes soils of a very different description. In the southern part of it, in Diss hundred, and some adjoining ones towards Norwich, there is much strong wet loam, where summer fallow and beans are found; and similar land is scattered in other parts; but the general feature is a good sandy loam, upon which turnips come in regular course: it is an old enclosed woodland country, which would not be noted as very famous for management; nor had Norfolk attained any great celebrity for her farming, from the practices (though, upon the whole, meritorious) which are found in this district. The natural fertility is considerable. Marle is found almost in every part.

It only remains to speak of the district of rich loam, which is certainly, in point of soil, one of the finest tracts of land that is any where to be seen: broads and marshes occupy too much of it; but the land, under the plough, is a fine, deep, mellow, putrid, sandy loam, adhesive enough to fear no drought, and friable enough to strain off superfluous moisture; so that all seasons suit it: from texture, free to work, and from chemical qualities, sure to produce in luxuriance whatever the industry of man commits to its friendly bosom. The husbandry is good, but by no means perfect.

The relative contents of these districts are found by measurement to be:

	Square Miles.
Light sand,	220
Good sand,	420
Marshland clay,	60
Various loams,	900
Rich loam,	148
Peat,	82
	1830

Or,

SOIL. 5

Or, in acres:

	Acres.
Light sand,	140,800
Good sand,	268,800
Marshland clay,	38,400
Various loams,	576,000
Rich loam,	94,720
Peat,	52,480
	1,171,200

LOCAL NOTES.

DISTRICT OF VARIOUS LOAMS.

In passing through the county, and discriminating the great divisions of the soils, some exceptions occurred, which were noted, and other circumstances not coinciding directly with the general features of the district.

Around Watton are various loams, some of them sandy; and some heavy, on a clay marle bottom: much land that is improved by draining.

Some of the black sand and gravel at Billingfold has proved very ungenial in cultivation; part of it broken up from the heath, and marled with 80 loads an acre, gave only *blind* ears; that is, the ears had chaff only, without grain; after this it was marled again; and cultivation has brought it to be profitable.

From Dereham to Wymondham, by Winborough, Yaxham, Garveston, and Kinderly, much of the country, especially towards Wymondham, is a thickly enclosed woodland; the soil is strong loam on clay marle, and very fine: I saw great crops of all sorts, and many *laid*. Around Wymondham, some light but good land. The two Morleys contain some very light, and some heavy. Deepham, heavy; and Wickle-wood, inclining to it. Fundenham, Ashwell, and Wramplingham, a strong loam.

From Wyndham, by Hethel, Wrenningham, and Flordon,

Flordon, very fine rich land, being a strong loam on marle; Mr. DAWES, of Wrenningham, has an immense crop of wheat, dibbled on clover ley. At Flordon, some land of the finest quality, deep, friable, and putrid. All good and strong land to Stratton; thence to near Harling (about which town there is some lighter land), by Wacton and the Pulhams, all strong land on clay marle: clover sown alone, a regular plant, and good; and a scattering of beans from Wymondham to Harleston, sufficient to shew (if this proof were wanted) what the soil is.

I met with few farmers in Norfolk who admitted that fern was a sign of good land; as I have an entire conviction that it is an excellent sign, I was pleased to hear Mr. BLOOMFIELD, the brother of the gentleman who lives at Billingford, observe, in riding over a heath in that parish, on my asking why they did not plough it—" this part, where there are brakes, would pay well; but some of it is poor."

Thelton, Rushel, Billingford, and to the back of Needham, are very wet and strong, as much so as any parts of Norfolk.

Earsham hundred has much mixed good land, some sharp gravel, and much heavy wet clay.

Diss hundred, good, though strong and wet. Rent 15s. to 20s.

Depwade has much strong land, but little meadow; nearly equal, however, to Diss.

The Pulhams are wet and heavy, and contain much pasture. Mr. DONNE's farm is broken up.

In Loddon hundred, Langley, Mendham, and Seething, strong land, but good; greatly improved by hollow-draining. At Langley, all I viewed, except the marshes, is a very fine and fertile sandy and gravelly loam; in various vas tracts on a fine clay marle, well worth 50s. an acre ; equally good for turnips and wheat. Twaite, strong. Hedingham, half strong, half mixed soil, none light. Broom has some heavy, but much gravel. Loddon, a little strong, but all good.

Chedgrave and Hardy, mixed; average rent, 20s.

Carleton and Ashby on loam ; and a little clay. Hillington, Claxton, and Houlston, better ; all good.

Brook divides the soils; on one side the land is as fine as in Fleg.

Kirstead has a good soil. Stoke has some heavy land.

There is a remarkable vein of strong clay soil, eight miles long, and two wide, extending from Brook to Tasborough turnpike at the Bird-in-Hand: taking in a little of Fladdon, and ending at Mulbarton ; this is all a strong clay land, with many beans cultivated; it was not worth above 8s. an acre ; but by draining, and other improvements, is now brought to be good land.

To the north of this line, towards Norwich, it is a country of good mixed soil ; the worst of it is at Dunstonhills, near Hartford bridges. To the south there lies also a range of mixed soil, till it meets the strong land of Thelton, Thorpe, Abbots, &c.

Clavering hundred, all good land.

Horsted and Belough, light. Wroxham, light.

The soil at Coltishal is various, but much a light sandy loam on a red running sand, three or four feet deep; then a bed of flints, then more sand ; another bed of flints, and then marle ; but little gravel.

About North Walsham a *mixed* soil, that is, a sandy loam ; some sand ; and the sub-soil sand : at the distance of three or four miles, much on brick earth, and some on clay marle.

The

The same at Scotter; a fine sandy loam opposite Mr. DYBLE's house, and much of it in the country.

At Oxnead, &c. a mixed sandy loam on a red sand, under which a layer of flints; then strong brick earth, and then white chalky marle, at the depth of 10 to 15 feet.

From Aylsham to North Walsham good land; better than to Holt. Around Wolterton good; and the first four miles to Cromer. It is bad to Causton; and much of it indifferent to Norwich. From St. Faith to Norwich enclosed within 30 years, and good.

From Causton to Reepham and Foulsham, a tract of good soil, but much wet, on which draining is a great improvement.

At Heveringland much sandy land, on a hard cold sand bottom.

Excellent sandy loam at Hackford and Reepham; much of it on a marley bottom.

Spixworth, Crostwick, Wickmer, and Wolterton, have very good land.

At Thurning there are sands of various qualities; some good; but much on a cold *clung* gravel; some stiff, even on the surface; so that water stands in the horses' foot-steps in turnip-feeding. Mr. JOHNSON remarks, as a general rule, that if the sand does not *wash* in the furrows, the land is bad, good neither for barley nor turnips. On many soils a thickish mud washes, which keeps the *plat*, the breast of the plough, from scouring. There is some good mixed soil on gravel, and some on clay; good strong land, with the sand washed in the furrows.

Gistick, as good land as any in this neighbourhood. Dawling better than Thurning. Briston part good, and part bad.

Croxton,

Croxton, Fulmerston, and Stibbard, good; also Wood Nowton, and Riborough.

The tract to the east of the Ouze, comprized in Downham, Winsbotsham, Stow, Crumplesham, Bexwell, Ruston, &c. is a mixed soil; sand, and sandy loam; some clayey, and a little moory; generally good land, upon a white clay marle.

At Watlington much sandy gravel, on a yellow clay; also strong silt clay, on a gault bottom, wet.

There is much very poor sand, gravel, and poor moory heaths, from Lynn to Swaffham; and poor shallow sand, on hard chalk, at Narborough.

The soil at Bestthorpe is strong, and much wants draining; but some very fine brown sandy loam is found. A hedge thrown down many years ago in Mr. PRIEST's turnip-field, and the rankness of the crop there, striking.

Hookering, and North and West Tuddenham, are a strong enclosure; mixed soil, being good turnip-land.

Part of Weston is sandy, and part good.

At Swanton Morley, Mr. EMMS's farm, 1000 acres, mixed land; but part of the parish heavy. Lyng is light, and Dillington various.

At Gressenhall there are three or four farms, of 1200 acres, to be sold, some of which might be watered. An advowson of 4 or 500l. a year, and manors; the price 50,000l.

Kempston is a fine mixed loam; no light land, but some heavy.

To Fakenham, all good.

Beccles to Loddon, a fine sandy loam, 20s. an acre in 1792.

Loddon to Norwich, and Norwich to Dereham, a fine sandy loam.

GOOD SAND.

At Great Dunham the soil is very fine land, and produces great crops. But in Little Dunham is some of the finest land I have any where seen: the crops immense; yet, if possible, inferior to the texture and appearance of the land. The herbage of the leys of a deep hue, that speaks their nourishment.

Along the coast, from Holkham, westward, towards Hunston, there is a tract between the marshes and the sand, from half a mile to a mile broad, of a singularly fertile sandy loam: it has tenacity sufficient to adhere into clods easily broken, and to produce great crops of vegetables demanding the richest soils; at the same time it has that dryness and friability which renders it excellent turnip-land. It may let at about 20s. an acre; whoever views it will not think the rent high.

At Burnham Westgate, some poorer sands, and chalk slopes. Choseley, middling sand. Summerfield, good mixed soil and sand. Thornham very good sandy loam. Holm equal, and part of it superior to Thornham. Tichwell is good, but has some light. Brancaster the same; and Hunston, very good.

Fring has some light, and some middling. All these parishes class as sand in general; the basis marle or chalk.

Snettisham, much good sand on chalk and marle, and some low and strong land. Hexham nearly the same.

At Ringstead a quarry of marble; the colour and veins something like that of Sienna. At Snettisham, a quarry belonging to Mr. STYLEMAN, of red sand-stone, 20 feet deep: it rises soft, but hardens in the air. The stables at Houghton are built of it.

Upon the sea-shore at Snettisham, Mr. STYLEMAN has
a large

a large tract of sea-shingle, which bears little more than eringo: he sowed some chicory seed on it for experiment, but it never vegetated.

The soil of Houghton, is sand on chalk and marle: that of Harpley better; the Rudhams good; Bagthorpe light; the Birchams ordinary; Anmer is better, but all sand.

At Hillingdon there is some black sand and gravel; a poor soil, but the hills are chalk: the sands all apt to be more foul than the chalk.

Much blowing sand at Riseing; and the evil of some of the soil there is, its being free from stones, and for that reason *burns* much; on which account Mr. BECK disapproves of picking stones. The remark is very judicious. To deep sands he thinks twitch so natural, that it is impossible to free them from it: if the field is made ever so clean, and lays two years, there will be some.

There is some sand that *burns*, at Grimstone. The soil of Massingham is much better, and never burns.

At Guyton the soil has a mixture of stones, and is the better for it.

At Morston, near the sea, there is some land so covered with stones, as to appear to the eye to contain little besides; but excellent for corn; 11 combs an acre of white vetches have been gained on it. Much good land is found at Cley.

Sarsingham, Dawling, Binham, are good sand; the loam on white clay marle.

Wighton, sandy loam; good mixed soil, on clay marle; but some is inferior.

In Sherrington, Hindringham, and Kilderston, there is much strong land, upon which hollow-draining is practised.

LIGHT

LIGHT SAND.

From Riddlesworth, by West Harling, there are poor thin sands on marle. At Quiddenham it improves, and continues better through Eccles to Snetterton. The crops in Harling-field, in this wet season, miserably poor, hardly yielding the seed again; yet a most luxuriant growth of spontaneous chicory and mellilot, which shews the profit to which the tract might be applied, if the hints offered by the beneficence of the Almighty were pursued.

From Attleborough to Euston, by Rowdham and Bretenham, a dreary country, beginning with rich, but sombre commons, and then crossing poor open heaths and sheep-walks, and open arable, that cries aloud for chicory: poor sand, on chalk and marle.

RICH LOAM.

One of the most interesting circumstances in the husbandry of Norfolk, is the soil of Fleg hundred; and much in Blowfield and Walsham hundreds is of the same quality: it is a sandy loam, from two to three feet deep, and much of it as good at bottom as on the surface; of so happy a texture, that almost any season suits it; subject neither to burn in droughts, nor to be wet with incessant rains. The basis most general, is a clay marle; but in several districts sand, both yellow and white. So fertile a soil I have very rarely seen of so pale a colour; it is a very light whitish brown in dry weather. The products are great: wheat from 6 to 14 combs; barley 9 to 16; oats 10 to 24; pease to 15. Yet these products do not altogether announce the merits of the soil, which perhaps is marked more by paucity of failures than by extraordinary crops. Ashby and Burgh were named to me as having extraordinary land, and at the latter I found it excel-

excellent. Mr. JAMES WIGG's is famous land, and his wheat this year great.

Mr. BROWN, of Thrigby, remarked to me a circumstance which well deserves noting; that at Ashby, Billaby, and Burgh, is land that, before marling, ran uncommonly to white clover; but after being marled from Thorpe and Wightlingham, will do it no longer, none coming at present without sowing.

At the Burlinghams and Linwood there is capital land.

Some of the finest land in Norfolk is at Acle, Moulton, Tunstal, and South Walsham; the crops of wheat were, in 1789, estimated, on an average of years, at ten combs an acre on their good land; they have, however, some that does not produce above six. At Mawby, near Castor, the average of the parish, for ten years together, ten combs: twelve, and even fifteen, have been had; but in 1802, they did not admit crops equal to these.

I examined with pleasure the fine loams near the respective houses of Mr. EVERITT, of Caistor, and Mr. FERRABY, of Hemsby, which class amongst the finest soils in Fleg; it ought to be termed a rich sandy loam; dry enough for feeding turnips, and rich enough for five or six quarters of wheat; equal to great crops of cabbage, beans, or any other production.

Southwood, Moulton, Lippenhoe, and Rudham, class high among the fertile parishes.

At Martham, much of the land is on a sand bottom, and some gravelly spots subject to *scald*, but towards the fens, brick earth.

At Catfield, &c. a pale, fine, sandy, loam, upon a sand bottom, esteemed thin skinned, but I found it to the eye the same, at a foot and eighteen inches deep.

Very

At Waxham the soil is very fine.

No clay for an under-stratum in Happing hundred, except at Happsborough, Walcot, and Bacton; generally sand or brick earth. Here are some of the finest lands in the county—equal to Fleg, but stiffer; yet the surface a fine friable sandy loam, and the bottom not too retentive of water. The best land in the Catfield district is at Stalham and East Ruston.

Happsborough, Walcot, and Bacton, again noted to me as the finest soils, perhaps, in the county; a rich, deep, mellow, friable loam, on a clay loam bottom, some on brick-earth and sand; all good. East Ruston, very good, deep, on brick-earth.

Mr. CUBIT, at Honing, has some very fine pale coloured sandy loam, resembling the Fleg soils, and worth 26s. an acre; yet, intermixed, he has some hills of sand and gravel of much inferior value.

MARSHLAND CLAY.

The whole district of marshland is probably a relict or deposition of the sea; it is a silt, or warp clay of great fertility, upon a sandy silt at various depths, but usually eighteen inches or two feet. The stiffer clays are the worst arable: the more mild and temperate ones, the best and easiest worked of course; but the strongest clay is the best for grass.

SECT. IV.—WATER.

NORFOLK is advantageously situated respecting navigation; for of its great circumference of 200 miles, there are but something more than thirty, from Thetford to Bungay, which do not consist of the sea, or of navigable rivers: to the north, the ocean; to the west, the great Ouze; to the east, the sea; to the south, the lesser Ouze, and the Waveney; and, exclusive of this beneficial boundary, the Yare and the Wensum penetrate from Yarmouth to Norwich, and the Bure and Thyrn, from the same port to Aylesham. With the last named river the Broads, and their communicating channels in the hundreds of Fleg and Happing, unite and connect the rich district with the sea and with Norwich, and the advantage is much felt in the conveyance of marle, &c. The navigation of the Nar reaches Narborough, and connects with the Ouze.

Smaller streams abound in every part of the county, and offer such opportunities of irrigation, as must excite the amazement of every farming traveller, at the utter neglect in which they have been suffered to run to waste for so many ages. At last, this important application has begun to receive a little attention; so that we may hope, that, in half a century more, these valuable treasures will be accepted.

Ponds, artificial.—Mr. COKE makes these ponds at Holkham to serve each four enclosures; they are set out 42 feet square; at bottom twelve, and seven deep. A bottom is worked with good clay, free from all stones, nor the least soil or sand in it, beating it as close as possible three inches thick; then three inches more are beaten in, and so on till a foot thick; then sand over it, to keep the drought out. The sides are made in the same manner as the bottom: within a yard of the top, the clay should be two feet thick. It is then paved with bricks set on edge. These ponds were made by men from Gloucestershire, who were paid 2s. 6d. per superficial yard, costing 28l. each,

besides

besides the bricks. To divide the ponds for four fields, a large stone with a hole wrought in it to receive a post, is necessary at the centre, with mortices to receive the rails.

CHAP. II.

PROPERTY.

ESTATES are of all sizes in Norfolk, from nearly the largest scale to the little freehold: one of 25,000l. a year; one of 14,000l.; one of 13,000l,; two of 10.000l.; many of about 5000l.; and an increasing number of all smaller proportions. When the larger properties are deducted, the remainder of the county will be found divided into moderate estates, and in the hands of gentlemen who pay a considerable attention to the practice of agriculture.

Seventy years ago, there was not, I believe, a great rental in the county, so that these considerable properties have been accumulated, first, by the most excellent of all causes, agricultural improvements, and, secondly, by additional purchases.

Estates sell now (1802) pretty currently at thirty years purchase.

In the Ovington and Sayham enclosure, the land sold by the commissioners to defray the expense of the measure, brought 43l. per acre, as it was assigned, waste and unenclosed. The average of all sales near Watton, 40l. an acre.

An estate lately sold at Fishley, near Yarmouth, contained

Arable land, good, - - -	250 acres.
Cars and marshes, worth 12s. an acre,	100
Marsh, worth 20s. an acre, - -	50
	400

Rent, 400l. a year, worth 500l. fairly, but 600l. a year offered for it; sold for 17,500l. to Sir EDMOND LACON, and 1500l. offered for the bargain.

Price of the estates sold at and near Happsborough, 30, 40, and 50l. an acre; much sold lately; and at this time, the best land would all sell at from 40l. to 50l. an acre.

Land worth not above 20s. an acre, between Coltishal and Norwich, has been sold at 50l. an acre.

In Marshland Smeeth, newly enclosed, at 50, 60, and 70l. an acre.

In Downham Westside, Denver, Welney, &c. fen farms, 10l. to 12l. an acre: to the east of Downham, at 24 years purchase.

In Upwell, some, not fen, to 50l. an acre, but the average 20l.

Mr. BAGGE, of Lynn, has land in Marshland which would now sell at 70l. an acre, which Mr. DIXON bought, 60 years ago, at 12l. 10s.

CHAP. III.

BUILDINGS.

SOME of the houses belonging to the proprietors of large estates in this county, have long been famous as objects of the attention of travellers, and deservedly so; for there are very few counties that rival it in this respect: the circumstance, however, is not interesting in an agricultural inquiry. The well-cultivated domain is here of much more consequence than the well-decorated palace.

In the species of building properly appropriated to an Agricultural Report, greater exertions have, I believe, been made in Norfolk than in any other county of the kingdom. One landed proprietor, Mr. COKE, has expended above ONE HUNDRED THOUSAND POUNDS in farm-houses and offices; very many of them erected in a style much superior to the houses usually assigned for the residence of tenants; and it gave me pleasure to find all that I viewed, furnished by his farmers in a manner somewhat proportioned to the costliness of the edifices. When men can well afford such exertions, they are certainly commendable.

One of Mr. COKE's barns at Holkham is built in a superior style; 120 feet long, 30 broad, and 30 high, and surrounded with sheds for 60 head of cattle: it is capitally executed in white brick, and covered with fine blue slate.

At Syderstone, he has built another enormous barn, with stables, cattle-sheds, hog-sties, shepherd's and bailiff's houses, surrounding a large quadrangular yard, likewise

in

in a style of expense rarely met with. In discourse with the men at work in this barn, they informed us, that to one man who *unpitched* the waggon at harvest, seven others were necessary on the *goff* to receive and dispose of the corn, after it was raised to some height; a great expense at a time of the year when labour is the most valuable. The farmers are, however, very generally advocates not only for barns, but for great barns. Another inconvenience is their not daring to *tread*, except lightly, in large barns: and the men complained that the corn threshed the worse for want of more treading. 140 acres were in this barn of Mr. SAVARY's. Floor, eleven yards; barn, nine wide.

In all Mr. COKE's new barns, and other offices, he has substituted milled lead for ridge tiles to the roofs, which is far more lasting, and the means of escaping the common accidents, in raising a heavy ladder on tiling, in order to replace a ridge-tile blown off.

For all locks, particularly in stables and other offices, Mr. COKE has found those with copper wards much more durable than any others.

The front edge of his own mangers are rollers covered with tin; the mangers themselves are plated with iron; and the bottoms of the stall fences are of slate. All these circumstances are found very economical in duration.

In building the walls around a new farm-yard for Mr. COKE, Mr. OVERMAN, after a certain height, draws them in to a brick's length at the top, a saving in these erections which merits notice.

Mr. COKE's *Method of making up and applying Lime-wash as a Preservative and Covering to Boarding, Walls, &c.*

In a tub of six or eight gallons, put of water a quantity sufficient

BUILDINGS. 21

sufficient to half fill the same; add thereto of clean sharp sand, and of lime *fresh burnt*, in about equal quantities, as much as will make, when well stirred up, a wash of moderate consistence. With this wash, as soon as made, pay over the boarding of any barns or buildings, keeping the sand constantly stirred up, so that the brush may take up the sand as well as the lime. As the quantity in the tub decreases, add by degrees, in small quantities, more lime and more sand, taking care to make up no more than will be immediately used. The quicker the lime the better, which, if good, will make the wash hot; and if it be required to make this wash particularly hard and durable, the same will be effected by making use of boiling water instead of cold, taking care to make it in such quantities that it can be laid hot on the boards.

Mr. COKE has, at Holkham, a brick manufactory, which ranks very high among the first in the kingdom; bricks in all sorts of forms are made, so that in raising an edifice, there is never a necessity for breaking a whole brick to have a smaller of a very imperfect shape, which takes time, and creates waste; cornice, round column, corner, arch bricks, &c. are made in great perfection.

Account in 1792	£.	s.	d.
Raising earth, making and burning per 1000 (white bricks 12s.), - -	0	8	0
[The brick-maker finds moulds, pails, barrows, rakes, boards, &c.]			
Duty, - - - -	0	2	6
A chaldron of coals, at 26s. burns 7000,	0	3	$8\frac{1}{2}$
A kiln, 32,000, - - -	0	14	$8\frac{1}{2}$

Red are burnt with coals; white with wood: the latter could be sold at 4l. 4s. a thousand; such as are stained,

at

at 3l. 3s.   8000 of the white are burned in the centre of a kiln of 32,000 red.

1802.—All now are burnt with wood.

Mr. SALTER, at Winborough, built a cart-lodge, and granary over it, in which are three circumstances that deserve noting: the main-posts that support the granary rest on brick-work square foundations, about two feet high, and these he has guarded by oak plank let into the ground; the posts are tied to the beams at top by knees, which add much strength.—[See *Plate* I. *Fig.* 1.] The stairs are without the building, at the end; and for the conveniency of loading waggons, there are two rollers fixed, over which sacks are slid down easily into the waggon.

Mr. ROBINSON, at Carbrook, sanded his cornice, window and door frames, and window soles. The method is, to paint white, and dash it immediately with *sea* sand from a dredging box: the effect, as I saw, is that of an exceeding good imitation of stone. It is said to be very durable.

Mr. COLLISON, of Dereham, has built at Bilney a very capital barn of brick, the walls 18 inches thick, and for three feet from the ground, 22; and 23 feet high: no cross beams to impede the filling.—[See *Plate* I. *Fig.* 2.] The porch, by being something lower at the point of the roof than the barn wall, and forming no junction in the roof, has no gutters. The lean-to sheds are joined to the walls by lead worked into the brick work. No ridge-tiles to any of his buildings, but milled lead. Stables, harness, and straw and hay houses, with two sheds, one on each side the barn porch: the whole covered with pan-tiles. A granary and cart-shed, and a double cottage, very good indeed, but with the universal error of the house-door opening into the keeping-room. Good gardens;

Plate 1.
to face Page 22

Fig: 2

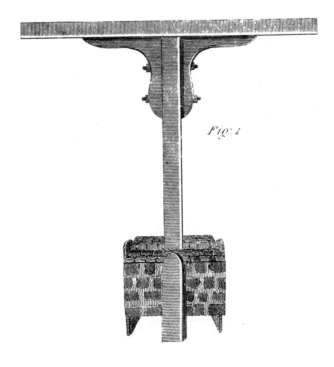

Fig: 1

Neele sc. 352 Strand.

dens; and the people seemingly more comfortable than common in Norfolk.

Mr. DRAKE, of Billingfold, near Scole, washes a fine white clay from a mixed soil, in order to make white bricks—*lumps*, as they are called—for barn-floors, &c. He has a boarded cistern, into which the earth is put, and water ladled or pumped from a contiguous pond: the earth is well stirred, and the sand, stones, &c. sinking, the suspended clay is let off into a broad shallow tank, and as soon as it is subsided, the water is let off into the same pond. But the clay thus gained is too strong to make brick alone—*flying* in the fire; it is therefore mixed with a yellowish loam.

At Thelton, to dig, turn over, water, make, and burn bricks, 9s. per 1000: every thing, as tables, boards, sand-barrows, &c. found. Some give 10s. 6d.

At Snettisham—	1770.	1803.
Bricks per thousand, -	20s.	- 30s.
Tiles, - - - -	3l.	- 4l.
Oak-timber per foot, -	1s. 6d.	2s. 6d.
Ash and elm, - -	1s. 2d	- 2s.
Soft woods, - - -	6d. to 8d.	1s. 2d.
Carpenter, per day, -	1s. 9d.	2s. 4d.
Mason and thatcher, -	1s. 9d.	2s. 4d.

Building a tolerable cottage at Hillingdon, costs 50l. 60l. and to 80l.; yet lime is burnt for 4d. a bushel; but it is bad, and the mortar decays.

An observation Mr. MAITLAND made, in shewing me a sluice in Governor BENTINCK's embankment, me rits noting. By the flooring of the arch being laid against the brick-work without due precaution, the water got through, and boiled up under the apron: he laid it afresh in Roman cement, stopt the water, and the evil ceased.

COTTAGES.

COTTAGES.

Mr. ROBINSON, at Carbrook, has built a double cottage of flint-work; the walls 18 inches thick, the workmanship of them 1s. the square yard. The rough-cast within is all of clay white-washed, which answers very well. The covering is pan-tiles. Cost 13ol.; but finished neatly, and with Gothic windows, perhaps may add 20l. or 3ol. extra: good gardens, and very well cultivated.

The two bed-rooms and wood-rooms are in a lean-to: a bed-chamber over each.—[See *Plate* II.]

Cottages are much wanted at Snetterton and the neighbouring parishes; if built, they would presently be filled with inhabitants. I wish they were erected, for the poor people are there very neat in their well-cultivated gardens; the land fully cropt, and in high order, and the hedges neatly clipt; but their gardens are much too small: they well deserve additions. All to Attleborough, &c. the same; and at Hingham, Mr. HEATH was certain that, if 20 or 30 were built, they would be all inhabited in three months. It is the same in the surrounding parishes.

It is a new practice in West Norfolk, to let cottages on leases of lives.

My guide into Marshland Smeeth, knows but three cottages built in consequence of cultivating 8000 acres: one by Mr. BAGGE, one by Mr. SILVERWOOD, and one by Lady TRAFFORD.

Rent of a cottage and a bit of garden at Walpole, 3l.

FARM-YARD.

The farm-yard of Mr. PURDIS, at Houghton in the Hole, lies on a slope, and runs into a ditch which conveyed

Plate 11.
To face Page 24.

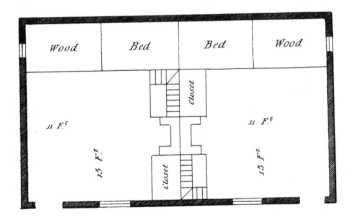

To face page 23.

A Cottage near Dorcham.

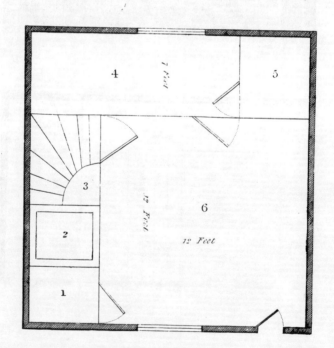

EXPLANATION.

1 — a Closet
2 — Chimney
3 — Stairs
4 & 5 — a Leanto
— one Chamb. over
6 — the Keep Room

Cost about £40. Rent 3.3.0.

veyed the drainings of it to the river: this ditch was cleaned out, and a great row of road-side mould raised; the declivity admitted the execution of a thought, beautifully, simple and useful. A trench is cut along the top of this row of earth, to receive the drainings of the yard; stops are made at a small distance from each other, to keep the liquor till the earth absorbs it; and thus the whole is impregnated. As the watered meadow is just below, Mr. PURDIS intends, when the earth is removed, to convey the drainings into his main carrier, for enriching his water of irrigation. When once men are led *to think*, nothing is lost: the mind is awakened to every hint; and circumstances which, in a sleeping age, would be passed by, are turned by active exertion into profit. In such a state of progressive improvement, every object becomes interesting, and all the faculties of the human mind are on the stretch to draw advantage, where our ancestors drawled on inactive and torpid.

I wish I had it in my power to add, that I saw a good farm-yard in the county, manifesting contrivance, and in which no building could be moved to any other scite without doing mischief. Where is such an one to be seen?

CHAP.

CHAP. IV.

OCCUPATION.

SECT. I.—FARMS.

UPON this subject we must divide the county generally into two parts—the dry soils and the wet ones. Upon the former the farms are large, and upon the latter much smaller.

The rich districts which, though dry enough for turnips, are strong enough for yielding great crops of wheat, possess some moderately-sized farms, such as from 400 to 600 acres; but many smaller. The wet land is more commonly held in small occupations. The poorer sands are usually in very extensive farms.

One near Holkham, 1792, 1000 acres, 450l. rent.

 50 acres wheat,   400 acres sheep-walk,
 150 ——— spring corn,   10 harvest-men,
 120 ——— turnips,   500 sheep—Norfolks,
 240 ——— layer,   14 farm-horses,
 40 ——— sundries,   40 cattle.

Mr. PURDIS, at Egmore, 1802.

 300 acres of turnips,
 300 ——— barley,
 600 ——— seeds,
 300 ——— tares,
 300 ——— wheat,
 100 ——— sundries.

1900

FARMS.   27

A finer farm is rarely to be seen, than Mr. MIL-
DRED's at Earsham, belonging to the Duke of NOR-
FOLK; as compact as a table; without a foot-path, near
a town and a navigation: a good soil, and a beautiful situation.

In the hundreds of Diss, Earsham, and Depwade,
more under than above 100l. per ann.

Average farm of Fleg under 200 acres; none very
large.

Mr. HEATH's, at Hemlington, of 500 acres, is one of
the largest farms, if not the largest, in South Walsham
hundred.

In Happing hundred, farms are very much in the
hands of the proprietors, the yeomanry being numerous;
hardly any so large as 4 or 500 acres, and not many under
50; from 50 to 300.

About North Walsham, few more than 2 or 300 acres.

Mr. PETRE's, at Westwick, 600 acres:

100 wheat,   25 horses,
180 barley and oats,   4 cows,
 12 pease,   50 bullocks,
 80 turnips,   340 sheep.
150 layer,

Mr. BIRCHAM, at Hackford, 840 acres:

720 arable,   120 pease, oats, and after-
120 turnips,   crop barley,
240 barley,   70 bullocks,
240 seeds,   600 sheep.
120 wheat,

His system is, to keep the greatest quantity of stock he
can; to keep his land clean, and then not to doubt of great
crops of corn.

Mr. JOHNSON, at Thurning, 880 acres:

200 grass,
                                      32 horses,

32 horses,
70 score sheep of all sorts,
95 beasts at turnips last year.

Mr. REEVE, at Wighton, 700 acres:
600 arable,
22 horses, and therefore not four to the hundred acres; all drilled.

Mr. STYLEMAN's, at Snettisham, 2000 acres:
1400 arable,
600 grass,
700 ewes, and their produce, or about 2000 in all

He has arranged his Ringstead farm in such a manner, in five divisions, each of six fields, like five distinct farms, that every sort of crop is scattered over the whole, which he considers as a great convenience.

Earl of CHOLMONDELEY:

650 acres arable,	120 turnips,
350 grass,	120 barley,
360 plantations, of which	240 seeds,
120 added by the present Earl;	120 wheat.

The size of farms in Marshland, will be seen by certain of Governor BENTINCK's:

695 acres,	264 acres,
438 another,	168 ditto,
231 ditto,	185 ditto,
497 ditto,	60 ditto,
324 ditto,	55 ditto.

About Downham, most are small; a few to 4 or 500 acres. About Wymondham, chiefly small: 200l. a year a large one; many 40l. 70l. and to 100l. per ann.

Around Hingham, and to Attleborough, farms are all small: Mr. GASS, at Hingham, has one of the largest.

In the richer lands of Attleborough, Hingham, and Watton,

Watton, the farms are generally small: about the latter place, 20ol. is a common size, and some of 100l. and even 60l. and 50l. Mr. BOUGEN, of Sayham, 700l. the largest in the neighbourhood.

Entering.—Mr. JOHNSON, of Thurning, entering 500 acres in sad order, he ploughed up and fallowed the seeds, for which he had given 30s. an acre.

Mr. OVERMAN, at Burnham, entered his new farm hired of Mr. COKE, with a most decisive energy. He paid per acre for liberty to fallow land, from which his predecessor had liberty to take a crop: he fallowed other lands left under seeds, and for sowing which he had also paid. He brought a ship-load of hurdles from Sussex, for defending his young setts: he marled such part of the farm as wanted it, the first year; remarking to me the great difference between enjoying the return from improvements for 19 or 20 years, rather than, by delay, possessing them only 14 or 16.

Stock.—The Rev. Mr. PRIEST, on 85 acres of very good land at Scarning, near Dereham, and 15 other acres at some distance, keeps 100 excellent South Down sheep, 6 cows, 5 horses, and has had a larger stock; yet 75 acres of it are arable.

Size.—In discourse with Mr. THURTELL, on the size of farms, he remarked that nothing could, in his opinion, be more absurd, than the prejudice against large farms. Wherever he has made any observation, large farms produce much more for the public consumption than small ones: that all improvements, if they arise at all, must be from large farmers, who are able, and now a-days willing, to make experiments. And he further observed, that whatever crop would produce most money, was of most public benefit; an observation perfectly coinciding with

with that of a celebrated writer, but whom this gentleman had not read, Mr. HARTE, in his *Essays on Husbandry*.

The size of farms is a subject upon which so much has been written, that to enter into a discussion on the subject in the Report of a single county, would be to swell a book with general subjects, which ought to be appropriated to local and particular ones. I shall only observe, that the vast improvements which have been made in Norfolk, by converting boundless heaths, sheep-walks and warrens, into well-cultivated districts, by enclosing and marling, are such as were never yet made by small farmers. Great farmers have converted in this county three or four hundred thousand acres of wastes into gardens: can any thing therefore be so grossly absurd, as to find fault with such divisions of the earth as have produced these effects? Little farmers have never, in any county that I am acquainted with, produced equal effects: if they have, let the Reporters of such counties explain it; it is my business to state what has taken place in Norfolk. In the eastern district of rich land, farms are moderate or small, yet the country is well cultivated; but natural fertility does the great business; upon such land, it is of much less consequence what the size of a farm may be.

When poorer tracts become highly improved, and very great exertions are not equally necessary, farms of extraordinary size may be profitably divided, since the invention of threshing-mills, which supersede the necessity of barns: as in this case, the interest of the money expended in new buildings, added to their repairs, may not be equal to the superior rent of a moderate farm over one of a great extent. The private interests of individuals may safely be trusted with all such arrangements, as much

more

more likely to produce a result nationally beneficial, than any of the laws which have been stupidly and absurdly called for.

Farmers.—The Norfolk farmers are famous for their great improvements, the excellency of their management, and the hospitable manner in which they live and receive their friends, and all strangers that visit the county. I have on various occasions found how well they merit their reputation.

In respect to their husbandry, the farming mind in this county has undergone two pretty considerable revolutions. For 30 years, from 1730 to 1760, the great improvements in the north western part of the county took place, and which rendered the county in general famous. For the next 30 years, to about 1790, I think they nearly stood still; they *reposed upon their laurels*. About that period a second revolution was working: they seemed then to awaken to new ideas: an experimental spirit began to spread, much owing, it is said, to the introduction of drilling; and as so new a practice set men to thinking, it is not unlikely: nothing can be done till men think, and they certainly had not thought for 30 years preceding. About that time also, Mr. COKE (who has done more for the husbandry of this county than any man since the turnip Lord TOWNSHEND, or any other man in any other county), began his sheep-shearing meetings. These causes combined (for what I know, the former sprung partly from the latter) to raise a spirit which has not subsided. The scarcities, and consequent high prices, brought immense sums into the county, and enabled the farmers to exert themselves with uncommon vigour. Experiments in drilling shewed that farmers might step out of the common road, without any danger of a gaol. South Down sheep came in about the same time. Folding was by many gradually given up.

These

These new practices operated upon the farming mind; ideas took a larger range; a disposition was established, that would not readily reject a proposal merely because it was new—the sleep of so many countries. Every thing is to be expected from this spirit. Irrigation is gaining ground, in spite of the dreams that have been ventured against it. And if the men who occupy, or rather disgrace so large a part of the light sand district, by steadily adhering to those *good old maxims* which have preserved it so long in a desart state, shall once imbibe a portion of this ardour, we shall see new plants introduced, and new practices pursued, to carry the county in general to the perfection of which its husbandry is capable.

Those who have visited Holkham as farmers, will not accuse me of flattery, if I assert of Mr. COKE, that he is *fairest where many are fair.* To name particulars, would be to detail the whole farm.

Mr. PURDIS, of Eggmore, is in the first class of excellent cultivators: his farm has many unequivocal signs of spirited exertion: 300 acres of tares; 3000 South Down sheep; and a watered meadow, are objects that speak for themselves.

The late Mr. MALLET, of Dunton, having, on coming to his farm of 2500 acres, nothing more than the stock, valued at 7000l. in thirty-four years acquired a fortune of 70,000l.

Mr. SALTER, of Winborough, is one of the most spirited improvers in the county: he hired 800 acres, in a state not far removed from a waste; and by ditching, draining, marling, and good husbandry of various kinds, has brought it to be one of the most productive farms in Norfolk.

The Rev. Mr. MUNNINGS, near Dereham, invented a method of preserving turnips, which he described in a late

late publication of merit. He drills successfully, and has various and useful implements.

Mrs. COLLISON, of Dereham, has made considerable improvements at East Bilney; drills successfully, and has built a capital barn, &c.

Mr. MONEYHILL keeps a farm of near 1300 acres, with a degree of neatness and attention which classes him among the first farmers of the county: the whole drilled. Fine South Down Sheep.

The Rev. DIXON HOSTE drills upon a very stiff and difficult soil, with success. Drains effectually.

Mr. JOHNSON, of Kempston, has the best threshing-mill I have seen in the county; Leicester sheep; and is a very good and attentive farmer.

Mr. FARROW, at Shipdam, 350 acres at Sayham; 200 new enclosure; very great and expensive new brick barn and a threshing-mill. STANTON of Weybread's drill: much drilling, and great crops.

Mr. ROBINSON, of Watton, has erected new cottages, with improved and economical circumstances in building: South Down sheep.

Mr. HEATH, of Hingham, after breaking up grass-land, has great crops.

Mr. FOWEL, of Snetterton, drills his whole farm: lucerne; soiling. An excellent farmer.

The Rev. Mr. PRIEST, at Scarning, drills all his arable land with barrow-drills, and horse-hoes the whole; three rows, at nine inches, on three-feet ridges: fine crops. Various implements; and a beautiful little flock of South Downs.

Mr. DRAKE, of Billingford, paring and burning; drilling and dibbling.

Mr. HAVERS, of Thelton, Devon cattle; piggery; beans; oxen.

Mr. PITTS, of Thorpe Abbots, drilling; beans.
Mr. KERRICH, of Harleston, manuring for beans.
Mr. MILDRED, at Earsham, beautiful farm.
Mr. BURTON, at Langley, general information; extensive knowledge, and excellent management · enclosures
Mr. THURTELL, near Yarmouth, manuring; tillage; building.
Mr. EVERIT, at Caistor, fine land, and good management.
Mr. FERRIER, at Hemsby, fine land, and great crops.
Mr. BROWN, at Thrigby, clay; marle; yard-dung; pease.
Mr. SYBLE, of South Walsham, intelligent; course; drilling; oil-cake feeding.
Mr. FRANCIS, at Martham, tillage; dibbling; manures; ploughs; products.
Mr. HORNARD, of Ludham, double ploughs; teathing wheat; manuring.
Mr. CUBIT, of Catfield, teathing wheat; potatoes; mixing dung; marle.
Mr. WISEMAN, of Happsborough, beans; pease for pigs; white clover.
Mr. CUBIT, of Honing, ploughing in buck-wheat; marle.
Mr. MARGATESON, of North Walsham, feeding and mowing clover; dibbling; marle.
Mr. PETRE, of Westwick, South Down sheep: beautiful improvements.
Mr. DYBLE, of Scotter, tillage; ploughing in green crops; good management.
Mr. PALGRAVE, of Coltishal, improvement of meadows; manuring; drilling.
Mr. REPTON, of Oxnead, courses of crops; products; drilling; steaming roots; Leicester sheep.

Mr.

Mr. REEVES, of Heverland, Norwich muck; improvement of wastes; drilling; threshing-mill; South Down sheep.

Mr. BIRCHAM, of Hackford, excellent management in the old school.

Colonel BULLER, of Haydon, steam-engine; drilled barley; enclosing.

Mr. JOHNSON, of Thurning, draining; improvement of meadows; drilling; tillage; South Down sheep.

Mr. ENGLAND, of Binham, drilling; South Downs.

Mr. REEVE, of Wighton, irrigation; draining; drilling; Leicester sheep.

Mr. H. BLYTHE, of Burnham, drilling; South Down sheep; sainfoin; white marle.

Mr. DURSGATE, of Summerfield, drilling; course of crops; manuring; South Down sheep.

Mr. RISHTON, of Thornham, drilling; South Down sheep; sea-weed.

Mr. STYLEMAN, at Snettisham, drilling; improvements; threshing-mill.

Mr. GODDISON (for Lord CHOLMONDELEY), manuring for wheat; marling.

Captain BEACHER, drilling; Wiltshire sheep.

Mr. BECK, of Riseing, South Down Sheep; drilling; marling; manuring; threshing-mill.

Mr. BENTINCK, a noble embankment.

Mr. SAFFORY, of Downham, fen-management.

Mr. PORTER, of Watlington, drilling; beans; manuring; course; Leicester sheep.

Mr. MARTIN, of Totterhill, drilling.

Mr. ROGERSON, of Narborough, drilling; manuring.

Mr. BURROUGHS, of Wymondham, good management.

Mr. PRIEST, of Besthorpe, drilling; lucerne; chaff-cutting.

Mr. TWIST, of Bretenham, drilling; Norfolk sheep.

Mr.

Mr. GALWAY, of Toffts, irrigation.

Mr. BEVAN, of Riddlesworth, considerable efforts in irrigation; marling; and various improvements; South Down sheep, &c.

SECT. II.—RENT.

THE circumstance which makes the rent of a county an object of any interest in an Agricultural Report, is its being a confirmation of the descriptions which are given of soils; the terms used in defining land, rich, poor, mixed, strong, light, &c. must in many cases be vague; but when sand is noted at 2s. 6d. an acre, or *rich* or *strong* at 20s. to 30s. the reader has more explicit information. When a man is told that sand produces six or seven quarters an acre of beans, the fact does not convey any knowledge; but if it be added, that the rent is 30s. an acre, it becomes easy to guess what the soil is. The minutes, however, of rent, are not numerous: on many occasions it is an inquiry prudently shunned, and on some, it is not an object of consequence.

Hundred of Shropham.—Two-thirds of Shropham hundred, containing 20 parishes, are sand, at 5s. an acre. The other third better land, at 12s.; average of the whole, 7s. 4d.

Around Attleborough, 20s. to 35s. Besthorpe, 23s. Shropham, 20s. but high let.

Wayland.—About Watton, small enclosures, up to 40s and 50s. an acre: all the country round, that is not light, 20s.

Gilcross.—Snare-hill, June 20, 1734. Valuation of stock: 15 neat beasts, 3 cart mares, 8 acres of grass, and all the other grass, and all the corn of the farm, 101l. 4s.

(Signed)   JOHN WARD.

In 1802, this estate sold for 15,000l. Memorandum with which I was favoured by Mr. SALTER, of Winborough.

Grimshoe and South Greenhoe.—Thetford to Swaffham, 7s. 6d. Some 3s. 6d. and 5s. Warrens, 5s. Lands that have been clayed, 7s.; if not clayed, 5s.

Laundich.—One of the most extraordinary instances of advance, is that of the farm at Rougham. Within the memory of many persons, it has been advanced from 500l. a year to 2500l.

Diss.—The whole of this hundred is good land, and much of it strong. At Thelton, in general, about 20s.; some higher. Rents in this hundred not raised more than 15 per cent. of late years.

Earsham and Depwade.—The former of these 20s. and not dear at that rent. Depwade equally good; but being further from navigation, is in reality worth 2s. or 2s. 6d. an acre less on that account. Rents have not been raised more than from 15 to 20 per cent. of late years.

Clavering.—The whole consists of rich land, rented at 20s. and cheap. Gillingham, 18s.; Stocton, 20s.; Alderby, 20s.; Whitacre better still.

Hensted and Loddon.—Both good, and let at near 20s. All the country for many miles around Langley, raised one-third in 25 years; and a great deal doubled. In that parish, 600 acres newly enclosed, at 12s. The river Yare bounds both these hundreds to the north; on its bank a line of marshes. At Wightlingham these are neglected, and do not let for more than 20s. an acre. At Surlingham and Rockland they are boggy, and let at about 8s.; but to Rudham-ferry at 20s.; thence, to Yarmouth, 23s. and to 30s.

Forehoe, Humbleyard and Mitford.—These are in the district

district of various loams, and let on an average at from 15s. to 20s. an acre.

For five miles around Wymondham, five years ago, the rents were not more than 20s.; now from 18s. to 25s.

From Attleborough to Hingham, 20s. to 35s. an acre. Two miles around Hingham, 18s. to 20s. Three miles around, 15s. or 16s.; but, if clayed and drained, 20s. to 25s.

East and West Fleg—These hundreds are famous for their excellent soil; and let on an average from 25s. to 27s. an acre; some up to 42s.; at which rate Mrs. MAPES, of Rollesby, has let some.

In the opinion of Mr. FERRIER, they have doubled in 20 years.

Happing.—Great tracts let at from 25s. to 30s.; and the whole at above 20s. on an average. The northern part of the hundred contains some of the finest soil in the county. But there is in other parts of it much low land, fen and commons enclosed, and some let in 1802, for the first time; the worst worth 12s. an acre, the better sort, 20s. The Bishop of NORWICH has let some as high as 26s.

Happsborough, Walcot, and Bacton, I have heard in conversations, valued too high to note: much lets at 30s., and well worth that rent. Sandy loam on a clayey-loam bottom; but sound.

Tunsted.—About North Walsham, raised one-third in 20 years. Westwick and its vicinity, 20s. to 25s.; but there is some much poorer. Scotto, 20s. to 25s.; some at 15s. The best soil in the hundred is in the maritime part.

North and South Erpingham.—Much of North Erpingham is in the better sand district. In the southern part of it, much at 20s. In South Erpingham great tracts very rich, and as high in rent. Around Coltishal, 16s.

Taverham

Taverham.—Consists of various soils, and varies much in rent: some up to 20s.; and some as low as 5s.

Horsted, Belough and Wroxham, 10s. Raised one-fourth, and some one-third, through most of this hundred in 20 years.

Blowfield and Walsham.—These hundreds are in the rich district of East Norfolk, and are in general let (with local exceptions) from 20s. to 25s. an acre.

Eynsford.—The whole of this hundred is included in the district of *various loams.*

For some miles around Reepham, Hackford, &c. rents 20s.; they have been raised one-fourth in 20 years.

Much grazing land at Gestwick lets at 30s. to 36s.; some at 20s.

Holt, North Greenhoe, Gallow, Brothercross, Smeethdon, Freebridge.—These hundreds are in the district of *good sand*, except only the marshland part of Freebridge.

From Holt to Burnham, 10s. to 20s. an acre.

Thirty-five years ago, I registered the rent from Holkham westward to Snettisham, and southward to Swaffham, at 2s. 6d. to 6s. per acre; some farms then newly let, at 10s.

Chosely, 10s. to 17s. tithe and rate free. Thornham, 21s. Some from Thornham to Snettisham, 8s. 10s. 12s.; much at the latter place, 20s. Five miles round Houghton, 8s. to 15s. Hillingdon, &c. 8s. to 16s.; average 10s. The rich level of marshes to the south of Lynn, and east of the Ouse, 2 guineas an acre; the acre something more than three roods.

The new *intakes* from the sea, in Marshland, by Governor BENTINCK, 2l. 2s. to 2l. 12s. 6d. an acre. The old lands 30s. Marshland in general, 28s. the statute acre; but much, near 30s. the short acre. It let 120 years ago at 15s.

Clackclose.—A considerable part of this hundred is fen;

much

much of it poor sand; but the rest is good, and lets at 17s.; some at 20s.

East of the Ouze, at Downham, Stow, Wimsbotsham, Crumplesham, Bexwell, Ruston, &c. 18s. Rise in seven years, one-fourth.

Recapitulation.—The light sand district, as marked on the Map, I conceive, lets, on an average, at 6s. an acre.

The various loams at 16s.

The better sand, 12s.

The rich loam, 26s.

The Marshland clay, 28s.

SECT. III.—TITHES.

So much has been written on the great national question of tithes, and their commutation, that any general observations are unnecessary. All that can with propriety be inserted here, are the notes taken of the compositions per acre.

At Harleston, 5s. an acre. In the hundreds of Loddon and Clavering, 5s.; some 6s.; and even 7s. an acre. Average of Fleg hundreds, 6s. marsh excluded. At Hemsby, taken in kind. Martham, &c. 7s. Happing hundred, 4s. to 6s. In Ludham, great tithe, 4s. 6d.; small, 1s. 6d. At Catfield, 6s. At Sutton, 5s. Some parishes more, few less. At Honing, great, 4s. 6d.; small, 1s. 6d. About North Walsham, 4s. to 6s. East Ruston and Happsborough, 7s. Around Westwick, 5s. At Oxnead, 4s. Heveringland free. At Causton, arable, 4s.; ordinary meadows and pastures, 1s. 6d. At Reepham, great and small, 4s. At Thurning, 3s. At Briston, sold by auction; and the buyer

buyer gathers. At Dawling, 3s. Binham, &c. 3s. to 4s. an acre. Snettisham, and the parishes around, average 4s. an acre. Houghton, &c. 4s. Gathered at Hillingdon: 2s. 6d. to 5s. the common payment in the vicinity. At Snetterton, and the neighbouring parishes in general, about 3s. to 4s. an acre. At Attleborough all gathered; and they gave this as a reason for not enclosing their immense commons. At Hingham, 4s. an acre. At Watton taken in kind. At Carbrook 5s. an acre. At Gursston 5s. and 1s. 6d. the vicarial.

Waterden something under 3s. per acre, grass included: in some parishes, 3s. 6d.

In the parishes around East Bilney, 3s. and 3s. 3d. an acre, on an average. At Goodwick, &c. 3s. and 3s. 6d. Some so high as 6s. said to be known near Holt.

At Terrington 4s. 6d. At Walpole 6s. arable; 1s. 6d. grass. Parishes around Downham 3s. At Wymondham 4s. 6d. Carleton 4s. Bunwell 4s.

North Walsham, 1782, rectorial 2s. 9d. and vicarial 1s. an acre, all round. North and South Reps about 3s. an acre for both. 1770, at Runcton, 20d. an acre round.

GENERAL AVERAGE of 37 minutes, 4s. 9d. per acre.

SECT. IV.—POOR RATES.

IT is to be regretted that returns similar to those made pursuant to an act passed in the 16th year of His present MAJESTY, have not been lately called for by Parliament. If the amount of the rates throughout the kingdom, during the late scarcity, were known, the necessity of some new system of provision for the poor would appear in a light so important, that measures would probably be had recousre course to, for preventing in future a return of similar burthens: not by way of lessening the comforts of the poor, but for increasing them; the grand objection to the present support being its insufficiency to answer the purposes for which it is given: the burden has been enormous, and the poor not provided for, by means which would produce in them industry and economy.

Snetterton in general 3s. in the pound. In Harfham, Wilby, Larling, and Eccles, in about the same proportion. The scarcity was met by different exertions. At Attleborough 3s 4d. for all parochial. At Hingham, in an incorporated hundred, 8s. to 10s.; once in the scarcity 14s. in the pound. At Watton 5s. 6d. land at rack-rent; houses at 3-4ths. In North Barsham, 4s. in the pound.

At Goodwick, in 1802, 2s. in the pound. At Repham 5s. in the pound, rated at 20s. an acre; some at 15s. or 16s.

The hundreds of Loddon and Clavering are incorporated; their house of industry at Heckingham, which has answered greatly. Five years ago, Mr. BURTON, of Langley, paid 16l. a year: the scarcity doubled, and even trebled it; but now it is reduced to 24l. and will come down again to 16l.

At Foncet and Tackleston, rates were 14s. in the pound, and now are 9s. to 10s. Edgefield 24s. in the pound, for two years running, now 16s. and the land not worth 20s. an acre.

The Fleg hundreds incorporated: rates no where high: about 3s. in the pound. Martham, &c. 2s. to 3s. an acre.

Happing and Tunsted hundreds incorporated; 41 parishes: income above 4000l. a year; rates in the scarcity rose, but they are now down again to 2s. in the pound; but not at full rent; about 2s. an acre: in winter between 3 and 400 in the house; in summer between 2 and 300. At

POOR RATES.

At North Walsham, not included in the incorporated hundred, 4s. in the pound; Westwick 1s. At Scotto, last year, 16s. in the pound, rack-rent: now 10s. Coltishal 1s. 6d. in the pound.

		s.	d.
Hevingham—			
Easter to Michaelmas 1795		6	6
To Easter 1796		7	6
The year		14	0
To Michaelmas 1796		10	0
To Easter 1797		6	6
		16	6
To Michaelmas 1797		4	9
To Easter 1798		5	6
		10	3
To Michaelmas 1798		4	9
To Easter 1799		5	6
		10	3
To Michaelmas 1799		5	9
To Easter 1800		13	6
		19	3

		£	s.	d.
Easter 1800 to July 1800		0	9	0
to Oct. 1800		0	6	6
to Jan. 1801		0	7	6
to Easter 1801		0	7	6
		1	10	6

Acres — — 1504
Rent — — — 994 ⎫
Tithe — — — 211 ⎭ 1205 l.

And

And this enormous rate on a new assessment, including tithe: the addition in the whole 35ol.

	l.	s.	d.
To July 1801	0	6	9
To Oct. 1801	0	6	9
To Jan. 1802	0	3	6
To April 1802	0	3	9
£.	1	0	9
April 1802, Rental assessed	1307	9	6
Disbursed	1360	9	8

The common enclosed was 1000 acres; it fed the cows of the poor, but they were greatly distressed to get winter food.

Total population	598
Deduct, not belonging to the parish,	83
	515

Disbursements 1360l. or 52s. a head.
Buxton rates 45s. in the pound, at 2-3ds rent.

	l.	s.	d.
Masham, one quarter	0	12	6
another	0	9	0
half year	£. 1	1	6

A new valuation, including tithe.

Before the scarcities, the rates at Causton were 4s. or 5s. in the pound. In the scarcity, 11s. or 12s. From Lady-day to Midsummer 1802, 1s. 6d. at rack-rent.

At Reepham and Hackford, 5s. an acre. At Thurning 5s. 6d. At Holt 10s. in the pound. At Burnham Westgate, from 1790 to 1800, 2s. 6d. to 3s. in the pound for the year. In 1801 they were 8s.; and this year (1802) they

they will not be lower than 5s. At Thornham—Easter 1797 to Easter 1798, 6s. 6d. in the pound, on half rent.

To Easter 1799, 4s. 6d.
To Easter 1800, 10s. 0d.
To Easter 1801, 17s. 6d.
To Easter 1802, 17s. 6d.
Rental 1265l.

At Holm, last year (1801), 9s. 6d. in the pound: this year 5s. 6d.

Rates, exclusive of scarcity, at Snettisham, &c. 4s. in the pound, rack-rent. In the scarcity, some were 7s.

In the parishes around Houghton, 2s. 6d. in the pound, on the average. At Hillingdon, 2s. in the pound on real rent. At Castle Riseing 1s. 6d. to 2s. in the pound; they were higher. At Lynn, they are now 10s. in the pound, and were lately 12s. and laid on stock; they would be near 20s. if on rent only: in the scarcity were 16s. besides great subscriptions. A gentleman in this town has paid 2l. 12s. in a year for poor-rates, and now (inhabiting the same house) 100l. At Terrington, in Marshland, 2s. in the pound: were in the scarcity 5s. to 6s. At Walpole, poor 2s. church 6d. surveyor 1s. in the pound. Dyke reeve 6d. Land-tax 1s. 4d. per acre.

Parishes around Downham 5s. in the pound. Besthorpe, last year, 10s. in the pound, rack-rent: now 5s. to 6s. Carleton 8s. last year.

RECAPITULATION.

	l.	s.	d.
Snetterton, in the pound - -	0	3	0
Harfham - - -	0	3	0
Wilby - - - -	0	3	0
Larling - - - -	0	3	0
Carry forward, -	0	12	0

Eccles

POOR RATES.

	l.	s.	d.
Brought forward, -	0	12	0
Eccles - - -	0	3	0
Attleborough - - -	0	3	4
Hingham - - -	0	9	0
Watton - - -	0	5	6
N. Barsham - -	0	4	0
Goodwick - - -	0	2	0
Reepham - -	0	5	0
Foncet - - -	0	9	6
Tackleston - - -	0	9	6
Edgfield - -	0	16	0
Fleg Hundred - -	0	3	0
Martham - -	0	2	6
Happing - - -	0	2	0
Tunsted - -	0	2	0
N. Walsham - -	0	4	0
Westwick - - -	0	1	0
Scotto - -	0	10	0
Coltishal - - -	0	1	6
Hevingham - -	1	0	9
Causton - - -	0	5	0
Hackford - -	0	5	0
Thurning - - -	0	5	6
Holt - -	0	10	0
Burnham Westgate -	0	5	0
Thornham - - -	0	17	6
Holm - - -	0	5	6
Snettisham - -	0	4	0
Houghton - - -	0	2	6
Hillingdon - -	0	2	0
Castle Riseing - -	0	1	9
Lynn - -	0	10	0
Carry forward, -	9	19	4

Terrington

			l.	s.	d.
Brought forward,	-		9	19	4
Terrington	-	-	0	2	0
Walpole	-	-	0	2	0
Round Downham		-	0	5	0
Besthorpe	-	-	0	5	6
Carleton	-	-	0	8	0
			£.11	1	10

Average of 40 minutes, 5s. 6d. in the pound.

SECT. V.—LEASES.

The great improvements which for 70 years past have rendered Norfolk famous for its husbandry, were effected by means of 21 years leases; a circumstance which very fortunately took place on the first attempt to break up the heaths and warrens in the north-west part of the county. These leases established themselves generally; and were, more than any other cause, powerfully operative in working those great ameliorations of wastes which converted that part of the county into a garden.

To explain generally the necessity of long leases, would at this time of day be an idle disquisition. I never heard any arguments against them which carried the least weight. Exceptions may, and will occur: in lands which are immediately around the mansion, it may be prudent to grant short tenures; and when a landlord is willing to take upon himself all those expenses which a tenant submits to merely because he has a term of 21 years, it is obvious that there is no *necessity* for a long lease; but, in general, it may be held for sound doctrine in Norfolk, that an estate can neither be improved, nor even held to its former state of improvement, without long leases.

Sorry

Sorry I am to perceive, that contrary ideas seem to be gaining ground in this county; that some landlords will give no leases, and others only for 7 or 9 years. That the agriculture of the country will suffer in proportion as these ideas prevail, I have not a doubt; and it is a very fortunate circumstance, that Mr. COKE, the possessor of the largest estate in it, adheres steadily to those principles which improved his noble property, never giving a shorter term than 21 years.

The views of landlords who act otherwise may easily be conceived; they have a quicker return of those opportunities for advancing their rents than occur with longer terms; and the late scarcities, among their other evils, have added much to this. The tenants' profits (supposed to be greater than in fact they were), glittered in the eyes of landlords, who were apt to think they had not a fair proportion of the product. But if such temporary fluctuations are to have weight in regulating the rent of land, the medium short of a *corn rent* will be difficult to find; and no leases at all are likely to be the consequence: what such maxims would produce *in Norfolk*, are easily conceived.

But in the main object of raising rents, confining myself to the county I treat of, I have great doubts whether an estate, in 43 years, will not be let for much more after two leases, than after six. Every sort of improvement, and what is of as much consequence, the common course of the husbandry, in points which no covenants can touch, will tend to improve the land in one case; while, in the other, the tenant will look to the duration of his term before he spends a shilling, or gives an order for a cart or a plough to move. The silent operation of such a constantly influencing motive, will gradually affect the farm in a manner that must be severely felt; and is a

perfect

LEASES.

perfect contrast to the spirit of animated exertion which pervades every part of the farmer's business, when he looks forward to a long period for his remuneration. The particular notes I made on the subject are few, but merit insertion.

There are more seven years leases about Holt than of any other term.

Mr. STYLEMAN, at Snettisham, gives leases of 10 or 15 years, of lands in the five-shift husbandry; that is, for two or three courses; and for 12 years, in the four or six-shift course.

Mr. M. HILL occupies two farms, one (Waterden) under Mr. COKE, on a 21 years lease, and another (Barsham) from a relation, on a 7 years tenure; the former land a lightish sand or gravel; sandy and gravelly light loams: the latter, a deep rich friable sandy loam, on marle, very fine land, and far superior to the Waterden soil; but he has upon it crops at least equal to the soil; fine corn, that is in its appearance to his credit as a farmer. I was much surprized to find the crops on the Barsham land very inferior; by no means equal to the soil: some very good barley; but some inferior, and no wheat comparable to the crops at Waterden. In such cases, I always expect to find some cross-cropping has been tampered with; and it turned out just so—wheat after barley; barley after wheat, &c. The circumstance is applicable not so much to the subject of courses of crops as to leases: it deserves the attention of landlords; for they never refuse 21 years leases without their farms suffering. It is true, the farmers suffer also, and nine times in ten lose by their calculation.

The New Covenants in Letting the Farms of T. W. COKE, *Esq. M. P.*

Supposing a farm to contain 540 acres arable land:

Shall, and will at all times, keep and leave ninety acres, part of the arable land, laid to grass of one or more years laying. Also ninety acres grass of two or more years laying—each to be laid down with a crop of corn, after turnips, and to continue laid two years at least; the time of laying to be computed from the harvest next after sowing the said seeds; and upon breaking up the same*, after January 1st, 1804, may be permitted to sow forty-five acres (part thereof annually) with pease, or tares, for seed, to be twice well hoed: other part thereof with tares, for green food, buck-wheat, or any leguminous or other vegetable plant, for ploughing in as manure, or summer tilling any portion of the remainder.

Shall not sow any of the lands with two successive crops of corn, grain, pulse, rape, or turnips, for seed, (except the above-mentioned pea and tare stubble), without the leave or consent of the said ——, his heirs, or assigns, being first had and obtained in writing.

Lands for turnips, four clean earths at least.

The turnips covenanted to be left in the last year, ninety acres to be mucked, so far as the same will extend, and to be paid for by valuation; at the same time a due regard to be had to the cleanness of the land upon which they grow.

Sheep, cattle, and all other live stock, to be lodged

---

* The Land intended to be sown with pease should not be till 4½ years after the commencement of lease, upon supposition that *new tenant* may not be so situated as to have the turnips (covenanted to be left by old lease) completely clean.

upon some part of the premises, when consuming the produce of the farm.

Straw, chaff, and colder, to be left without allowance.

Incoming tenant to carry out the crop of corn, not exceeding the distance of ten miles, gratis.

Rent payable forty days before St. Michaelmas (wherever a threshing machine is, or shall be erected), if demanded, by notice in writing being left at the farm-house to that purpose.

CHAP. V.

IMPLEMENTS.

For more than half a century, the implements of Norfolk remained without alteration or addition; but of late years many and great improvements have been introduced.

PLOUGHS.

The common Norfolk wheel-plough varies from other wheel-ploughs in three circumstances: 1. By the high pitch of the beam. 2. By the wheels being, when in work, brought so near to the point of the share. 3. By the general lightness of the tool, when compared with some others.

The Hertfordshire plough has a very long beam, with a low pitch, a circumstance thought in that county to be essential to steadiness of draught; and when Mr. Arbuthnot made a wheel-plough for gaining 18 inches depth for the culture of madder, he adopted a long beam, as essential to a great power.

Whether the second circumstance in the Norfolk plough, that of the points in the periphery of the wheels which touch the ground, being so near to the share point, is really a benefit, remains a question, and greatly merits experiment to ascertain.

The general lightness of the plough is probably a merit, when the work is easy; but when we find it not an uncommon practice to load the body of the tool with a great stone,

stone, to keep it steady, doubts will suggest themselves, that the mathematical construction is erroneous.

But let me recur to the notes.

I found wheel-ploughs common through Earsham, Loddon and Clavering hundreds, but a sprinkling of short swing ones, such as are in general use, about Thetford, Brandon, &c.

In discourse with Messrs. THURTELL and EVERIT, near Yarmouth, they both insisted on the propriety of having the share point and the wheels as near together as possible, as the ease of draught depended much upon it; but admitted that a greater distance would make the plough go steadier, in case of difficulties.

Examining the ploughs of Mr. FRANCIS, at Martham, from perceiving the beams not mounted so high as in common, he remarked, that the wheelwright made his upon his own plan: he has ploughed much with his own hands, and knows that when they are very high, the plough is apt not to cut a flat furrow, nor to go close at heel, he therefore lowers the beam, and the share is two feet from the points where the wheels touch the earth; and the beam-ring being in the centre hole, the plough will then go alone without holding.

Mr. JOHNSON, of Thurning, has his beams lower than common, and rather further from the share point to the wheel points of contact with the ground; nor does he find that his ploughs are at all less easy in the draft.

Mr. ENGLAND, of Binham, thinks that the nearer the share is to the draft, the easier for the horses; the reason for mounting the beam.

Mr. REEVES, of Wighton, is of the same opinion, and that the points of wheels and share being near, does not make the plough go unsteady. When he wants to *whelm* a layer well, he uses ploughs with the *plat* rather longer,

and

and the beam a little longer, as well as the share point a little farther from the wheels.

Mr. HILL remarks, that there are three sorts of plough-wheels used in West Norfolk; first, all of cast-iron; second, wooden boxes, spokes, and fellies shod with iron; third, wooden boxes and spokes, with rims of hammered iron. The first are good and safe while at work, but very apt to break as they move through stony lanes; the third are light, and do well in dry weather; but the second are the most durable; he thinks they last out ten sets of the cast-iron, and two sets of the iron-rims. He uses the cast shares of Messrs. GURNEY and Co.; No. 6 of these he much approves. Cast-iron wheels are 10s. 6d. a pair, wooden ones 30s.

They do not use wheels of unequal height in Norfolk, common in Hertfordshire, &c. because in one-furrow work (a species of half-ploughing), the wheels would be reversed, the high one run on the land, and the low one in the furrow, as the plough turns the contrary way.

The line of traction, from the tuck of the collar to the heel of the plough, passes through the axletree.

The ploughs are a foot wide at the heel.

In the south-west angle of the county, and from Wymondham, swing-ploughs only are used. On different farms I followed several of them, most of which were ill constructed; they *ride on the nose*, to use the farming term, not going close at heel, a defect arising from the shortness of the beam.

It would be improper to omit noticing the high opinion which the great agriculturist, Mr. COKE, has of the Norfolk wheel-plough; which goes so far as to induce him to be always ready to bet it against any other, on any soil; and he has in several trials been successful. But for want of minutes being kept of such trials, and, above all, for

for want of the force exerted being accurately ascertained, these experiments have not been attended with the conviction which might have been the consequence.

From the preceding minutes it appears, that some doubts have actually been entertained, by very able and practical farmers, on the usual structure of this plough; and Mr. REEVE, varying the form in the length of beam and the mould-board, when he wants to *whelm* the furrow well (that is, turn it completely over), confirms the propriety of our considering the structure of this plough as not well ascertained.

In addition to this observation I have to remark, that in passing through almost every part of the county, I never omitted any opportunity of following ploughs at work, and noticing their steadiness and other circumstances; and I remarked two very deficient points—they do not *generally* go close at heel; and when I desired the men to quit their hold, and let the plough go alone, not one in twenty would do it, even for a single rod; now, a *wheel-plough* must be badly constructed that will not stand this trial, which, for a short distance, is perhaps the best criterion of a swing-plough.

I have an high opinion of the Norfolk plough, when well constructed, and offer these remarks merely to instigate the gentlemen of the county who take any pleasure in rural mechanics, to ascertain these circumstances by experiments which are not difficult to make, and would tend powerfully to give the wheelwrights and blacksmiths more certain rules to work by, than they possess at present.

HARROWS.

Mr. JOHNSON, of Thurning, thinks that it is common to put too many teeth in harrows: he chuses to have his harrows

harrows *snatch* in moving; and always trots the horses when finishing turnips.

Mr. M. Hill, of Waterden, has improved the light harrows of the country, by making them three four-baulked (or rows of teeth), instead of two six-baulked; the division fits them better to the lands. The teeth all round, as he thinks that square teeth gather more as they move, and impede the work. He inclines the position of the teeth to the iron hooks, by which the whipple-trees are fastened.

ROLLER.

Mr. Priest, at Besthorpe, uses a roller divided in two parts, rising and falling in the centre, for rolling the slopes of ridges. I have seen the same useful tool at the Rev. Mr. Hill's, in Suffolk.

Mr. Coke has the most powerful roller for grass-lands I have seen: it was cast at the Carron foundery; it is 5 feet 6 inches high, and 5 feet 6 inches long; weighs $3\frac{1}{2}$ tons, drawn by 4 horses, and cost 60l. It leaves the surface of grass-land in the order it ought always to be in.

WAGGONS.

Mr. Denton, of Brandon, has found a considerable saving by the use of light caravan waggons for two horses abreast, with which he carries a chaldron and half of coals, and other loads proportioned. Every man who reduces the teams of any country, will be sure to do this till he arrives at perfection in a one-horse carriage.

Very few waggons are used in Fleg, except for road-work; chiefly carts and *wizzards*.

CARTS.

Mr. Overman, of Burnham, made an improvement in his carts, of beautiful simplicity: instead of the toe-stick, as in the common, drawing out to let the buck tilt up, and deliver the load, it turns in the centre on a pivot, and the hooks which confine it at the ends, being each in a position the reverse of the other,

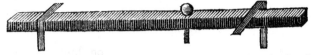

the least motion laterally, frees it, and permits the buck to rise.

A thought of no small value also, is that of chaining the tail-board to the cart. It is not uncommon elsewhere to see the carter, in unloading, leave his board, and have to go many steps for it; not to mention carelessness in mislaying, and time lost in searching.

Yarmouth, from the singular arrangement of the streets, will not admit common carriages for the domestic purposes of the town. It consists of two large streets, but all the cross-ones (called *rows*) are so narrow, that a cart more than 3 feet or 3½ broad, cannot enter them. Necessity, therefore, has happily driven the inhabitants to the best of all vehicles for carriage, those of a single horse or ass; they are not well constructed, as too great a weight rests on the back of the animal, from the load being all *before* the axletree: however, the loads they carry are considerable; seven coombs of wheat are a common load; a hogshead of sugar the same; one man and cart have moved ten score of corn in a day.

DRILL

DRILL ROLLER.

I have at different periods made many inquiries for the inventor of this tool, but could not ascertain it thirteen years ago. Mr. SILLIS, of Hartford Bridge, near Norwich, was mentioned to me as a person who had improved it. It is a cylinder of iron, about seven feet long, around which are cutting wheels of cast-iron, that turn, each independently of the others, around the common cylinder, weighing from a ton to $1\frac{1}{2}$, drawn by four horses, and heavy work. The cutting wheels, being moveable, may be fixed by washers, at any distance, commonly at four inches. By passing over a fresh-ploughed layer, the soil is cut into little channels, four inches asunder; the seed is then sown broad-cast, and the land bush-harrowed in the direction of the drills; thus the seed is deposited at an equal depth. GEORGE Earl of ORFORD gave the Writer of this Report one, but the soil was too heavy for it: for breaking clods in a dry season, no tool I ever beheld comes near to it.

They are much in use in Loddon hundred. Mr. BURTON, of Langley, put in a great deal of corn thus, and approves the method so much, that hitherto he has drilled little; but thinks dibbling a vast improvement.

The implement was more commonly used in the county ten or twelve years ago than it is at present, for the drill machine has been adopted by many who formerly had a good opinion of this tool.

DRILL MACHINE.

Mr. COOKE's drill is very generally used in Norfolk, and I found it every where highly approved.

Mr.

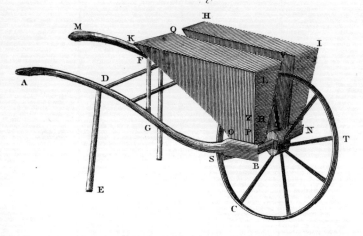
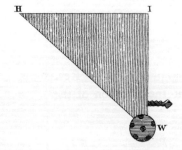
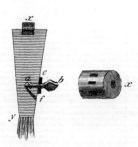

Mr. FARROW, of Shipdam, uses the drill of Mr. STANTON, of Weybread, the shares of which swing separately; but on his land it executes very badly; chokes at 5½ inches, and the delivery very unequal.

Mr. REEVES, of Heverland, has a drill machine made by Mr. ASHBY, of Blyborough, in Suffolk, which executes to his satisfaction: he has had it three years; price 5l. 5s.

The Rev. Mr. MUNNINGS, at Gorget, near Dereham, uses only the barrow-drill, which he had originally from the Rev. Mr. LINDSEY, of Lincolnshire, at present Lord Bishop of KILLALOE. He finds it excellent for all purposes, and especially adapted for little farmers, who are weak in horses. He has himself only 48 acres of arable land, and 14 of grass; by means of it, on this poor land, he gets 5 comb of wheat, 7 of barley, and 5 of pease; and once had 12.

FIXED HARROW.

This implement, newly invented by Mr. COOKE, for attaching to his drill machine, has been used this year (1802) by Mr. REEVES, of Wighton, who thinks it the best tool he ever saw of the kind.

HORSE-HOES, &c.

The following tools, invented or improved by the Rev. Mr. PRIEST, of Scarning, merit the attention of the reader.

No. I.

Is an attempt to delineate the *Barrow* I used to drop my wheat. [See *Plate* III.]

A B
M N } the two handles are 4 feet long.

D E, one

D E, one of the feet 13 inches.

The diameter of the wheel, viz. S T, 22 inches.

The dimensions of each hopper:

$\left.\begin{array}{l}\text{K Q}\\ \text{L V}\end{array}\right\}$ each $6\frac{3}{4}$ inches.

$\left.\begin{array}{l}\text{K L}\\ \text{or}\\ \text{H I}\end{array}\right\}$ $18\frac{1}{2}$ inches.

$\left.\begin{array}{l}\text{L P}\\ \text{or}\\ \text{I W}\end{array}\right\}$ 12 inches of which, x y, makes 4 inches; and the length of the hairs of the brush is one inch.

X Y, is the form of the brush, made moveable upon a hinge at Z, so that the hairs of the brush may act by pressing against a cylinder at the bottom of the hopper, and suffer more or less seed to drop from it, according as it is pressed against the cylinder by the screw a b. This screw is 3 inches long, fastened to the brush x y, and by acting through a female screw e f, fixed at the end of the handle A B, viz. at c, presses the brush against the cylinder.

The cylinder x is in length two inches; its diameter $2\frac{3}{4}$ inches. It has twelve cups indented in its surface; each cup $\frac{1}{4}$ of an inch in diameter. The cylinder is fixed upon the axis of the wheel, and revolves with it at the bottom of the hopper, so as to carry the wheat out of it by the cups on its surface.

The hopper is supported upon the handle A B, at P O, and by the stay F G.

No. II.

Is an attempt to delineate my *Double Barrow*—[See *Plate* IV.] It is only the Barrow No. I. with two wheels instead of one; the two wheels and the axle-tree are

Plate IV
to face Page 60.

The Brush

a b c

A View of the Cylinder

End View of the Cylinder

View of the Rev.d Mr Priest's Double Barrow.

Plate V.
to face Page 61.

The Rev.d Mr Priest's Scuffler.

Fig. 1

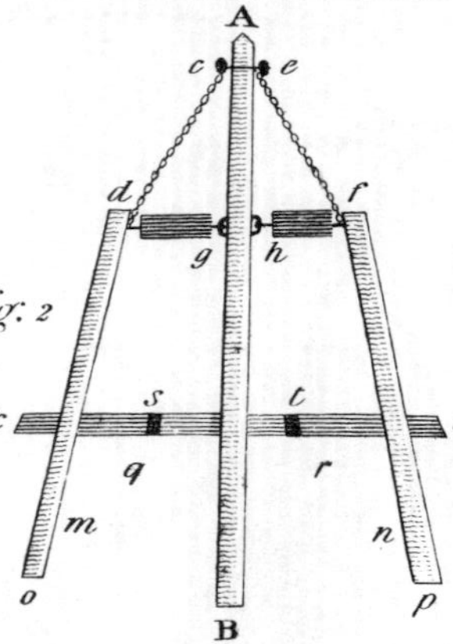

Fig. 2

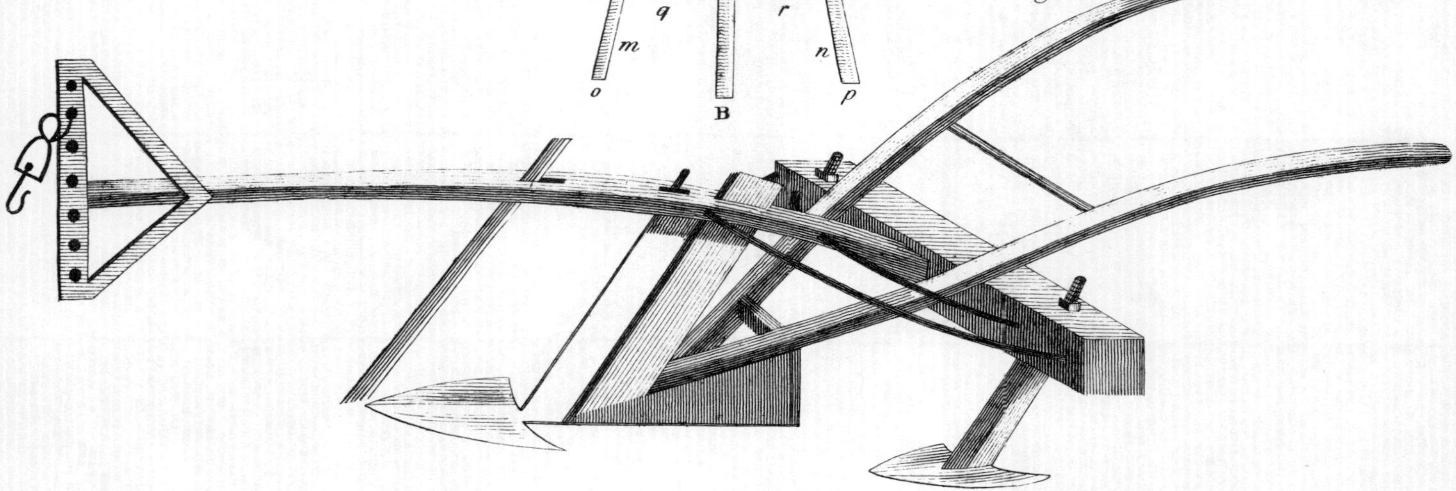

Neele sc. 352 Strand

IMPLEMENTS.   61

are united like the wheels of what are called *Yarmouth carts.*

a b c are different views of the springs A B, against which the lids of the hoppers fall, and are fastened; a is a side view, b the back, and c a front view. This barrow, and No. I. I formed myself from a single barrow, which I saw when I accompanied my friend MUNNINGS into Lincolnshire, to visit the Rev. and Hon. Mr. LIND-SEY.

Note.—The flat piece of iron k l, lies parallel, and the plates d g, f h, perpendicular, to the horizon.

No. III.

My *Scuffler* [See *Plate* V. *Fig.* 1.] is an instrument formed from a double-breasted foot-plough: thus—I took off the breasts of the plough, and had a share larger and flatter than the original share made. I then fastened, at the end of the beam of the plough, a cross beam of wood, 3 feet long, 4 inches broad, 4 inches thick, and at the distance of $12\frac{1}{2}$ inches each way from the centre of this cross-beam, inserted two coulters, each 12 inches long, 3 inches broad, and $\frac{3}{4}$ of an inch thick on the back, but reduced to 3-8ths in front; and into these coulters, at the bottom, I rivetted two shares, of nearly the same size as the first share, which was 9 inches broad, but these two only 8 inches. The cross-beam I strengthened by two iron reins, or bars, fixed to the cross-beam, and also to the beam of the plough, thus:

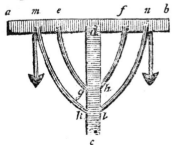

c d, a part

c d, a part of the beam of the plough.

a b, the cross-beam.

e g
f h
m k
n l
} iron reins, or bars to strengthen the cross-beam.

The coulters which are fixed into the cross-beam, do not stand perpendicularly, but inclining, as the coulter at the head of the plough, thus they are fixed into the cross-beam by a screw, a b, and a nut, c d.

The advantage of this scuffler above any that I have seen is, that it is used with *two* horses only. It does the work of more than two ploughs, as the three shares cut nearly the width of 30 inches, whereas two ploughs would cut only 24 inches. My farm consists of heavy land, as well as land of a mixed soil.

No. IV.

Next I am to describe my *Hoe-Plough*, which I formed purposely to scarify and hoe wheat on four-furrow work. At the time when I visited Mr. LINDSEY, I saw a double mould-board foot-plough, which he sent to Mr. MUNNINGS, and from it I had one made like it. Upon the beam of this *hoe-plough* I have fixed two scarifiers (or hoes, as I want them), [See *Plate* V. *Fig.* 2.]

A B represents a part of the back of the beam of the plough; A the head, and B the tail.—c e is a small iron bolt, with a nut and screw at one end, fixed into a cap upon the beam, to be taken out at pleasure. At each end of this bolt is fixed a chain, c d, e f, each two feet long, and hooked to plates of iron, d g, f h. These plates, d g, f h, are flat, and 8½ inches long, 2¼ inches broad, and moveable upon staples fastened into the beam: upon the

ends

ends of these flat plates of iron are fixed two pieces of wood, d o, f p, parallelopipeds, 2 feet 3 inches long, $2\frac{1}{2}$ inches broad, and $2\frac{1}{4}$ thick; moveable at d and f by hooks (upon the wood of the same breadth as the iron plates), and eyes (upon the iron plates). The points q and r mark the distances to which the expanding mould-boards are opened, viz. about 10 inches from one to the other.

At m and n, distances beyond the points q and r, are cut grooves through the wooden parallelopipeds, for the scarifiers and hoes to be fixed in.

k l, is a flat piece of iron, 3 feet long, and 2 inches broad, $\frac{1}{2}$ an inch thick, passing through grooves in each parallelopiped, and in the grooves through which it passes are screws to fasten it. This iron does not communicate with the beam of the plough. At s and t, points on each side the beam, are hung weights, such as may be necessary to make the scarifiers or hoes do their work properly.

I do not describe my scarifiers and hoes, because they are nearly the same as are used upon all instruments of this kind.

RAKE.

The hand-wheel rake of Fleg is an excellent implement for both hay and corn: it is to answer the purpose of the common hay-rake; about four feet long in the rake, and the two wheels of nine inches diameter, so fixed that the teeth are kept in any position, at the will of the holder.

THRESHING MILL.

Mr. JOHNSON, at Kempston, was so obliging as to order horses into his mill, that I might view it. To me it appears to be one of the best I have seen; if not the very best of the larger machines. The movements are uncommonly smooth. It requires from six to eight horses, six men, and one woman; threshes, without any question, much cleaner than the flail, and, without any doubt, cheaper. To bring it to its present perfection cost Mr. JOHNSON much time, attention, and expense; but as he was determined to carry his point, he never stopped till it worked to his mind; and having completed it, the repairs have been quite trifling. He attributes the common complaints of their being always out of order, to original errors or inattention in the construction. The arrangement is excellent, for disposing of the chaff, colder, straw, and corn, at once, in their respective places, without any confusion or removals; and it takes up a very small part of a barn. It was built by Mr. Wigful, of Lynn.

Mr. DROZIER's, of Rudham; built by Messrs. WIGFUL, 1799:

	£
Machine - - - - - -	70
Blacksmith } including repairs - -	25
Carpenter }	25
Workmen's wages and board, 25 weeks, 4 men	105
Timber and deals - - - - -	40
	£. 265

Including a granary over the wheel, of 26 feet by 24.

The threshing-mill, with dressing addition, would have cost about 120l.

A repair now in hand will cost 10l. at least.

EXPENSE OF THRESHING.

	£	s.	d.
Four strong horses, at 5s. - - -	1	0	0
One boy to drive - - -	0	1	0
Carry forward - - £.	1	1	0

Three

IMPLEMENTS.

	£.	s.	d.
Brought forward	1	1	0
Three women, to hand and untie sheaves	0	2	0
One man with ditto	0	1	6
One man to feed the machine	0	2	0
One man clears the machine, and throws the foul corn to the riddle	0	2	0
One man to carry away straw	0	1	6
	0	9	0
Repairs and oil, calculated at 10 per cent. on 120l. 12l. At 60 days work	0	4	0
	£.1	14	0

The mill will thresh 40 comb of wheat in eight hours, 50 comb of barley, 60 of oats and pease, and threshes pease better than any thing. The same men, while the horses bait, dress the corn with the winnowing machine.

40 comb of wheat, 34s. at 10d. a comb.
50 comb of barley, 34s. at 8d.
60 comb of oats and pease, 34s. at 6¼d.

He is confident that the machine threshes much cleaner than any flails. Every man that has examined the straw, admits this without a shadow of doubt; and barley equally with any other sort. One objection however, is, breaking the straw for thatching; but this is none for cattle. It makes the chaff good; but leaves no corn in the colder—not a grain. He has not tried threshing with the mill any straw from flails; but that there is a saving, he has no doubt.

Common threshing was never less than 1s. per comb; may be reckoned for wheat, on an average 1s. 3d. for this

NORFOLK.] neigh-

neighbourhood; for barley, 6d. the general price, but 7d. the average now; oats and pease, 6d. On an emergency, there is a power of getting corn out much quicker than without a machine.

Mr. WHITING, of Fring, has a large threshing-mill, built by an engineer from Scotland, Mr. FORDYCE. It cost him 200l.; is worked by 6 horses; threshes 24 comb of wheat in the day, 55 of barley, and from 63 to 84 of oats. It has five beaters on the drum-wheel, and the fluted segment of a cylinder which covers the drum in two parts, with an unfluted plate between them, which is raised or sunk by a short lever; this as a guard against stones getting in. In another circumstance also it is singular; there is a long platform, with a rolling cloth bottom; the whole raised or sunk at pleasure, for delivering the corn, across the floor space of the barn, from the *goff* in which the corn is stacked, to the other end in which the mill is built; which saves much labour, and works to his satisfaction.

EXPENSE OF THRESHING.

	£.	s.	d.
Four men; one to feed, one to hand the corn, two at the straw, and one to drive	0	7	0
One boy at the straw	0	0	10
One woman	0	0	8
Six horses	0	12	0
	1	0	6
Repairs have been considerable, but 10 per cent. a large estimate, or 20l.: if it works 80 days, this is	0	5	0
	£. 1	5	6

This may be called 1s. a comb for wheat, 5½d. for barley, and 4d. for pease and oats. As to its performance, Mr.

IMPLEMENTS.

Mr. WHITING is well satisfied with it; no flails in common management equal it for cleanness, and with barley as well as all other sorts of corn.

The horse-wheel is upon a different construction from the common ones, working by a cogged-wheel of small diameter below, instead of above the horses; and the communicating spindle under their path; but Mr. WHITING says it is hard work.

Mr. STYLEMAN, of Snettisham, has a very capital mill, which cost 300l. It is worked by 8 horses, and threshes in a day of 8 hours 120 comb of barley, pease, or oats, and 80 of wheat. It is afterwards dressed in the hand machine.

EXPENSES.	£.	s.	d.
Two men to feed, alternately	0	4	0
One at the chaff	0	2	0
One at the straw	0	1	6
Four women	0	4	0
Two drivers	0	2	0
	0	13	6
Eight horses	1	4	0
Repairs, 10 per cent. 30l.: if 80 days work, it is, per diem,	0	7	0
	£. 2	4	6

The wheat is therefore threshed at 6d. per comb, and the other sorts of corn at 4¼d. As to clean work, it does every sort of corn far cleaner than the generality of tasking, and barley equally with the rest.

Mr. COKE, of Holkham, has a very large machine, which cost about 600l. Besides threshing, it grinds corn, works two chaff-cutters, and breaks oil-cake.

EXPENSES.

IMPLEMENTS.

EXPENSES.	£	s.	d.
Two men to feed - - - -	0	3	0
Three men to remove straw - -	0	4	6
One sacking the corn - - -	0	1	6
Three to hand the sheaves - - -	0	4	6
Two drivers - - - -	0	2	3
One looker-on, to oil, and see all in order	0	2	6
Eight horses - - - -	1	4	0
Interest of 600l. at 10 per cent. 80 days -	0	15	0
	£. 2	17	3

Its work has not been tried with sufficient accuracy with any thing but wheat, of which it threshes 64 combs a-day: it has sacked 13 combs in an hour; but no rule, as it was a mere exertion: 64 combs, at 52s. 3d. is 9¾d. per comb. The common price is 1s.; here, therefore, is a saving; no flails equal it for cleanness in every sort of corn; and it dresses better than any method whatever that has been yet tried. It occupies too great a space in one of the finest barns in England: it prevents the stowing of near 300 quarters of barley. The more of these machines I see, the more I am convinced of the saving that would attend the moveable stacks I proposed in the *Annals*.

Mr. BEVAN, of Riddlesworth, built one by a person from Leith, in Scotland: it cost 100l. The following is Mr. DAY, the bailiff's, calculation of its work, compared with common flails.

IMPLEMENTS.   69

TO THRESHING BY FLAILS.

	£.	s.	d.
Forty combs barley, at 7d. come to —	1	3	4
To dressing of ditto —	0	2	0
£.	1	5	4
Fifty combs oats, at 6d. — —	1	5	0
Dressing ditto, at 1s. per score —	0	2	6
£.	1	7	6
Forty combs rye, at 9d. — —	1	10	0
Dressing ditto, at 1s. per score —	0	2	0
£.	1	12	0
Forty combs wheat, at 1s. —	2	0	0
Dressing ditto —	0	2	0
£.	2	2	0

THRESHING BY MILL.

	£.	s.	d.
Forty combs barley take eight horses, at 2s. 6d. — — — —	1	0	0
Ten men, at 1s. 6d. each — —	0	15	0
To dressing ditto, five men one day	0	7	6
£.	2	2	6

Barley costs more by 17s. 2d.

	£.	s.	d.
Fifty combs oats take eight horses, at 2s. 6d.	1	0	0
Ten men, at 1s. 6d. each —	0	15	0
To dressing ditto — —	0	7	6
£.	2	2	6

Oats cost more by 15s.

IMPLEMENTS.

	£.	s.	d.
Forty combs rye take eight horses, at 2s. 6d.	1	0	0
Ten men, at 1s. 6d. each	0	15	0
To five ditto, at 1s. 6d. one day dressing	0	7	6
£.	2	2	6

Rye costs more by 10s. 6d.

	£.	s.	d.
Forty combs wheat take eight horses	1	0	0
Ten men, at 1s. 6d. each	0	15	0
Five ditto to dressing	0	7	6
£.	2	2	6

Wheat costs more by 6d.

The flails thresh much cleaner, as the thatching of the stacks proves; so that if this article were brought to account, it would go much further against the mill. Nor is here any thing reckoned for repairs, which have always amounted to from 6l. to 8l. a year: to which add 5l. interest of capital, and the worthlessness of bad threshing mills will be sufficiently obvious.

1802.—Mr. BEVAN has had the machine eight or ten years; and thinks the material object is the power of getting out a crop to sell for seed, or to employ the men and horses in a wet day; but is clear that if he had it not, he would not build one, unless he was certain of its threshing barley cleaner than his does.

Mr. FARROW, of Shipdam, has built one of these mills on his new farm at Sayham: Mr. WIGFUL, of Lynn, was the engineer: it is worked by four, five, or six horses, requiring also three men and three women and a boy. It threshes with that strength 40 combs of pease in a day; 40 of oats; 30 of barley; and 20 of wheat; but the dressing is not more than half effected, so that the common

mon machine is necessary to finish it, by which a man and woman will dress 60 combs a day. It threshes all grain very clean; barley as well as the rest; and in a year and quarter's nearly constant work, has not demanded any repairs. Yet upon the whole Mr. F. is not well satisfied with it. On examination, I found it very aukwardly placed: the delivery of the corn is in too confined a space, close to a wall, so that men must attend to take it away; the straw is thrown out against a door into a yard, when it cannot all be wanted, and no sufficient receptacle for the chaff: the appearance is as if the machine was hooked into a building, and not the building raised for the machine. One is never to see an end to ill adapted farm buildings.

Mr. GOOCH, of Quiddenham, has a threshing-mill built by BURREL, of Thetford. It works with two horses; and threshes all sorts of grain to his satisfaction, barley included.

Mr. REEVES, of Heverland, has a threshing-mill, which is, I think, nearer to perfection than any other I have seen: it is made by Mr. ASSBY, of Blyborough, in Suffolk; works with two or three horses, and cost 100 guineas. I found it at work, threshing oats: it does for barley, as well as for any other grain, threshing 32 combs in a day, of $7\frac{1}{2}$ hours; more of oats; 40 of pease, and 30 of wheat: its day's work of wheat, threshed the day before I was there, was 31 combs, standing sacked in the barn. It varies considerably in the beating-drum cylinder from the others I have seen, it being of a much larger diameter, and has 11 beaters. Mr. REEVES is perfectly well satisfied with it; and the men all agreed that it does the work much better than the flail. It has had no repairs in the threshing one crop; nor does he, from the simpli-

city

city of the movements, expect that it will be in that respect the least expensive.

Mr. BECK, of Castle Riseing, has a threshing-mill built by Mr. WIGFUL, of Lynn; he works it with four, five, or six horses, three men and three women. It threshes 32 combs of wheat, 64 of barley, or 80 of pease, in a day; cost 200 guineas, and has had very trifling repairs in three years, not 3l. and threshes barley as clean as any other grain. I saw it at its work, and clean done.

CHAFF-CUTTER.

Mr. BURTON, of Langley, has so high an opinion of cutting hay into chaff, that he gives his horses nothing else; and finds that a bushel weighing 14lb. will go as far as 30lb. given in the common way.

Mr. KERRICH, of Harleston, has attached one of SALMON's chaff-cutters, which cost him twenty-two guineas, to his malt mill; and as he had to fix it in a chamber near to the stable, for the chaff to fall at once into the chaff-room, which joins it, and the whole building detached from the mill, he very ingeniously contrived a communication of the power, under the pavement of the yard, by a universal joint. The engine cuts 40 bushels an hour, with no other expense than feeding; and the addition of labour to the horse so small, that a man's force will take the whole pressure off. He finds the saving, by giving hay in chaff only, to amount, at the lowest computation, to one-fourth. He also applies the engine to cutting green rye, mixed with an equal quantity of hay, which makes a most fragrantly sweet food, with attention not to keep it after cutting more than six or seven hours. Two acres of rye thus used in soiling, last twenty horses

horses three weeks: succeeded by winter tares, and these by summer tares.

STEAM-ENGINE.

Mr. GOOCH, of Quiddenham, in Norfolk, having a water-mill which was sometimes unemployed for want of water, erected a steam-engine contiguous, at the expense of about 50ol. The stove which heats the boiler, is so contrived as to burn coal to coke for his malt-house. One man attends both the engine and the cinder oven. It was, in the drought of 1800, of singular use to the whole country, for wind and water having both failed in a great measure, corn was brought from ten miles distance, to be ground by this engine: he has two pair of stones to the water-wheel, and two pair to the engine. The power, that of twelve horses.

The first steam-engine erected in Norfolk for merely agricultural purposes, and, for what I know, in England, is one now erecting at Haydon, by Colonel BULLER. He has contracted for the sum of 600l. It is to do the work of ten horses; to work a threshing-mill that shall thresh and dress six lasts a day: it is to grind corn also, and cut straw; to grind nine bushels of wheat with one bushel of good Newcastle coals, of 84lb. weight, and this with all the other works going on at the same time: the Colonel to find timber. Last year his hay and straw cutting cost above 70l. therefore little doubt can be entertained of the plan answering.

Under the head Implements, I must not conclude without mentioning a person of most extraordinary mechanical talents. Mr. JEX, a young blacksmith at Billingford, at 16 years of age, having heard that there was such a machine as a way-measurer, he reflected by what machinery the result could be produced, and set to work to con-

trive

trive one: the whole was his own invention. It was done, as might be expected, in a round about way, a motion too accelerated, corrected by additional wheels; but throughout the complexity such accurate calculations were the basis of his work, that when finished and tried, it was perfectly correct without alteration. His inventive talents are unquestionable. He has made a machine for cutting watch pinnons; a depthening tool; a machine for cutting and finishing watch-wheel teeth, of his own invention; a clock barrel and fuzee engine, made without ever seeing any thing of the kind. He made a clock; the teeth of the wheels cut with a hack saw, and the balance with a half round file. He has made an electrical machine, and a powerful horse-shoe magnet.

Upon being shewn by Mr. MUNNINGS a common barrow-drill, the delivery by a notched cylinder, he invented and wrought an absolutely new delivery; a brass cylinder, with holes, having moveable plugs governed by springs, which clear the holes or cups, throwing out the seed of any size with great accuracy; and not liking the application of the springs on the outside of the cylinder, reversed the whole; and in a second, now making, placed them most ingeniously within it. He has not yet failed in any thing he has undertaken: he makes every thing himself: he models, and casts them in iron and brass, having a powerful wind-furnace of his own invention. It is melancholy to see such a genius employed in all the work of a common blacksmith. However, he is only 23 years of age, and I am mistaken greatly, if he does not ere long move in a much higher sphere. This is not a country in which such talents can long be buried: a mind so occupied has had no time for vicious habits; he is a very sober honest young man, and bears an excellent character.

CHAP.

CHAP. VI.

ENCLOSING.

THE number of parliamentary enclosures that have taken place of late years in Norfolk, and the remarkable improvements which were known to have flowed from them, made it an object of considerable importance to ascertain the result, as far as it could be procured by visiting the respective places, or obtaining information from the Commissioners, or other persons interested in the work thus effected. With this view I visited many of them, and gained the best intelligence to be procured concerning the rest. The following alphabetical table contains the result of these inquiries, with such additions, not immediately relative to enclosing, as circumstances induced my attending to.

ACLE, 1797.

Quantity.—About 350 acres of common, 300 of it grass; the great object laying lands together: no half-year land.

Rent.—Now 20s. and upwards.

Corn.—Increased.

Sheep.—None before, nor at present: not 100 in the parish.

Cows.—The same; but few; they might keep more: perhaps more bullocks.

Course.—Now, 1. Turnips; 2. Barley; 3. Seeds, one year; 4. Wheat; 5. Barley or oats; a middling crop of barley,

76   ENCLOSING.

barley, 10 comb an acre; wheat, average 9 comb; oats 20 comb sometimes, average 14; turnips, all for bullocks, and generally in stalls.

Improvement.—The chief improvement is the bringing marle from Thorpe, near Norwich, by water; notwithstanding which conveyance it costs 4s. 6d. a cart-load, and to 5s. or more on the land: they lay on 8 or 9 loads an acre, which has more effect than 40 or 50 of what they call clay, though that has a good effervescence with acids. The marle works sooner, and mixes better. The marshes are very extensive, and have for centuries yielded much *stuff*, as rushes, &c. for making dung, and thus contributed to the great fertility of much of this country. More than 8 or 9 loads of marle is prejudicial for a time; a farmer in the neighbourhood tried 12 loads, and did mischief at first.

POPULATION.

1780.—Males, 3	1785.—Males, 9
Females, 5	Females, 6
— 8	—15
1781.—Males, 7	1786.—Males, 4
Females, 7	Females, 3
—14	— 7
1782.—Males, 10	1787.—Males, 8
Females, 5	Females, 7
—15	—15
1783—Males, 7	1788.—Males, 12
Females, 11	Females, 8
—18	—20
1784.—Males, 5	1789.—Males, 5
Females, 8	Females, 5
—13	—10
	135

   Males - - - 70
   Females, - - - 65

               1790.

ENCLOSING.

1790.—Males,	7		1795.—Males,	4
Females,	4		Females,	2
	—11			— 6
1791.—Males,	12		1796.—Males,	8
Females,	9		Females,	6
	—21			—14
1792.—Males,	13		1797.—Males,	6
Females,	13		Females,	12
	—26			—18
1793.—Males,	2		1798.—Males,	11
Females,	5		Females,	12
	— 7			—23
1794.—Males,	3		1799.—Males,	10
Females,	13		Females,	8
	—16			—18
				160

Males,   -   -   -   76
Females, -   -   -   84

BURIALS.

1780.—Males,	9		1785.—Males,	6
Females,	10		Females,	10
	—19			—16
1781.—Males,	4		1786.—Males,	5
Females,	9		Females,	3
	—13			— 8
1782.—Males,	6		1787.—Males,	0
Females,	9		Females,	5
	—15			— 5
1783.—Males,	5		1788.—Males,	7
Females,	5		Females,	5
	—10			—12
1784.—Males,	8		1789.—Males,	5
Females,	7		Females,	6
	—15			—11
				124

Males,   -   -   55
Females, -   -   69

1790.

ENCLOSING.

1790.—Males,	6	1795.—Males,	11	
Females,	8	Females,	6	
	—14		—17	
1791.—Males,	6	1796.—Males,	3	
Females,	2	Females,	3	
	— 8		— 6	
1792.—Males,	8	1797.—Males,	5	
Females,	3	Females,	6	
	—11		—11	
1793.—Males,	4	1798.—Males,	6	
Females,	3	Females,	4	
	— 7		—10	
1794.—Males,	4	1799.—Males,	3	
Females,	2	Females,	4	
	— 6		— 7	
			97	

Males, - - - 56
Females, - - - 41

First period.   Baptisms, - - - 135
      Burials, - - - 124

      Increase, - - - 11

Second period. Baptisms, - - - 160
      Burials, - - - 97

      Increase, - - - 63

ASHILL.—ENCLOSED 1785.

Soil.—A very good *mixt* soil; scarcely better corn-land in Norfolk.

ENCLOSING.

	Acres.	Value.
Quantity.—Whole year land	893	£. 610
Half year ditto	819	468
Land in dispute	329	190
Commons	933	550
	2974	1818
Roads	20	
	2994	

Rent.—Quality price 18l. 18s.

Fifty-four Proprietors.

Course.—Now, 1. Turnips; 2. Barley; 3. Clover, one year; 4. Wheat; 5. Sometimes Barley.

Cows.—Have diminished.

Poor.—39 acres of common for the poor; fed by all whose occupation is under 5l. a year; and 32 let, and applied in coals. The former of no benefit, from abuses. They take in stock, and call it their own, but none of the real poor enjoy the benefits.

Tithe.—Remains subject, and pays 5s. and 5s. 6d. an acre.

Corn.—The produce of corn has increased to an extraordinary degree. They raise upon the best land wheat of 8 combs an acre; oats 20 combs; and barley 14.

Improvement.—Has been in general great: the land, which before enclosing was worth but little, has been sold so high as 40l. an acre.

	Baptisms.	Burials.		Baptisms.	Burials.
1770	15	6	1785	23	14
1771	14	4	1786	13	14
1772	7	7	1787	18	6
1773	12	8	1788	14	7
1774	12	9	1789	16	5
1775	15	9	1790	18	9
1776	13	5	1791	11	12
1777	19	7	1792	14	4
1778	10	16	1793	26	11
1779	16	15	1794	11	7
1780	15	9	1795	19	9
1781	14	9	1796	23	8
1782	10	17	1797	17	5
1783	15	14	1798	16	8
1784	8	13	1799	18	10
	195	148		257	129

First period, 15 years before the enclosure,
 Baptisms - - - 195
 Burials - - - 148

   Increase - - 47

Second period, 15 years since the enclosure,
 Baptisms - - - 257
 Burials - - - 129

   Increase - - 128

BANHAM.—ACT 1789.

Quantity.—About 1000 acres: 600 of heath and common; no open arable.

Soil.—Very good strong land; clay bottom.

Rent.

ENCLOSING.

Rent.—Quality price from 15s. to 25s.

Corn.—All turned to arable.

Sheep.—6 or 700 sheep were kept; now not more than 300; but much better: some farmers breed, and some graze. The common was much subject to the rot.

Cows.—Nearly the same as before.

Tithe.—Remains subject.

Rates.—Heavy, and always were so: 3s. 6d. or 4s. in the pound.

Expenses.—There was much road to make, and cost above 2000l.

Poor.—An allotment of 30 acres of turf fen, for fuel. No complaints of any injury. The herbage of the 30 acres is given among them.

Improvement.—All clayed; and where wet, hollow-drained.

POPULATION.

	Baptisms.	Burials.		Baptisms.	Burials.
1778	36	11	1789	28	18
1779	25	16	1790	38	14
1780	28	21	1791	39	9
1781	31	21	1792	41	15
1782	27	26	1793	50	25
1783	31	27	1794	41	29
1784	30	24	1795	34	19
1785	32	16	1796	39	34
1786	29	17	1797	35	16
1787	40	21	1798	37	13
1788	29	19	1799	44	18
	338	219		426	210

First period. Births - - 338
              Burials - - 219
              Increase - - 119

Second period. Births - - 426
               Burials - - 210
               Increase - - 216

BARTON.—ENCLOSED 1774.

Quantity.—Open fields, 2626 acres; commons and wastes, 784; total, 4087.

Soil.—Much of it strong land; but some on chalk.

Rent.—Now about 2000l.

Tithe.—Land assigned, about one-seventh.

Corn.—Increased much.

BINTREY AND TWIFORD.—ENCLOSED 1795.

Soil.—Mixed soil; turnip and wheat land.

Quantity.—About 1950 acres; open fields 661 acres; 309 commons; the rest old enclosures, or whole year land.

Rent.—Before enclosing about 1200l.; quality rent, 1560l. The commons let at 20s. which were not worth 2s. 6d. before.

Course.—On part of it five shift husbandry; on part six.

Corn.—A very great increase from the common.

Sheep.—Certainly increased.

Cows.—Not diminished; as some part remains in pasture.

Improvement.—The commons drained and clayed.

Tithe.—Remains subject.

Poor.—There were 26 acres allotted for fuel, let by the parish. There were 46 commonable rights; the whole divided according to value: very few little proprietors; but small *occupiers* suffered.

Expenses.—1900l.; paid by rate.

BIRCHAM

ENCLOSING.

BIRCHAM (GREAT).—ENCLOSED 1740.

I attempted to procure information here, but every body was dead who lived at the time, except the old Parish-Clerk, who remembered nothing more than working at the fences. Corn must be increased.

	Baptisms.	Burials.		Baptisms.	Burials.
1680	10	11	1700	11	8
1681	9	8	1701	9	9
1682	12	13		196	155
1683	7	5			
1684	8	5	1702	13	6
1685	12	5	1703	13	7
1686	5	10	1704	4	12
1687	6	6	1705	9	8
1688	6	6	1706	2	
1689	7	7			
	82	76	A Gap in the Register.		
			1722	7	9
1690	8	4	1723	13	9
1691	7	5	1724	7	3
1692	12	7	1725	6	7
1693	9	3	1726	12	5
1694	8	4	1727	12	10
1695	6	11	1728	8	12
1696	13	10	1729	7	7
1697	9	9	1730	8	7
1698	9	4	1731	10	3
1699	13	5	1732	15	9
			1733	20	9

1734

ENCLOSING.

	Baptisms.	Burials.		Baptisms.	Burials.
1734	8	7	1760	4	4
1735	17	10	1761	7	8
1736	9	11	1762	3	6
1737	15	9	1763	6	11
1738	12	2	1764	4	1
1739	19	5	1765	3	6
1740	9	14	1766	7	8
			1767	8	8
	214	148	1768	6	8
			1769	6	6
1741	23	5	1770	8	7
1742	13	8	1771	5	5
1743	17	7	1772	7	6
1744	9	2	1773	5	5
1745	5	8	1774	5	5
1746	8	5	1775	6	5
1747	9	2	1776	12	6
1748	7	7	1777	15	8
1749	8	2	1778	7	6
1750	12	11	1779	7	6
1751	6	5	1780	15	7
1752	8	1	1781	6	12
1753	8	6	1782	7	8
1754	12	7	1783	15	8
1755	2	5	1784	14	9
1756	8	4	1785	16	8
1757	6	9	1786	12	7
1758	7	7	1787	12	7
1759	5	0	1788	9	4
			1789	14	8
	173	101		251	203

1790

	Baptisms.	Burials.		Baptisms.	Burials.
1790	5	6	1797	9	5
1791	13	11	1798	9	6
1792	11	5	1799	11	2
1793	13	6			
1794	10	12		102	62
1795	12	4			
1796	9	5			

Baptisms in 19 years immediately preceding
  the enclosure   -   -   -   214
    Burials   -   -   -   148

            Increase   -   -   66

Baptisms in 19 years immediately following
  the enclosure   -   -   -   175
    Burials   -   -   -   101

            Increase   -   -   74

Ten years from 1780 to 1789:
  Baptisms   -   -   -   120
  Burials   -   -   -   78

            Increase   -   -   42

Ten years from 1790 to 1799:
  Baptisms   -   -   -   102
  Burials   -   -   -   62

            Increase   -   -   40

The whole of this account is unfavourable: the enclosure lessened Baptisms so far as these data extend; and in the last 10 years there are 18 fewer baptisms than in the 10 preceding. These are circumstances not uncommon in in this part of Norfolk, and they deserve an attentive investigation. I know knothing to which to attribute it, except the parishes being occupied by one farmer; or perhaps by two, or at most three, who unite and prevent the settlement of poor, by employing hands only from more populous places. In such a case the local deficiency is made up elsewhere.

BRANCASTER.—ENCLOSED 1755.

Quantity.—About 2350 acres, besides salt-marshes, of which 960 were *breaks*; 1020 common field; 210 half-year land; 160 old enclosures.

Soil.—Good sandy loam.

Rent.—What it was before the enclosure is unknown; now about 2500l. a year. In 1792, there were 1570 acres sold, containing 940 arable, 75 fresh marsh, and 555 salt marsh, which were then rented at 500l. a year.

Poor.—Very well off; Barrow-hills, a common of 65 acres, allotted to them; and each dwelling-house has a right to keep two cows or heifers; or a mare and foal; or two horses; and also to cut furze.

Rates.—Disbursements for the poor:

1760	-	£.91 0 0
1770	- -	119 0 0
1787	- -	180 0 0
1794	- -	203 0 0
1799	- -	337 0 0

A rate of 2s. in 1l. raises 20l.; this last year the rates therefore were above 3s.

Tithe.—Remains subject.

Corn.—Before the enclosure it was in an open, rude, bad state; now in five or six regular shifts.

ENCLOSING. 87

POPULATION.

	Baptisms.	Burials.		Baptisms.	Burials.
1780	7	11	1790	25	10
1781	14	17	1791	22	12
1782	14	8	1792	16	26
1783	8	13	1793	16	5
1784	15	13	1794	19	8
1785	10	10	1795	24	10
1786	26	6	1796	18	7
1787	14	21	1797	13	9
1788	20	16	1798	11	5
1789	14	8	1799	18	10
	142	123		182	102

Tenements 1789, 136—544 souls.
     1797,   560 souls.

First period.	Baptisms	-	-	142
	Burials	-	-	123
	Increase	-	-	19
Second period.	Baptisms	-	-	182
	Burials	-	-	102
	Increase	-	-	80

In the last ten years, 1 in 31 born; 1 in 56 died.

BRESSINGHAM AND FERSFIELD.—1798.

Quantity.—About 800 acres of common.
Rent.—Will be 18s. some worth 30s.
Corn.—All corn.
Sheep.—Will lessen.
Cows.—More will be kept.

                Poor.

Poor.—Never cut any fuel on the common: no allotment.

Rates.—About 5s.

BRINTON.

	Baptisms.	Burials.		Baptisms.	Burials.
1780	6	3	1790	13	3
1781	10	3	1791	6	7
1782	3	3	1792	9	2
1783	7	7	1793	9	3
1784	5	8	1794	7	5
1785	8	5	1795	5	4
1786	7	7	1796	7	6
1787	6	2	1797	8	5
1788	9	8	1798	8	1
1789	9	6	1799	7	3
	70	52		79	39

First period.	Baptisms	-	70
	Burials	-	52
	Increase	-	18
Second period.	Baptisms	-	79
	Burials	-	39
	Increase	-	40

BROOK.—1800.

Quantity.—200 acres of common; no open field; all will be ploughed, except 40 acres; the value will be above 20s.; quality price, 15s.; never paid any thing but to jobbers and idle fellows.

OLD BUCKENHAM.—ACT 1790.

Quantity.—Near 900 acres of poor common, some wet, the rest sandy and gravel; a little good, that the act operated upon. No open field arable.

Rent.—Quality, price 5s. to 25s. Now 15s. and much raised since enclosing. All now, 15s.

Corn.—All, except 100 acres, converted to arable.

Sheep.—Very few kept before; certainly increased.

Cows, &c.—Lessened, perhaps two-thirds; but as much butter as before; other kinds of cattle increased.

Improvements.—The soil not adapted to the common ones.

Tithe.—Free before.

Rates.—About 2s. except in a year of scarcity.

Expenses.—Roads included, above 1500l.

Poor.—Had 100 acres of fen allotted for fuel; the herbage lets for 12l. a year; had, as at Carleton, a double portion, and set out near their houses; nobody suffered or complained; very few under 5l. a year kept cows; they have not been injured in any degree whatever: for fuel, are better off than before; as no draining was done to the fen, it is not under any regulations.

BURNHAM NORTON.

This parish was instanced to me, as a proof that the nominal number of cows kept before enclosing, was a mere fallacy. There is a salt-marsh common, of 15 acres, which, if embanked, would not let at 15s. There are 24 cows kept on it by the common-right cottagers; they have no other land; no hay; no turnips; no means of winter support; no enclosures belonging to them. It is not known that they do, or can buy hay or turnips. How the cows are supported is an enigma; but, as in summer, there is not feed hardly for a sheep, it may easily be supposed what the 24 cows yield.

Such was the account I received in the neighbourhood; but the circumstances seeming extraordinary, I went to the place itself for further information, and there I was told a rather different account: that the number of cows was 17; that they had another small common besides that alluded to; that Mr. FOLEY, the farmer of the parish, sold them hay and straw for their cows; that he did not disapprove their keeping cows, and that the poor were better contented, and better off here, and at Brancaster, than in any parish near them. This the second attempt; I shall make a third, by seeing Mr. FOLEY himself.

Called; but not at home.

CANTLEY AND HASSINGHAM.—1800.

Quantity.—600 acres of common; no open fields; 100 acres of it to till; the rest marsh.

Rent.—Increased from 5s. to 16s.

Sheep.—No sheep; but will have some.

Cows.—Will be increased; and grazing also.

Corn.—Will be greatly increased.

CARLETON.—ENCLOSED 1777.

Quantity.—About 3000 acres in all: enclosed about 1200, chiefly common.

Soil.—Sandy loam; good mixt soil, upon a clay bottom.

Rent.—Old enclosed land, 15s. to 20s. before the enclosure; very little open field; new rent of enclosed common, 15s. to 20s. quality price.

Corn.—Greatly more corn produced since than before; it is the best corn parish in the neighbourhood: the 1200 acres all converted to corn, and very little has been laid down: nearly half more corn than before.

Sheep.

Sheep.—About 1000 sheep were kept on the common before; now not above half; they buy lambs and fatten them, having changed their system to grazing: the sheep before were much subject to rot, and were very ordinary; now very good.

Cows.—As many as ever, or rather more, and have a considerable return of fat beasts, as turnips are now largely cultivated.

Horses.—Are increased a third, in consequence of the increase of tillage.

Improvement.—The whole 1200 acres have been marled, and all the parts that were too wet, have been hollow-drained. In general produce, it may fairly be estimated that the parish produces double what it did before. It gave me very great satisfaction to view this parish, with Mr. GOOCH, of Quiddenham, who was a Commissioner in the enclosure, and to whom I am obliged for these particulars. The crops, very generally, great, and the whole face of the parish bore every appearance of a most prosperous cultivation.

Tithe.—Remains subject to tithe; they would not take land.

Rates.—The rental increased one-third; and till the last year, not more than 2s. in the pound.

Expenses.—The whole about 2500l.

Poor.—40 acres allotted for fuel; not *turf,* but *flag*; and, therefore, the land let at 15s. an acre, and the money applied to buying fuel. In the allotment for common-rights, every poor man who had a 5l. or 6l. a year tenement, had equal to 10l. given him, in case he was a poor owner; but not *for* the tenement, if belonging to a large proprietor. Several poor owners, received for 10l. occupancy, two acres. Almost all kept them; and several kept cows, others turned their attention to hemp, turnip seed,
&c.

ENCLOSING.

&c. and have acknowledged that they are better off than before. And very few indeed who had received an advantage from the common before, but what had some thing better, and to their content, in the enclosure.

POPULATION.

Twenty-three years before and 23 after the enclosure.

	Baptisms.	Burials.		Baptisms.	Burials.
1754	30	30	1777	28	16
1755	14	16	1778	20	9
1756	25	6	1779	26	20
1757	14	23	1780	19	19
1758	28	11	1781	19	11
1759	15	16	1782	17	20
1760	27	18	1783	20	14
1761	25	19	1784	17	13
1762	17	33	1785	16	9
1763	26	18	1786	18	5
1764	27	29	1787	16	11
1765	19	17	1788	17	12
1766	27	21	1789	12	6
1767	26	14	1790	19	14
1768	25	17	1791	8	24
1769	26	28	1792	14	12
1770	21	23	1793	14	6
1771	22	21	1794	18	10
1772	24	18	1795	11	10
1773	26	22	1796	21	12
1774	24	15	1797	12	15
1775	27	6	1798	17	6
1776	22	19	1799	10	13
	537	440		389	287

First period, 15 years before the enclosure:

Baptisms	537
Burials	440
Increase	97

Second period, 15 years since the enclosure:

Baptisms	389
Burials	287
Increase	102

The difference marked in this account is far inferior to the fact; for the number of Anabaptists have multiplied so of late years, as to form a considerable proportion of the whole population. The parish is crowded with inhabitants, who have increased uncommonly since the enclosure.

CAUSTON.—ENCLOSED 1801.

Thirteen hundred acres of common and warren; the warren subject to sheep-walk, and the rent only 20 guineas a year, to Colonel BULLER.

Remains subject to tithe.

The mode of improving has been, to plough it up, and leave it for a year, then coleseed sown, for sheep to manure and tread it; then marle, by lease 25 loads an acre for the first course; and in every after course 10 loads, mixed with dung.

Colonel BULLER has arranged the divisions in such a manner, that the roads open into all the pieces.

The act, roads, commons, &c. cost Colonel BULLER 130ol.: probably the other proprietors' shares might have made the total 2000l.

By

By means of small allotments let by him to the poor, cows will increase, as they exceed the rights that were actually exercised.

Sheep will increase after seven years; but during that period excluded by the act.

The whole parish above 4000 acres; and the valuation two years ago for the rate, 2800l. besides 400l. tithe.

The rent of the new enclosed lands, on 21 years leases, for 10 years, 7s. to 8s.; the tenants doing every thing except raising the buildings; but at the end of 10 years, 2s. an acre more.

The Colonel reserved between 70 and 80 acres for small occupiers, to enable them to keep cows; and he has planted 321 acres of the parts where the common had been most pared for fuel, with all sorts of trees.

CRANWORTH, REMIRSTON, SOUTHBOROUGH.—ENCLOSED 1796.

Quantity.—Commons to four parishes, Cranworth and Letton, Remirston and Southborough; the two latter the largest. 743 acres in all. Remirston 306, Cranworth 171, Southborough 272 acres.

Rent.—Will be 20s. an acre.

Corn.—Greatly increased: the whole under it.

Cows.—Few or none were kept by the poor. Now more cattle of all sorts.

Sheep.—None to be kept for seven years in the new enclosures, by the act: they will be lessened; but very often used to be rotten.

Poor.—They kept geese on the commons, of which they are deprived. But in fuel they are benefited: an allotment not to exceed $\frac{1}{20}$ let, and the rent applied in coals for all not occupying above 5l. a year: this is to the advantage of those at Southborough, having enough allowed for

for their consumption; at Cranworth the poor are more numerous, and the coals of little use. The allotment for the poor at Letton, is 17 acres. Ten are left as a pasture for the cows of those who rent under 5l. per annum. Seven are let to the highest bidder, and the money is applied to the purchase of fuel. The number of those who keep cows, is five.

All encroachments within thirty years to be divided as common; but to be allotted to persons in possession, if entitled to any.

Tithe.—The commons to pay 1s. 6d. an acre the first year; 2s. the second; 2s. 6d. the third; 3s. 6d. the fourth; and after that remain subject.

Rates.—Four shillings to five shillings in the pound.

CRANWORTH AND LETTON.

POPULATION.

	Baptisms.	Burials.		Baptisms.	Burials.
1780	8	6	1790	21	9
1781	9	6	1791	9	9
1782	10	11	1792	15	3
1783	9	11	1793	14	7
1784	12	15	1794	16	10
1785	11	11	1795	13	12
1786	17	9	1796	19	8
1787	10	5	1797	19	7
1788	12	3	1798	22	11
1789	9	8	1799	19	4
	107	85		167	80

First period.	Baptisms	-	-	-	107
	Burials	-	-	-	85
	Increase	-	-	-	22

Second

Second period. Baptisms   -   -   -   167
                Burials   -   -   -   80
                                       ———
                Increase   -   -   -   87

SOUTHBOROUGH.

	Baptisms.	Burials.		Baptisms.	Burials.
1780	4	5	1790	8	1
1781	5	4	1791	6	3
1782	6	2	1792	3	1
1783	4	8	1793	4	4
1784	4	8	1794	4	0
1785	6	2	1795	3	0
1786	3	1	1796	9	2
1787	3	5	1797	4	3
1788	3	3	1798	4	3
1789	4	0	1799	3	1
	42	38		48	18

First period.   Baptisms   -   -   -   42
                Burials   -   -   -   38
                                       ——
                Increase   -   -   -   4

Second period. Baptisms   -   -   -   48
                Burials   -   -   -   18
                                       ——
                Increase   -   -   -   30

REY-

RIMERSTONE, ENCLOSED.

	Baptisms.	Burials.		Baptisms.	Burials.
1780	6	5	1791	11	2
1781	11	5	1792	5	3
1782	4	4	1793	6	3
1783	4	1	1794	7	3
1784	6	5	1795	5	3
1785	10	8	1796	6	2
1786	3	7	1797	7	2
1787	6	1	1798	1	7
1788	8	6	1799	6	2
1789	5	5	1800	7	2
	63	47		61	29
Increase	16		Increase	32	

DERSINGHAM.

Rent.—Doubled.

Corn.—At least twice as much corn raised as before the enclosure.

Sheep.—Increased.

Cows.—About the same number as before.

Tithe.—Remains subject.

Poor.—The poor are not affected, except by the increase of employment. The common remains common, but stinted by the Commissioners.

ENCLOSING.

DOCKING.

	Baptisms.	Burials.		Baptisms.	Burials.
1780	24	18	1790	20	11
1781	22	12	1791	17	8
1782	20	21	1792	24	16
1783	19	14	1793	20	14
1784	12	22	1794	23	17
1785	23	11	1795	16	7
1786	20	15	1796	15	16
1787	29	12	1797	18	10
1788	13	9	1798	19	17
1789	24	10	1799	22	15
	206	144		194	131

First period. Baptisms - - - 206
              Burials - - - 144
              Increase - - - 62

Second period. Baptisms - - - 194
               Burials - - - 131
               Increase - - - 63

DOWNHAM.

	Baptisms.	Burials.		Baptisms.	Burials.
1780	26	44	1790	57	35
1781	38	54	1791	42	23
1782	27	45	1792	39	25
1783	49	42	1793	34	27
1784	30	42	1794	45	31
1785	36	27	1795	39	40
1786	31	43	1796	53	26
1787	39	36	1797	44	28
1788	42	46	1798	40	32
1789	35	33	1799	42	25
	353	412		435	292

Burials

Burials in the first ten years - - 412
Baptisms - - - - 353

Decrease - - - 59

Baptisms in second ten years - - 435
Burials - - - - 292

Increase - - - - 143

LITTLE DUNHAM.—ACT PASSED 1794.

Quantity.—About 1800 acres in the parish, of which 360 are common.

Improvement.—Had great effect on the common, and by exchanges; and half-year lands were much improved.

Poor.—Thirty-five acres of common let for 46l. a year, to be distributed in coals, instead of their cutting whins (not turf). Last winter each family had 22 bushels of coals. Very few kept cows.

Tithe.—Remains subject.

Expenses.—126ol.

POPULATION.

	Baptisms.	Burials.		Baptisms.	Burials.
1780	5	8	1790	11	5
1781	8	6	1791	8	9
1782	5	3	1792	10	3
1783	6	2	1793	7	5
1784	7	6	1794	7	3
1785	10	7	1795	11	1
1786	9	3	1796	5	1
1787	8	4	1797	11	8
1788	9	4	1798	7	3
1789	12	8	1799	9	5
	79	51		86	48

Eighteen

Eighteen years, 1562 to 1579:
- Baptisms — 68
- Burials — 59
- Increase — 9

Twenty years, 1580 to 1599:
- Baptisms — 135
- Burials — 72
- Increase — 63

Twenty years, 1720 to 1739:
- Baptisms — 116
- Burials — 81
- Increase — 35

Twenty years, 1740 to 1759:
- Baptisms — 131
- Burials — 72
- Increase — 59

Twenty years, 1760 to 1779:
- Baptisms — 102
- Burials — 78
- Increase — 24

Twenty years, from 1780 to 1799:
- Baptisms — 165
- Burials — 99
- Increase — 66

Baptisms.
- First period — 68
- Second — 135
- Third — 116
- Fourth — 131
- Fifth — 102
- Last — 165

The account is remarkable; but in whatever light viewed,

viewed, it proves the superior population of the present period.

Survey taken in 1792:

Men (above 16) - 55 ⎫
Women - - 62 ⎭ 117
Male Children - 45 ⎫
Female ditto - - 44 ⎭ 89

Souls - - - 206

Forty-four families: 39 married couples.
Males 100.
Females 106.
Forty-one houses.
In 1763 there were 172 souls.

If the average number be 190, deaths will be 1 in 38; births 1 in 21.—*(Note in the Register.)*

EATON.

	Baptisms.	Burials.		Baptisms.	Burials.
1780	6	6	1790	11	6
1781	16	8	1791	6	4
1782	9	5	1792	11	5
1783	5	6	1793	7	4
1784	9	8	1794	13	4
1785	9	6	1795	7	3
1786	6	6	1796	8	5
1787	13	4	1797	10	5
1788	6	8	1798	14	1
1789	8	4	1799	6	3
	87	61		93	40

First

First period. Baptisms - - - 87
                Burials - - - 61
                Increase - - - 26

Second period. Baptisms - - - 93
                Burials - - - 40
                Increase - - - 53

ELLINGHAM.—ENCLOSED 1798.

Quantity.—About 2000 acres, old enclosed land.
800 common divided by the act.

2800

Rent.—About 14s. or 15s. old rent: quality price, 15s. to 17s.

Soil.—A loam on clay; some light soils.

Poor.—The allotment for fuel by the act, as the Commissioners shall appoint. There are 64 common-rights; they have allotted 60: there were scarcely any cows kept by the poor, as they would have been starved by the sheep.

Cows.—More will be kept in consequence of the enclosure.

Sheep.—Will be decreased.

Corn.—All will be in course under corn, and the increase very great.

Tithe.—For five years from the award. 1s. an acre, if not broken up; and 2s. on such as is ploughed: then to remain titheable.

Rates.—Are about 4s. in the pound.

FELTHORPE.—ENCLOSED 1779.

Quantity.—About 1500 acres; old enclosure, about one-third, or two-fifths; common, three-fifths.

Soil.

Soil.—Old enclosures, red sand; the new, a grey sand on red and white sand.

Rent.—Before enclosing, 600l.; at present not 800l.

Poor.—The common was so valuable to the poor, or thought to be, that the farmers could not get their work done; they cut fuel, whins, &c. for sale, and the place harboured poachers, &c.—Twenty-five small occupiers; only four above 40l. a year: many very small, and generally owners. They are comfortable, though they work harder than day labourers; they were not well treated respecting pasture; but they have a common of 50 acres for fuel, which they feed.

Corn.—There may be a little increase; but questionable: perhaps none at all.

Cows.—Perhaps more cows; but doubtful: the poor kept before, and do so now.

Sheep.—Little difference; for several years fewer.

Tithe.—Remains subject to tithe.

Expenses.—The enclosure did not pay them.

Rates.—This year about 6s; in general about 4s.

Improvement.—Mr. WRIGHT, on a summer fallow, drilled an acre with sainfoin, July 1799; in 1800 it did nothing: soil, a red sand bottom. In 1800, six acres carrots of his did well, though a bad season for hoeing; these were on red but good sand, worth 12s. an acre. He had lived at Sutton, near Woodbridge, where carrots are a common article of culture, and did not observe the husbandry in vain: I hope he will fully establish it on these sands, some of which are well adapted to it.

The husbandry which should be adopted, is to pare and burn for turnips, fed with sheep; then plough well for buckwheat (of which there is much in the parish), and with it lay down to grass for a sheep-walk; sowing burnet, chicory,

cocks-

cocks-foot, yarrow, &c.: this would be a great and lasting improvement, and would, hereafter, give good corn.—I do not know any where, after an enclosure, a parish that carries so unimproved a countenance as this, unless it be the heaths at Kelling Little or no use is made of the allotments of common: they generally lie in their waste state: they say, for want of marle or clay, without which turnips are *anburied*; and yet some small pieces have been pared and burnt, but being exhausted by repeated crops of corn, the land and the husbandry both are abandoned. Some pieces have been sold since the act passed, at 40s. an acre; at 20s. and even at 14s. as it is said: yet these poor grey sands do exceedingly well for potatoes; many are cultivated, and in 1799, as many were produced on one acre, as would have paid the fee simple of 10.

Population.—The register most irregularly kept, and births and burials so jumbled together, that even for the last eight or ten years (all I could get at) they are not easily ascertained:

	Baptisms.	Burials.
1792	8	4
1793	8	10
1794	11	5
1795	11	4
1796	9	5
1797	20	1
1798	10	2
1799	6	1

I have little faith in the account: no resident clergyman; the case of half the county.

FINCHAM.

FINCHAM.—ENCLOSED 1772.

Quantity.—2953 acres; by the award, divided amongst 29 proprietors, viz.

No. 1.	-	1288	No. 16	-	21
2	-	358	17	-	12
3	-	307	18	-	13
4	-	141	19	-	5
5	-	39	20	-	5
6	-	90	21	-	10
7	-	55	22	-	16
8	-	66	23	-	10
9	-	121	24	-	10
10	-	74	25	-	7
11	-	27	26	-	7
12	-	20	27	-	7
13 poor	-	51	28	-	61
14	-	17	29	-	5
15	-	6			

Commons	- -	618
Open field summer fallow	-	297
Crop - -	-	891
Cottage houses	- -	45

Valuation of the common houses - 261*l*. 18s.

Soil.—Strong good land; some wet, and tenacious.

Rent.—Commissioners' valuation, 1040*l*.

Course.

Course.—Before, 1. Fallow.
2. Wheat.
3. Barley.
4. Pease or oats.
Now, 1. Turnips.
2. Barley.
3. Clover.
4. Wheat.

Corn.—The common, except 30 or 40 acres, is under corn, and has been uncommonly productive; though for two or three years at first it did not well, as they ploughed without burning. The product far more than ever.

Sheep.—There were two large flocks; not so many kept at present.

Cows.—Dairies much lessened.

Tithe.—Remains subject to it.

Rates.—By an old rate-book it appears,

	£.	s.	d.
that at Easter, 1729, there was disbursed for the poor	21	0	1
1730 ditto	39	6	5
1731 ditto	27	11	4
1742 ditto	48	4	3
1747 ditto	75	10	0
1750 ditto	84	2	4
1753 ditto	42	5	3
1754 ditto	68	14	3
1758 ditto, collected for it 1s. 6½d. on 913l.	96	1	0
1762 ditto, rate 1s. 9d. on 920l. rent	78	1	7
1767 ditto, rate 1s. 10d. on 733l. rent	67	6	2
1769 ditto, rate 2s. 6d.	89	18	4
1771 ditto, rate 3s. 3d.	119	15	9
1782 ditto, 3s. 4d.	198	18	0
1783 ditto, 3s. 9d.	200	10	6
1784 ditto, 2s. 7d.	176	18	11
1785			

ENCLOSING.

1785 ditto, on a rental of 2258l. 1s. 11d.   214 10 10
1791 ditto, on 2252l.   -   -   159 18 3
1794 ditto,   2s. 3d.   -   -   259 2 10
1795 ditto, on 2303l. at 3s. 7d.   -   412 13 10
1796 ditto,   3s. 6d.   -   -   408 7 0
1798 ditto,   2s. 7d.   -   -   304 0 6
1799 ditto, on 2340l. at 2s.   -   240 0 5

Poor.—As much land allotted for their use by the act, as should produce 36l. a year, to be distributed to poor persons not receiving relief or alms; which has been applied in the purchase of coals: the land is now worth 50l. a year. In fuel, they are not deficient; but not quite equal to the former privilege. As to cottage cow-keepers they are all over: many before the enclosure, but the allotments all thrown to the farms, and in this respect they are much worse situated, though many had no right, and paid when the common was driven. Stredgit has a common, and better. But take the poor here in general, employment has multiplied; so much, that they are in a better condition; better fed, and better clothed.

POPULATION.

Twenty-eight years before the enclosure.

	Baptisms.	Burials.		Baptisms.	Burials.
1744	15	6	1754	14	13
1745	16	10	1755	11	10
1746	11	23	1756	11	8
1747	11	13	1757	14	13
1748	10	26	1758	11	11
1749	10	17	1759	14	10
1750	11	9	1760	9	7
1751	15	5	1761	12	21
1752	16	3	1762	15	12
1753	11	9	1763	12	10

ENCLOSING.

	Baptisms.	Burials.		Baptisms.	Burials.
1764	10	10	1770	12	9
1765	16	12	1771	16	13
1766	8	11	1772	11	6
1767	20	12			
1768	10	10		366	328
1769	14	9			

Twenty-eight years since the enclosure.

	Baptisms.	Burials.		Baptisms.	Burials.
1773	12	11	1788	16	10
1774	13	11	1789	10	9
1775	10	9	1790	15	3
1776	13	3	1791	17	6
1777	15	5	1792	11	7
1778	18	8	1793	14	3
1779	15	10	1794	16	9
1780	17	16	1795	16	4
1781	16	9	1796	24	9
1782	12	13	1797	9	9
1783	16	23	1798	12	8
1784	9	8	1799	16	8
1785	19	13			
1786	13	11		396	242
1787	22	7			

First period.   Baptisms   -   -   -   366
                Burials   -   -   -   328
                Increase   -   -   -   38

Second period. Baptisms   -   -   -   396
                Burials   -   -   -   242
                Increase   -   -   -   154

FOULDEN,

FOULDEN.

Baptisms and Burials for twenty years.

	Baptisms.	Burials.		Baptisms.	Burials.
1780	5	13	1790	8	6
1781	12	8	1791	8	5
1782	10	9	1792	12	7
1783	11	10	1793	11	7
1784	12	19	1794	10	4
1785	14	6	1795	11	4
1786	13	9	1796	11	6
1787	6	3	1797	11	6
1788	21	6	1798	12	4
1789	6	5	1799	14	5
	110	88		108	54

First period.	Baptisms			110
	Burials			88
	Increase			22
Second period.	Baptisms			108
	Burials			54
	Increase			54

Enumeration—1782.

Inhabitants	367
Houses	61
Per House	6
Families	89

Baptisms

110 ENCLOSING.

Baptisms and Burials for seven years,
from 1780 to 1786   -   -   77——74
Average   -   -   -   11——10
One in 33 born.
One in 36 dies.

FRING.

	Baptisms.	Burials.		Baptisms.	Burials.
1780	3	1	1790	7	3
1781	5	5	1791	9	5
1782	7	4	1792	6	1
1783	9	5	1793	8	1
1784	5	3	1794	8	3
1785	2	5	1795	4	2
1786	10	3	1796	4	2
1787	3	6	1797	5	1
1788	10	7	1798	8	1
1789	8	3	1799	7	1
	62	42		66	20

First period. Baptisms   -   -   62
               Burials   -   -   42
               Increase   -   -   20

Second period. Baptisms   -   -   66
               Burials   -   -   20
               Increase   -   -   46

HARDING-

HARDINGHAM.

The register so kept, or rather so rotted with damp; and the entries made in such a scattered topsy-turvy manner, that the authority not too good.

	Baptisms.	Burials.		Baptisms.	Burials.
1783	2	4	1791	10	10
1784	12	2	1792	11	4
1785	7	5	1793	10	7
1786	17	3	1794	16	6
1787	8	6	1795	10	4
1788	14	5	1796	12	3
1789	10	3	1797	14	8
1790	10	8	1798	16	6
	80	36		99	48

HARLESTON AND REDENHALL, 1789.

Enumeration.—
Widowers	14
Married	357
Unmarried	25
Widows	41
Children	499
Servants	185
Lodgers	223
Total	1344
Families	240
Deduct, as living in separate apartments in same house	9
	231
Houses being tenements	235

Nearly

ENCLOSING.

Nearly six to a family, consisting of

Husbandmen	-	38	Mason -	1
Spinners	-	26	Surgeons -	3
Carpenters	-	8	Grocers -	4
Farmers	-	12	Barbers - -	3
Painter	-	1	Knitter -	1
Shoemakers	-	15	Thatchers - -	3
Tailors	-	8	Heelmaker -	1
Millwrights	-	2	Brickmaker -	1
Blacksmiths	-	5	Brewer - -	1
Butchers	-	5	Ditto Servants -	9
Bakers	-	6	Midwife -	1
Schoolmasters and Mistresses	}	5	Carrier - -	1
			Glaziers -	4
Gentlemen and Gentlewomen	}	5	Shop general -	1
			Sawyer -	1
Fruiterers	-	2	Brazier -	1
Chaise-driver	-	1	Gelder - -	1
Sadlers	-	3	Woolcombers -	2
Watchmaker	-	1	Milkwoman -	1
Drapers	-	2	Nurse - -	1
Gardener	-	1	Ironer -	1
Coopers	-	2	Bookseller -	1
Weavers	-	6	Milliner -	1
China and earthenware		2	Wheelwright -	1
Attornies	-	3	Mole-catcher -	1
Publicans	-	12	Rector -	1
Ostlers	-	2	Clerk -	1
Whitesmith	-	1	Sundries -	4
Breeches-makers	-	2		
Drovers	-	2		231
Hosier	-	1		

Wortwell, a hamlet, besides, not included.

ENCLOSING.

In 1798—Men between 15 and 60	-	284
Incapable of active service	-	27
Men above 60; women and children		1153
		1464
In 1789	-	1344
Increase	-	120

Cows	-	119
Oxen	-	63
Young cattle	-	129
Sheep	-	414
Pigs	-	475
Riding horses		32
Cart ditto	-	139

In Wortwell, in 1798—Men	-	112
Women	-	98
Children	-	124
		334
Harleston	-	1464
Total	-	1798

	£.	s.	d.
Rates.—Poor-rates, in Harleston, in 1780,	461	2	5
Ditto, in 1680	108	5	10
Increase in 100 years	352	16	7
In 1783	542	5	10
1784	473	17	5
1785	454	10	3
In 1796	1115	16	3
1797	696	14	3¼
1798	736	18	10½

Midsummer 1799, to Midsummer 1800, 8s. 5d. in the pound.

REDENHALL, WITH HARLESTON.

	Baptisms.			Burials.	
1780.—Males,	36		1780 Males,	24	
Females,	28		Females,	23	
	—64			—47	
1781.—Males,	28		—— Males,	21	
Females,	38		Females,	25	
	—66			—46	
1782.—Males,	19		—— Males,	9	
Females,	28		Females,	18	
	—47			—27	
1783.—Males,	42		—— Males,	24	
Females,	46		Females,	26	
	—88			—50	
1784.—Males,	21		—— Males,	19	
Females,	21		Females,	21	
	—42			—40	
1785.—Males,	26		—— Males,	22	
Females,	29		Females,	25	
	—55			—47	
1786.—Males,	24		—— Males,	24	
Females,	34		Females,	13	
	—58			—37	
1787.—Males,	31		—— Males,	14	
Females,	20		Females	10	
	—51			—24	
1788.—Males,	41		—— Males,	16	
Females,	27		Females,	17	
	—68			—33	
1789.—Males,	27		—— Males,	13	
Females,	44		Females,	21	
	—71			—34	
	610			385	
				1790	

ENCLOSING.

Baptisms.			Burials.	
1790.—Males,	25	1790 Males,	22
Females,	32	Females,	17
	—57		—39
1791.—Males,	38	—— Males,	11
Females,	25	Females,	17
	—63		—28
1792.—Males,	41	—— Males,	23
Females,	33	Females,	27
	—74		—50
1793.—Males,	48	—— Males,	14
Females,	30	Females,	10
	—78		—24
1794.—Males,	38	—— Males,	20
Females,	30	Females,	20
	—68		—40
1795.—Males,	39	—— Males,	24
Females,	35	Females,	23
	—74		—47
1796.—Males,	38	—— Males,	10
Females,	29	Females,	13
	—67		—23
1797.—Males,	32	—— Males,	25
Females,	36	Females,	26
	—68		—51
1798.—Males,	36	—— Males	16
Females,	49	Females,	18
	—85		—34
1799.—Males,	33	—— Males,	20
Females,	30	Females,	21
	—63		—41
	697		377

First

ENCLOSING.

First period.	Baptisms	-	-	610
	Burials	-	-	385
	Increase	-	-	225
Second period.	Baptisms	-	-	697
	Burials	-	-	377
	Increase	-	-	320

THE SEXES.

	Males born.	Males died.		Males born.	Males died.
1780 -	36 -	24	1790 -	25 -	22
1781 -	28 -	21	1791 -	38 -	11
1782 -	19 -	9	1792 -	41 -	23
1783 -	42 -	24	1793 -	48 -	14
1784 -	21 -	19	1794 -	38 -	20
1785 -	26 -	22	1795 -	39 -	24
1786 -	24 -	24	1796 -	38 -	10
1787 -	31 -	14	1797 -	32 -	25
1788 -	41 -	16	1798 -	36 -	16
1789 -	27 -	13	1799 -	33 -	20
	295	186		368	185

Born	-	-	663
Died	-	-	371
Died elsewhere	-		292

ENCLOSING.   117

	Females born.	Ditto buried.		Females born.	Ditto buried.
1780	28	23	1790	32	17
1781	38	25	1791	25	17
1782	28	18	1792	33	27
1783	46	26	1793	30	10
1784	21	21	1794	30	20
1785	29	25	1795	35	23
1786	34	13	1796	29	13
1787	20	10	1797	36	26
1788	27	17	1798	49	18
1789	44	21	1799	30	21
	315	199		329	192

Born   -   -   644
Buried   -   -   391
Ditto elsewhere -   253

Men born   -   663   Died elsewhere -   292
Women ditto   -   644   Women ditto   -   253
Excess of men -   19   Excess   -   39

The number of people in 1798, being 1798; and the average of ten years baptisms 69. There is born annually one in 26; and the average burials being 37, there dies annually one in 47.

HARPLEY.

ENCLOSING.

HARPLEY.

	Baptisms.	Burials.		Baptisms.	Burials.
1780	11	11	1790	10	3
1781	7	6	1791	8	5
1782	10	10	1792	13	9
1783	4	14	1793	13	2
1784	6	12	1794	13	5
1785	11	9	1795	11	5
1786	5	3	1796	9	4
1787	11	11	1797	14	1
1788	10	8	1798	13	4
1789	5	5	1799	11	4
	80	89		115	42

First period. Burials - - - 89
     Baptisms - - 80
     Decrease - - 9

Second period. Baptisms - - 115
     Burials - - 42
     Increase - - 73

This parish joins Massingham; and offers in the last ten years a complete contrast to it.

HETHERSET, 1798.

Quantity.—About 480 acres of common, good land; and about 270 of open field arable: let very high in small parcels.

Rent.—The whole parish, of 2100 acres, improved about 5s. an acre: the common is at 20s. and to 25s.

Sheep.—Decreased. Many were kept; but many rotted.

Cows—And cattle will be increased: the common was so fed with sheep, that cows could get nothing. The poor now keep some.

Corn.

ENCLOSING. 119

Corn.—Immensely increased.

Course.—In 1800 broke up and dibbled with oats and pease; but moderate crops; and turnips after oats; and wheat after pease. Those who scaled shallow, had very bad crops: one who ploughed deeper, a very tolerable one.

Poor.—There were fifty or sixty small allotments. Only one has been sold. Twenty guineas for half an acre and to be at the expense of the measure. An allotment was made for the poor's fuel, let at 50s. an acre, 34l. 10s. a year, including a public-house at 14l.

Expenses.—These were very heavy; amounting to 6l. an acre, on the common allotments, free from all exchanges (2700l.); which were numerous, and for which the charge was 25s. for each piece, whether great or small.

Rates.—Were on an average, before the scarcity, 5s. in the pound. In the scarcity 10s.

Tithe.—Remains subject.

POPULATION.

	Baptisms.	Burials.		Baptisms.	Burials.
1783*	10	15	1790	20	7
1784	28	7	1791	20	9
1785	15	14	1792	19	14
1786	24	12	1793	23	10
1787	25	14	1794	17	11
1788	24	17	1795	18	11
1789	24	7	1796	21	5
	150	86	1797	22	7
			1798	23	12
Average	21	12	1799	17	9
				200	95
			Average	20	9

* I could not see the preceding, Mr. EDWARDS being absent.

First

First period, adding two years at the average:

Baptisms	192
Burials	110
Increase	82
Second period. Baptisms	200
Burials	95
Increase	105

HEVENINGHAM, 1799.

Quantity.—In Heveningham 1553 acres of arable and meadow, and 1000 acres of common.

Rent.—Better than 800l. on the old rate. Valuation 1024l. Tithes 212l.

Corn.—Will be considerably increased. Four hundred acres will yield good barley and wheat.

Sheep.—Will be increased considerably. Six hundred acres will be turned into sheep-walk. Some so light they would be otherwise unprofitable.

Cows.—Will not probably be increased.

Poor.—They had allotments to common-right houses. The poor that had no rights, have no benefit. Others will be benefited in proportion to their properties. The common was the source of all sorts of immorality, poaching, smuggling, &c. &c.

Tithe.—Remains subject to tithe.

Rates.—The Norwich manufactories are here both spinning and weaving, and shawls. The poor complain that they are forced to lay out half their earnings with those at Norwich, who supply them with work.—*For further particulars see the Chapter on Poor Rates.*

Population.

Population.—In 1760,

Men	110
Women	121
Under 16 years	175
	406

In the year	1787	1791	1795	1796
Males { Married	90	93	101	112
Males { Unmarried	35	45	38	32
Females { Married	90	93	101	112
Females { Unmarried	37	38	43	26
Widowers	8	6	12	7
Widows	12	8	17	12
Males under 20 years	128	113	121	140
Females under 20 years	91	116	126	126
	491	512	559	567

Increase from 60 to 85, average increase
of nine years   -   - $28\frac{1}{3}$
Increase from 1787 to 1796   -   -   76

Additional increase in the last nine years — $47\frac{2}{3}$

	Baptisms.	Burials.		Baptisms.	Burials.
1780	22	18	1790	10	13
1781	18	9	1791	21	7
1782	9	14	1792	21	9
1783	16	6	1793	21	5
1784	14	12	1794	20	7
1785	18	16	1795	19	14
1786	24	11	1796	17	10
1787	16	11	1797	23	6
1788	17	6	1798	26	18
1789	19	7	1799	18	9
	173	110		196	98

First

First period.   Baptisms   -   -   -   173
                Burials   -   -   -   110
                                       ———
                Increase   -   -   -   63

Second period. Baptisms   -   -   -   196
                Burials   -   -   -   98
                                       ———
                Increase   -   -   -   98

Population in 1795   -   -   -   559
Average baptisms from 1790 to 1799 -   19
  One in 29 therefore born annually.
  Average burials in the same period, ten.
  One in 55 dies annually.
  The Rev. Mr. ALDERSON, the Rector, has great merit for the regular manner in which this register is kept, and the people numbered.

January 1, 1800.

Males {	Married	-	114
	Unmarried	-	39
Females {	Married	-	114
	Unmarried	-	44
Widowers	-	-	7
Widows	-	-	14
Males under 20 years	-		118
Females ditto	-		145
			595

January 1, 1801.

Males married	113
Females ditto	113
Males unmarried, above 20 years	30
Females ditto	31
Males under 20, and above 14	26
Females ditto	34
Males under 14	104
Females ditto	132
Widowers	8
Widows	19
	610

January 1, 1802.

Males married	109
Females ditto	109
Males unmarried, above 20 years	40
Females ditto	48
Males between 14 and 20	18
Females ditto	22
Males under 14	113
Females ditto,	121
Widowers	7
Widows	11
	598

The decrease has been occasioned by the enclosure of the common, which has lessened the temptation to getting settlements in the parish: poor-rates had risen to such a height, that this parish, Buxton, and Marsham, have united

united for building a work-house, under Mr. GILBERT's act, which has carried off some to settle elsewhere.

Woods.—Memorandum in possession of ROB. MARSHAM, Esq. of Stratton Strawless, made by Mr. PLUMBSTEAD, Rector of Heveningham: it appears that the Spanish chesnut, now standing in the church-yard, was planted in 1610.

It girted in	-	1742	-	12 feet	7 inches.
		1778	-	14	8¼
		1782	-	15	0½
I measured it in		1802	-	15	11

All the measures were taken at the same place, four feet 11½ inches from the ground, on one side; and four feet four inches on the other. There are knots at five feet.

HEACHAM.—ENCLOSED 1780.

Quantity.—Three thousand three hundred and twenty-nine acres, of which 400 salt-marsh; now worth about 5s.

Rent.—Now above 15s. (deducting marsh) an acre, which is more than double what it was before the enclosure.

Soil.—Fine loamy sand, on marle, or a chalky bottom.

Course.—Before the enclosure they were in no regular shifts, and the field badly managed; now in regular five shift Norfolk management.

Corn.—The produce of corn is increased by the enclosure very considerably.

Sheep.—More and larger sheep are kept, and the crop of wool more considerable.

Cows.—More cows are kept; for the common was not divided, only stinted by the act.

Tithe.

Tithe.—Remains subject.

Rates.—Of late years 1s. 9d. to 2s. in the pound.

Expenses.—1174l.

Poor.—There are fifty-five commonable right-houses, of which none belonged to poor people; but many to little tradesmen and small occupiers. The really poor and distressed people had no stock on the fields or common, further than geese, and could suffer by the enclosure to no other amount; abundantly made up to them by an ampler and better paid employment. The common-rights themselves were worth very little before the enclosure, which gave two head of large cattle per right to feed on the common of 209 acres. As to fuel, the poor had no right to cut flag, &c. on the common before, nor of course since: they burn coals, supplied by the parish. To the common-right houses were assigned for each right, two acres of middling land, or one and half of good, for open field shackage and feeding; the right to the stinted common remaining: there are from twelve to fifteen little and very comfortable proprietors and renters of small plots, from two to ten acres; who have cows and some corn, and what they like to cultivate. A remarkable instance, and I cordially wish it was universal. Most of them have two cows; some more.

Population.—This register, like that of Snettisham, kept previous to 1784 so ill, with such a gap of years, as to be useless.

1780

	Baptisms.	Burials.		Baptisms.	Burials.
1780	19	18	1790	19	9
1781	8	5	1791	19	11
1782	16	21	1792	22	8
1783	15	10	1793	28	16
1784	17	15	1794	19	16
1785	14	20	1795	19	23
1786	22	14	1796	17	16
1787	18	17	1797	17	9
1788	17	12	1798	15	15
1789	18	11	1799	19	16
	164	143		194	139

First period. Baptisms - - - 164
              Burials   - - - 143
              Increase - - -   21

Second period. Baptisms - - - 194
               Burials - - - 139
               Increase - - -   55

HILLBOROUGH.—ACT PASSED 1769.

Soil.—Sand: some mixed land.

Quantity.—The parish contains 3020 acres: of which 420 commons.

Rent.—In 1791 the parish was valued.

Farms—No	Acres	Rent
1.	572	£.272
2.	535	250
3.	356	117
4.	244	120
5.	407	190
6.	199	80

Farms

ENCLOSING.

```
Farms—No  7.—109 Acres    £.  30 Rent
          8.—  54             50
          9.—  90             50
         10.— 100             40
         11.—  36             30
              ————           ————
              2702           1229
Sundries       318             71
              ————           ————
              3020          £.1300
```

Tithe.—Glebes, &c. 300l.

Corn.—Produce greatly increased.

Sheep.—The number lessened: before the enclosure there was a large flock of ewes for folding; now they buy and sell for fatting; under 1000.

Cows.—Lessened; there are not above 60 at present; some few cottagers have ten or twelve among them.

Rates.—For three or four years past, average 3s. in the pound. In 1800, 6s.

Poor.—Twenty acres were directed by the act to be sown with whins; and six others for turf and whins, and feeding their cows; but the crop of whins was so badly sown or managed, that the produce is trifling; and the people suffered.

Population.—Baptisms, &c. for thirty years before the enclosure.

	Baptisms.	Burials.		Baptisms.	Burials.
1740	5	1	1747	7	4
1741	5	5	1748	3	2
1742	5	9	1749	6	4
1743	10	5	1750	6	6
1744	8	1	1751	3	4
1745	7	7	1752	9	8
1746	6	7	1753	2	4

1754

128   ENCLOSING.

	Baptisms.	Burials.		Baptisms.	Burials.
1754	5	1	1764	3	8
1755	5	3	1765	7	7
1756	9	4	1766	11	7
1757	4	7	1767	9	6
1758	4	3	1768	12	3
1759	5	3	1769	7	2
1760	4	2			
1761	4	4		186	136
1762	5	6			
1763	10	3			

Thirty years since the enclosure.

	Baptisms.	Burials.		Baptisms.	Burials.
1770	8	6	1786	6	2
1771	11	4	1787	12	3
1772	18	6	1788	10	7
1773	10	6	1789	10	5
1774	13	6	1790	14	6
1775	13	16	1791	9	10
1776	12	4	1792	8	7
1777	13	9	1793	14	9
1778	10	4	1794	7	9
1779	10	7	1795	10	2
1780	12	7	1796	13	9
1781	6	15	1797	9	1
1782	8	10	1798	14	7
1783	10	6	1799	13	6
1784	9	8			
1785	8	4		320	201

First period.	Baptisms	-	-	-	186
	Burials	-	-	-	136
	Increase	-	-	-	50

Second

ENCLOSING. 129

Second period. Baptisms - - - 322
            Burials - - - 199
            Increase - - - 123

In 1792 the Rev. Mr. NELSON, the Rector (brother of the ever-celebrated Admiral Lord NELSON), numbered the inhabitants.

Males - - 152
Females - - 163
          ———
          315

Baptisms and Burials in seven years,
  from 1786 to 1793 - - 83——49
Average per annum - - 12—— 7
One in 26 born. One in 45 dies.

That year the parish was inoculated, and of 98 children and 17 adults, not one died. In 1800 there was another inoculation, and of 91, none died.

GREAT HOCKHAM.—ENCLOSED 1795.

Quantity.—About 1000 acres of common, and some open field arable.

Rent.—About 7s. valuation: some will let at 18s.

Corn.—Much left in grass for sheep-walk, perhaps half. But corn greatly increased.

Sheep.—Increased.

Cows.—Not increased; nor much decreased.

Poor.—Have an allotment of 40 acres; 20 acres of flags and 20 of furze. Some kept cows, and so they do now on their allotments.

Tithe.—Land allotted.

Rates.—About 5s. in the pound.

HOLM HALE,

Has an inter-commonage with Necton and West Bradenham; there are in the three commons 1000 acres: Hale has 300; Bradenham 200; and Necton 500. The soil in the low commons on a gravel bottom, with good loam surface, but for want of draining, is much injured by water in the winter. Hale has 36 common rights, which would sell now for 30l. each. They turn on what horses and cows, and geese they please, but no steers, sheep or hogs. There is not one cottager that has a right of his own; all farmers and tradesmen. Some so low as 5l. or 6l. a year. All have a right to cut flags. The same nearly occurs at Hale and Bradenham; it is a great injury to the common, and if it continues long, will render it hardly worth enclosing. The cattle are all turned on by the proprietors, and not by the cottagers to whom the houses are let: they have not a cow in the town; but keep geese, and cut flags.

Sheep.—Very few sheep kept; many bullocks grazed.

Cows.—Not 200 cows in the parish.

Land.—About half grass and half arable.

NECTON.

The above unenclosed common of 1000 acres, with Bradenham and Hale.

Baptisms from 1780 to 1789	246
Burials	158
Increase	88
Baptisms from 1790 to 1799	223
Burials	140
Increase	83

Baptisms

ENCLOSING. 131

Population of Holm Hale.

Baptisms from 1780 to 1789	119
Burials	85
Increase	34
Baptisms from 1790 to 1799	122
Burials	48
Increase	74

These parishes enjoying the right of commonage over 1000 acres of good land, and joining Ashill, whose commons are enclosed; I was solicitous to compare the results; and it is to be noted, that for the last ten years, the poor in the two former have been permitted to pare turf for fuel, before which time they only cut whins.

	Baptisms.	Increase.
First period. Necton	246	88
Hale	119	34
	365	122
Average of the two	182	61
Second period. Necton	223	83
Hale	122	74
	345	157
Average of the two	172	78
First period	182	61
Second ditto	172	78
	354	139
Average of the two	10	17

Baptisms in the second period less by ten.
Increase in the second period, more by seventeen.

Ashill

	Bap.	Bur.	Inc.
Ashill—10 years: two-thirds of the period of 15 years before enclosing	130	99	32
Ditto of that since — —	172	86	86
Superiority of the last —	42	13	54

With commons open at Hale and Necton, baptisms have *diminished.*

With commons enclosed at Ashill, they have *increased* considerably.

With commons open, the increase, that is, the superiority of baptisms to burials, equals one-tenth of the baptisms.

With commons enclosed, this increase equals one-third.

KEMNINGHALL, 1799.

Quantity.—About 1000, or 1200 acres of common: 2500 in all the parish.

Rent.—Will be 18s. an acre.

Corn.—Will be all under corn.

Sheep.—Will be decreased much.

Cows.—Will be increased.

Rates.—About 5s.

The little commoners, as the common laid at a distance, complained that they should be obliged to sell their allotments: Mr. GOOCH proposed that the whole parish should be included in the act for exchanges, to lay every man's ground near his house; they consented, and done to general satisfaction.

KETTERINGHAM.

	Baptisms.	Burials.
Twenty-nine years, from 1700 to 1728 —	118	92
Twenty-nine years, from 1729 to 1757 —	89	74
Twenty-nine years, from 1758 to 1786 —	102	62

Inhabitants

ENCLOSING. 133

  Inhabitants in 1787 - - 166
  Houses charged to window tax - 20
Fourteen single, five double, one triple.
   Acres - - - 1200
    Pasture - - 200
    Wood - - 150
Soil, from sandy loam to wet clay.
Rent.—12s. 6d.
Wheat.—175 to 200 acres; one half dibbled.
Course - 1. Turnips,
      2. Barley,
      3. Clover,
      4. Wheat.

LANGLEY.—ENCLOSED 1800.

Quantity.—Five hundred and fifty acres of low common; none of it for the plough; no open field.

Rent.—Quality price 7s. an acre. It yielded no rent before. Before enclosing, the people were told that to 200l. a year, 10l. for common would be given; and they agreed, that if they continued at their old rent, they would have nothing to do with the common, for they valued their rights at nothing. Rent now 12s.

Corn.—No increase.

Cattle.—Six times as many as ever. Three hundred may be kept, instead of 40.

Sheep.—None before or since.

Poor.—There were 24 cottagers; 40 acres allotted to the commoners for firing, in regular cuttings: a very good improvement for them, as it is under direction by the act. They used to beg for straw, &c. and often lost their cows; and went begging to get others; deriving very little benefit. Sir T. BEAUCHAMP, the landlord, has agreed also to mark out some lots of marsh land for such as can keep cows,

cows, by which means they will be in much easier circumstances; and all tenants to himself; none pay more than 2l. 2s. rent for their cottage: nor are they charged with any expense of the act or enclosing, if their rents did not exceed 5l. before the enclosure.

In examining this enclosure, Mr. BURTON pointed out many cottages, with good gardens annexed, and various small grass fields enclosed, to all who kept cows; Sir T. BEAUCHAMP's orders being at all times to furnish land to such as are able to get a cow. They have each a piece near the river, assigned for mowing fodder for their cows—too much cannot be said in favour of that system.

EAST LEXHAM AND GREAT DUNHAM.—ENCLOSED 1795.

Quantity—2500 acres in the two parishes. Whole year and half year land.
500 common
—
3000 in all.

Rent.—Before enclosing, Dunham 14s. or 15s. and Lexham 10s. Quality price, Dunham 18s. Lexham 14s.

Course.—Five shifts. Now the same in Dunham. Corn after wheat.

Corn.—Greatly more.

Sheep.—Increased.

Cows.—Diminished for many small proprietors; there were 51 rights.

Improvement.—Clay and draining.

Tithe.—Remains subject.

Expenses.—Two thousand two hundred pounds, including every thing.

Poor.—Twelve pounds a year allowed for fuel; an allotment ploughed, and sowed with whin seeds: part let. Employment much increased.

LYTCHAM.

LYTCHAM.—ENCLOSED 1758.

Quantity.—Above 1700 acres half year land.
Farms.—No. 1.   -   1000
           No. 2.   -   200
           No. 3.   -   300
           No. 4.   -   100
           No. 5.   -   100

                        1700

Besides some smaller.

Poor.—The commons, except a small one given in lieu of sheep shackage, were not enclosed, but left as they were. They are fed by tradesmen and small occupiers, but not by the poor, whose chief benefit is fuel.

Population.—Forty-two years before the enclosure.

Baptisms in 42 years before the enclosure -	367
Burials - - -	328
Increase - - -	39
Baptisms in 42 years since the enclosure -	523
Burials - - -	326
Increase - - -	197

LUDHAM.

The commons were enclosed in 1801: all cottagers that claimed, had allotments; and one for fuel to the whole; but the cottages did not belong to the poor; the allotments in general went to the larger proprietors, and the poor consequently were left, in this respect, destitute: many cows were kept before, few now. All the poor very much against the measure.

Mr.

Mr. CUBIT, of Catfield, thinks, that where enclosing has converted land to arable, the poor get as much by gleaning as they did by the common: the high price of cattle has lessened the number of their cows: geese and fuel have been of late their chief advantage; nor does he remark, that those who support themselves by commons, are richer, or better off than others depending on the labour of the farmers.

MARHAM.—ENCLOSED 1793.

Quantity.—Near 4000 acres in the parish; 1000 turf and fen; not 500 acres of old enclosures.

Rent.—Before the enclosure, between 1100l. and 1200l.; now twice as much.

Corn.—Very greatly increased.

Cows.—Lessened considerably; for there were many renters of 10l. 12l. or 15l. a year, who loaded the commons with cattle.

Sheep.—As many now as before.

Poor.—An allotment of 205 acres of turf moor for fuel, and feeding cows, &c. for which they pay, as a regulation, 2s. per annum for a cow, and 3s. for a mare; and in this manner 30 are supported better than they were before the enclosure. This measure has been so favourable that the poor have not suffered at all.

Rates.—Before the enclosure they were from 1s. to 2s. Since, 6d. to 8d.; except the late scarcity, in which 3s.

Improvement.—In the Fen the improvement has been very great, by means of paring and burning; upon which they sow cole for sheep feed; but some have seeded it. Then oats, producing 10 to 14 combs per acre: after this, some have sown wheat; and others gone on with two or three more crops of oats, and have, by over cropping, hurt the land, and done themselves no good. The fen soil is not good and solid, but rather loose and *frothy*.

Much

Much of the upland is on a chalk bottom, on which sainfoin has been tried: Mr. WINEARLS says it has not answered; if forced by manuring, the same efforts would make the land produce corn.

Expenses.—2870l.

MARSHLAND, SMEETH, AND FEN.

The act passed about seven years ago, for the drainage of the Fen, and the division and allotment of both.

	A.	R.	P.
The Smeeth common	1585	1	0
The Fen	4757	3	0
	6343	0	0

Belonging to 528 common-rights. Besides this, there were drained, also of private property belonging to

	A.	R.	P.
General BROWN	592	1	10
Mrs. PATRICK	40	0	0
Mrs. ALLEN	28	3	17
Dr. HORNE	15	2	24
Mr. HARDY, &c.	11	2	0
	688	1	11

And in another Fen, Well-Moor:

	Acres.
Mr. TOWNLEY	163
Mr. HOWEL	73
	236

The Lords of Manors had one-twentieth, and the clergy half tithe, or 2s. 6d. an acre for five years: then subject to tithe.

	£.	s.	d.
There has been raised by the Commissioners	29,000	0	0
And it will demand 10l. a right more, or	5,280	0	0
	£34,280	0	0

Sixty-five pound a right in all; 55l. a right for drainage; and 10l. for act, Commissioners, &c.

For each right the allotment in the Fen is from 12 to 15 acres, in the distant and lower part; and $6\frac{1}{2}$, for a mile and a half from the mill; and on the Smeeth, from $2\frac{1}{2}$ to 3.

Some of the latter has sold for 50l. an acre; and much would sell for 70l. and even more. Rights sold three or four years ago, from 100l. to 125l. each.

The Smeeth would, under the hammer, let at 3l. an acre; and the Fen, at 25s.

This great tract of land was, in its former state, worth little: the Fen not above 1s. an acre in reed, being two or three feet deep under water: the Smeeth was often under water, in parts to the amount of half; and then at the Midsummer after rotted the sheep that fed it.

The first crop taken on the Smeeth, a strong clay soil on silt, has been chiefly mustard; the crops great (see that article); wheat by some; and oats by others: the last were great crops, of 16 or 18 coombs an acre; and would have been 20, or more; but the bulk of straw too great. I found the country in a blaze, burning the oat stubbles, to sow cole for seed, about Michaelmas, on one earth.

Above 30,000l. a year is added to the produce of the kingdom, by this most beneficial undertaking.

The poor people who turned cows, geese, and ducks upon the common, without possessing *rights*, have suffered, as

in

in so many other cases; and it is to be regretted, that some compensation is not in all such cases provided by the act. There cannot be a doubt, that the immense system of labour created, is worth far more than such practices; still many individuals are injured, and without any absolute necessity for being so.

Mr. COE, of Islington, is one of the greatest undertakers in the cultivation of this Fen: he has 400 acres of it, and has already cultivated the whole of that tract. His method has been to pare and burn, at 30s. an acre for some; but paid too much; now 21s. an acre paring, and 6s. burning. But after this operation he found it too soft and rotten to go on with horses; he therefore breast-ploughed, and sowed cole seed; the crop very great, but killed by the frost: he then breast-ploughed again, and sowed oats; both these ploughings at 21s. an acre: the crop from 15 to 18 combs an acre: he intends on 40 acres of this stubble to sow rye-grass to feed, that the land may be trodden, and consolidated: on the rest he will plough, and sow oats: some intend wheat, at Candlemas: he this year sowed 12 acres at that time, and got 9 or 10 coombs an acre.

I viewed a fine piece of cole in the Fen belonging to Mr. J. THISLETON, which had been gained without burning, by tillage; but examining the surface, I found it in so loose and puffy a state, and so like bears' muck, that if it be not very severely trodden, it will yield bad corn: another piece near it, after burning, more solid.

Crossing by the six-and-twenty foot road, a bridge now building, I came to a piece which had been broken up by the Rev. Mr. ASHMOLE, by paring and burning, and sown to mustard: this failed; it was then tilled, and sown again; but again failed; and presented only a furrow of bears' muck; but examining the bottom, found a much

more

140   ENCLOSING.

more solid peat under the furrows: this furrow should have been burnt, and the next would then have yielded good crops of any sort.

A vast improvement on such a Fen, would be to pare and burn, and sow chicory, and feed, and trample well for two years, to consolidate; then to break up for cole, and after it corn.

The fen of Downham, Outwell, Wimsbotsham, and Stow, under an act for draining, four years ago, extending to Old Podyke; and they are now at work.

The tract on to which Stowbridge leads, and marked in the new map as fen, is under cultivation.

MARSHAM.

Enclosed in the same bill with Heveningham.

Quantity.—The common 5 or 600 acres; and so bad as not to be a great object; but the half year lands several hundred acres.

Tithe.—Remains subject.

POPULATION.

Baptisms from 1780 to 1789	152
Burials	195
Increase	43
Baptisms from 1790 to 1799	196
Burials	90
Increase	-106

LITTLE MASSINGHAM.

Baptisms from 1780 to 1789	37
Burials	29
Decrease	8

Baptisms

ENCLOSING.

Baptisms from 1790 to 1799	-	-	26
Burials	-	-	18
Decrease	-	-	8

This account appeared to be so singular, that I looked back in the register, and finding the period from 1740 to 1749, well kept, and entered, I noted it.

Baptisms	-	-	34
Burials	-	-	30
Increase	-	-	4

Mr. MORDAUNT, the rector, was not at home, nor Mr. GODFREY, one of the three farmers in the parish, or I should have inquired into the circumstances which have had this effect. It is an arable parish of dry land, and very well cultivated.

MATTISHALL, 1801.

Nine hundred acres of common: average 20s. They sold land to pay the expenses, 35l. per acre: subject to tithe; but for five years to half tithe.

METHWOLD UNENCLOSED.

It is a vast parish, containing, probably, not less than 12,000 acres. There is a very great common fen; but situated at such a distance, that many poor people who would otherwise use it do not, except for fuel. Here is an immense chalk-pit, of hard building chalk, called *clunch*; in and about which are many cottagers; but there are only three cows kept among them. I know no parish that calls for enclosure more than this.

MILGHAM.

MILCHAM.

Arable	- -	1372 Acres.
Pasture	- -	770
Wood	- -	90
Common	- -	300
		2532

Rent, arable and pasture, 15s. in 1788.
Land-tax - £.181 8s.

Houses	-	50
Double	- -	13
Tenements	-	63
Inhabitants	-	315

Maintenance of the poor to Launditch House of Industry £.79 3s.

Course.—1. Turnips,
  2. Barley,
  3. Seeds, one or two years,
  4. Wheat; and some add,
  5. Barley.
   Produce, wheat 5 to 6 coombs.
    Barley and oats, 8 coombs.

NORWICH.

Poors rate in St. Andrew's parish, which is very nearly in the same ratio as all the rest: it is one half the rent. The more wealthy persons are rated for stock, in which 50l. is accounted equal to 1l. of rent.

	In the pound.			In the pound.	
	s.	d.		s.	d.
1790 -	14	11	1796 -	22	4
1791 -	13	10	1797 -	22	7
1792 -	8	4	1798 -	19	7
1793 -	18	2	1799 -	18	1
1794 -	18	8	1800 (3 quarters)	15	7
1795 -	18	9			

There

ENCLOSING.

There appears, however, to be some difference in the parishes: it is calculated that 6d. in the pound, in the whole city, raises 500l. Now, in the Annual General State of the Court of Guardians for 1799, the rates are set down at 18,405l., deducting from which 1722l. for arrears and empty houses, there was raised 16,683l. which makes, at 6d. producing 500l. the rates to be that year 16s. 8d. in the pound, instead of 18s. 1d.

POPULATION.

Births from 1781 to 1789, average	1138
Ditto from 1790 to 1799	997
Decrease	141
Deaths in the first period	1203
Ditto in the second	1056
Decrease	147
Average deaths, first period	1203
Ditto births	1138
Annual loss	65
Average deaths, second period	1050
Ditto births	997
Annual loss	53

Here is almost every mark (as far as these accounts are authority) of depopulation; and some other circumstances confirm it: the state of the manufacture is certainly that of great declension; as I am assured by persons of every party and description. There is not a single person of any consequence bringing up to the business there is only one house of any great consideration in the city; the introduction of cotton languishes; shawls the only new article that

that flourishes. They have a large yarn trade, however; which marks the transfer of the fabrics elsewhere.

Account of empty houses returned to the Court of Guardians, and arrears of rates:

Year ending April 1, 1791	£.1325
1792	1090
1794	1776
1796	1866
1797	2126
1798	2361
1790	2279
1800	1722

This circumstance does not explain the difficulty.— Rated houses are not probably in question: I applied to several persons who knew the city well; but they were not acquainted with such a number of little houses untenanted, as would account for a great decline in population.

There is one great encouragement to manufacturing; which is the cheapness of labour: I was assured by several persons, by one in particular, remarkable for accuracy (Mr. SWIFT), that on an average of all the trade of Norwich, a weaver does not earn more than 5s. a week the year round; yet, in some works, and at some seasons, he will get a guinea or more. This is very low; nor does any thing enable him to support his family, but his wife, if she weaves, earning 4s. In flourishing fabrics no such wages are heard of.

Proportion of the Sexes.

Baptisms of males from 1781 to 1789, eight years, average	587
Ditto females	551
Baptisms of males from 1790 to 1799, average	524
Ditto females	478

Deaths

ENCLOSING. 145

Deaths of males from 1781 to 1789, eight years,
average — — — 583
Ditto, females — — — — 619
Deaths of males from 1790 to 1799, average — 511
Ditto, females — — — — 539

Average baptisms in the first period:
   Males — — 587
   Females — — 551
   Less by — — — 36
Second period. Males — — — 524
   Females — — — 478
   Less by — — — 46

These are or may be the number of weavers that enlist or go elsewhere, to leave the sexes equal.
Average deaths in the first period:
   Females — — — 619
   Males — — — 583
   Less by — — — 36
Average, second period:
   Females — — — 539
   Males — — — 511
   Less by — — — 28

If many young weavers had emigrated, or enlisted, in a greater proportion than females go to London, &c. it might be seen in this comparison. If 583 give a diminution of 36, then 511 should diminish 31, but it is only 28; had they emigrated more than common, this number would have exceeded 31.

But there is one circumstance which throws much uncertainty on the inquiry, which is the number, an increasing

146   ENCLOSING.

ing evil every where, of sectaries: at Norwich they are very numerous; several of them have burying-grounds. There will be no judging of any of these circumstances with tolerable accuracy, till returns are made from all descriptions of persons. Mr. EASTLAND does not conceive that Norwich is less populous than it was, excepting the immense number of recruits it has furnished; and substitutes for the militia. The amount of arrears and empty houses returned to the Court of Guardians, he thinks may be probably a tolerable rule for marking this point; as there is a practice of rating houses of very low rent, down even to 20s. or 30s. if they stick out a broom, or a plate of apples to sell, considering them as shops.

NORTHWOLD

Act passed, 1796.—In 1800, the award, map, and Commissioners accounts, dated November 1, 1798, not given in.

Quantity.—Five thousand acres in the parish: of which 1500 fen, 550 ancient enclosure, not subject to the act; the remainder, about 3000, open arable fields.

Lord's Allotment.—One-twentieth of the commons.

Tithe.—Remains subject.

Soil.—The open field arable, a sort of sandy loam. much sand, and some gravel.

Rent.—Before the enclosure, average not exeeding 9s. the statute acre: now it is 13s.; some up to 15s.; of the 1500 acres of fen the worst at 6s.; the best of the common up to 16s. Valued rent of the whole parish, houses included, 3374l.

Corn.—The enclosure will undoubtedly increase the product very greatly.

Course.—1. Fallow for turnips, or wheat; 2. Barley;

or oats; 3. Land for shackage; no seeds: now four shifts.

Improvement.—Much has been and will be clayed, and the success very great: clover introduced and wheat on it, and the crops bear quite a new face: Dr. HINTON has had seven comb an acre of wheat. In all respects the improvement very great indeed. The fen is not cultivated, for the drainage depends on Bedford Level; and it is not dry enough to pare and burn. In 1799, they were drowned so late as May.

Sheep.—There were once 3000 very bad and ill-fed sheep, with numbers dying for want of food; now there are not more than 1100 kept, but they are far more valuable, from being better fed, though no change in breed.

Cows.—There were about 240 or 250 kept in a wretched manner, wandering the day through half-starved: now much lessened; perhaps half.

Rates.—Average of three years, to Michaelmas 1799, 2s. 3d. in the pound. In 1800, 4s.

Poor.—There were 72 commonable messuages. They are supposed to have received land to the amount of 8l. 10s. per annum, for each right. The value of a common messuage, about five years before the enclosure, was 110 guineas. Now, the house and allotment would sell for 340l. These are freehold; but for copyholds, the value less. They may, on an average, be reckoned, rent of habitation, 3l.; allotment, 8l. 10s.; total, 11l. 10s.; and value, 300l. Besides this allotment, each has 1¼ acre of fen set out for fuel, supposed to yield 12,000 turfs yearly, for one hearth, the calculated consumption. The ancient cottages, of 20 years standing, that had no common rights, were favoured with a right of turbary over 113 acres, subject to the controul of the fen reeves, to cut each 800 turfs. There might be about 20 cottagers, who without a right

kept

kept half-starved cattle, geese, &c. which were subject to be *driven*, and fined. These are all at an end; and the loss is not what might be supposed, for they were not profitable speculations by any means. The allotments of 8l. 10s. are from four to six acres of fen, near the town, subject to inundations; the remainder in value in high fen, $4\frac{1}{2}$ miles distant from the village; and this circumstance proves a great hardship to them: the lands which would have suited them have been allotted to the great proprietors; all they can do is to mow it: thus in the winter they have no dry land on which to put their cattle. The allotments of fen ground for fuel, to common-rights, cannot be separated from the houses, neither let nor sold from them, by a clause in the act of parliament.

POPULATION.

Baptisms from 1780 to 1789	224
Burials	142
Increase	82
Baptisms from 1790 to 1799	232
Burials	112
Increase	120

OVINGTON.—(SEE SAYHAM.)

Quantity.—Six hundred acres of common.
Rent.—Was worth 8s. an acre; will be 20s. to 22s.
Sheep.—Decreased.
Cows.—Decreased.

OXBOROUGH.—ENCLOSED 1723.

Quantity.—About 2000 acres in the parish.

Rent.

ENCLOSING. 149

Rent.—Now near 900l. a year. The award and all the documents relative to the enclosure being lost, the Rev. Mr. WHITE, the rector, to whom I applied, and who was so obliging as to give me such information as was in his power, could not satisfy various of the inquiries.

Population.—For 20 years past:

Baptisms from 1780 to 1789	67
Burials	65
Increase	2
Baptisms from 1790 to 1799	74
Burials	53
Increase	21

The people are certainly very much increased, and more than any account of this sort would shew: for the last seven or eight years the Roman Catholics are much increased; these bury, but do not baptize in the church.

In 1782 an enumeration taken:

Inhabitants	233	
Houses	35	
Charged to window tax	9	
Families	43	
Souls, per house	$6\frac{2}{3}$	
Baptisms and burials for 7 years, 1780 to 1786,	44	45
Average	6	6

One in 39 born—one in 39 buried.

PLUMBSTEAD.

Quantity.—About 600 acres of common. No half year land.

Rent.—Common worth 12s. and will be worth 25s.

Corn.—Greatly increased.

Sheep.

Sheep.—Decreased.
Cows.—Much the same.

PORINGLAND AND FRAMINGHAM, 1800.

Quantity.—One thousand one hundred and forty acres of common. Known for many years to be fed only by asses, and the worst of live stock: never paid (except by fining) one shilling. It is bad land; not worth more than 10s.

Tithe.—Commuted by a corn rent; which, on the commons and waste lands, shall equal *one-tenth part of their improved annual value.*

Sheep.—Decreased.
Cows.—A decrease.

RANWORTH.

Baptisms from 1780 to 1790	58
Burials	45
Increase	13
Baptisms from 1790 to 1799	88
Burials	38
Increase	50

RINGSTEAD.—ENCLOSED 1781.

Quantity.—Two thousand six hundred and ninety-seven acres.

Soil.—Sandy loam.

Rent.—Now near 15s. an acre round, which is considerably more than double what it was before.

Course.—Five shifts at present; before irregular.

Corn.—The produce of corn more than doubled.

Sheep.—As many; not more; but they were wretchedly kept before the enclosure.

Cows.

Cows.—More; and more cattle of all sorts.

Tithe.—Remains subject.

Rates.—Of late about 2s. 6d. in the pound.

Poor.—Fifty acres were allotted for cutting whins, and they are restricted to cut in regular crops; much more than wanted; so that others take for various uses. A fair equivalent was given to common-right houses. Very few little proprietors, like those at Heacham, adjoining.

Improvement.—Mr. STYLEMAN, after grubbing the whins, ploughed 100 acres of heath, quality at 4s. an acre, and sowed cole, which killed many wethers by the murrain, which arises sometimes from feeding cole. After this, he summer-fallowed for wheat, which gave seven coombs an acre, and sold for 93l.

POPULATION.

Burials from 1762 to 1780	147
Baptisms	137
Decrease	10
Baptisms from 1781 to 1799	233
Burials	159
Increase	74

SALT HOUSE AND KELLING.—ENCLOSED 1780.

Quantity.—About 1200 acres of heath.
1500 half year land.
2700

Soil.—Much black sand on a gravelly bottom; but some better land.

Rent.—The heath yielded little or nothing; and the half year land was from 7s. to 10s. an acre; it is now 15s.

Improve-

Improvement.—The rent now made by the heath is chiefly by cultivating parts of it, for the assistance of the rabbits, that are upon the remainder: before the enclosure the poor cut whins and flags every where; since, they are restricted to their own allotments. Mr. GIRDLESTONE, who is lord of the manor, and has a considerable property here, manages his warren by thus improving parts. He ploughs, and leaves the furrow two years to rot; then clays it 50 loads an acre, and 10 loads of muck, works it for turnips, which are good, worth 40s. an acre; these are sowed with sheep and cattle. Sows oats next. which, in a season not too dry, yield ten coombs: with these he lays down to grass for as many years as it will stand, for the rabbits; these new lays enable them to give milk, and bring up their young; when grown, they feed upon the ling, thus giving value to all the rest, and in this way is worth 20s. an acre: but would not let to a farmer, by itself, for more than 5s. or 6s.

Corn.—Very much increased.

Sheep.—Not so many kept as before the enclosure.

Cows.—Much more cattle kept. Cows are increased, yet the farmers graze more than dairy; and the tread of great cattle is wanting in feeding off the turnips.

Poor.—To every commonable right house under 10l. a year, there is a right allotted of keeping a cow, and two heifers, or a mare and a foal, and of cutting fuel; between three and four hundred acres being allotted for this purpose. This common is not overstocked, and the poor are much better off than they were before, as they have it to themselves, all great commoners being excluded; but being invested in the church wardens and overseers, as well as in the lord of the manor and the rector, there are some abuses which might have been avoided.

Tithe.—Remains subject: pays 5s. an acre.

Expenses.

Expenses.—About 1500l.

Present General State.—A very large part of the parish, from being left in warren, has, to the eye, the same dreary, uncultivated, barbarous state, as so many other common heaths in the neighbourhood. Those animals are never found but in deserts, and it seems to have been a strange exertion to have gone to parliament for powers to leave any part of a parish in such a state. The soil is certainly not good; but turnips worth 40s. and ten coombs an acre of oats, are proofs that the land might be profitably cultivated in an alternate husbandry of sheep-walk and corn. Their manner of breaking up and leaving to rot two years, explains the failure; this has been tried in various parts of the kingdom; and almost every where, whether it fails or not, proved unprofitable. It should be pared or burnt for turnips or cole, and laid down to grass; burnet, chicory, cocksfoot, Yorkshire white, and a little ray, and being well loaded with sheep as long as it would last, and clayed or not, would prepare for one crop of corn to lay down again. But the notion that the land is good for nothing but rabbits, makes it so.

POPULATION.—SALT-HOUSE.

Baptisms twenty years before the enclosure	164
Burials	128
Increase	36
Baptisms twenty years since the enclosure	121
Burials	103
Increase	18

This is a singular instance; for population has unquestionably declined, as far as this document proves any thing, and considerably too, whether the increase or the baptisms be confided in.

KELLING.

KELLING.

Burials twenty years before the enclosure	55
Baptisms	52
Decrease	3
Burials twenty years since the enclosure	66
Baptisms	61
Decrease	5

The two parishes together.

First period.	Baptisms—Salt-House	164
	Kelling	52
		216
	Burials—Salt-House	128
	Kelling	55
	Increase	183
Second period.	Baptisms—Salt-House	121
	Kelling	61
		182
	Burials—Salt-House	103
	Kelling	66
		169
	Increase	13

Whatever the cause be, it has, therefore, operated on both parishes.

SAYHAM AND OVINGTON.—ENCLOSED 1800.

Quantity.—One thousand six hundred acres of common; no half year land.

Soil.—A marley clay bottom at 12 to 18 inches, over it a vegetable mould. Very superior land: friable loam.

Rent.—The value nothing; for it was so overstocked, that though the land is very fine, yet it would have answered much better to have paid a good joist price for putting cattle out, than keeping them gratis on the common: notwithstanding this circumstance, however, and the fact that the common was of no value to the public, yet it certainly yielded a rent to the landlords of the enclosed property, probably to the amount of 8s. an acre, or 640l. a year. After enclosure, worth 21s. an acre. The Commissioners, in setting out 50l a year directed by the act for the poor, finding that the rents of land were rising, sold off a part they had intended for that purpose, for about 250l. and letting the remainder by auction, it produced, on a 21 years lease, 98l. 13s. As this rent arose from the competition of the farmers, it offers a curious anecdote touching the value of land; not, however, to be separated from the price of corn.

Course.—Mr. CROWE, who has a large portion of the common, ties his tenants to either, 1. Turnip; 2. Barley; 3. Clover; 4. Wheat; or, 1. Turnip; 2. Barley; 3. Grass seeds two years; 4. then two crops of corn or pulse.

Corn.—An immense increase.

Sheep.—There will be fewer sheep kept; they often rotted before; now none. There was a walk on one of Mr. CROWE's farms for 36 score ewes, on this common and that of Ovington, and the two tenants, for the last eleven years, did not make a shilling profit, such were the losses. The tenant

tenant of the park farm has bought 300 ewes in one year, to make up these losses.

Cows.—Very few cows were kept, except by copyhold tenants: but it is supposed that there will be fewer; but a very great increase in grazing beasts.

Tithe.—Remains subject.

Rates.—About 3s. in the pound.

Expenses.—3600l. on Sayham only, which was 1000 acres of the 1600: land was to be sold for paying it: it fell short 500l. collected by a rate.

Poor.—An allotment of not less than 50l. a year, for distributing to the poor in coals, was ordered by the act; it let for 98l. There were 100 commonable right houses. They used to sell a cottage of 3l. a year, with a right, for 80l. For each, four acres were allotted: and the cottage, with this allotment, would now sell for 160l. And what is very remarkable, every man who proved to the Commissioners, that they had been in the habit of keeping stock on the common, was considered as possessing a common-right, and had an allotment in lieu of it. Nor was it an unpopular measure, for there were only two people against it from the first to the last.

Fences.—A ditch, five feet wide and four feet deep, at 2s. for seven yards; with a dead hedge and quick. Sixteen rod, at seven yards, cost 20s. for dead fence: quick, 5s. a 1000, and 80 laid into a rod. The best fence is white thorn and sweet-briar alternate; the latter protects the former, while it is young, and the quick killing it when grown, remains an excellent fence at a good distance.

Improvement.—The husbandry of breaking up the common, is to plough once for pease, oats, or cole; the two former all dibbled. Then clay 60 loads an acre of 24 bushels; and fallowed for turnips; and then the common husbandry.

husbandry. It may be calculated that the first year there were 600 acres of oats, and 600 of pease.

	£.	s.	d.
600 acres of oats, are 15 coombs, some broke up this year 25. This is 9000 coombs at 10s.	4500	0	0
600 acres of pease at 5, are 3000 coombs at 14s.	2100	0	0
For the first year	£.6600	0	0

No cole would be sown, but they cannot get all ready in time for oats and pease.

Second year.

400 acres of cole, at 6 coombs, 2400 coombs at 30s.	3600	0	0
600 acres of wheat, after pease, at 8 coombs, 4800 at 24s.	5760	0	0
600 acres of turnips, at 3l.	1800	0	0
	£.11,160	0	0

Third year.

400 acres of wheat, 6 coombs, 2400 coombs at 24s.	2880	0	0
600 acres of turnips, at 3l.	1800	0	0
600 acres of barley, at 10 coombs, at 12s. after turnips fed	3600	0	0
	£.8,280	0	0

And this prodigious product will arise in three years, from the culture of a common which most certainly never produced 500l. in any one year, reckoning at the highest which such indefinite returns as that of an unlimited common can be estimated at.

Births

POPULATION.

Baptisms from 1780 to 1789	182
Burials	178
Increase	4
Baptisms from 1790 to 1799	256
Burials	126
Increase	130

SEDGFORD.—ENCLOSED 1795.

Quantity.—About 4000 acres.

Soil.—Sand; loamy sand; and some sandy loam on marle.

Rent.—About 11s. an acre, which is more than double what it was before the enclosure.

Course.—Before enclosing, the management was quite irregular. Now it is in five shifts: 1. Turnips; 2. Barley; 3. and 4. Seeds; and 5. Wheat, pease, or oats, according to soil.

Corn.—A great deal more than ever.

Sheep.—More, and better than before.

Cows.—Fewer.

Tithe.—Remains subject.

Rates.—Are 2s. to 2s. 6d. in the pound.

Poor.—The real poor did not suffer by the enclosure; but the allotments to common-right houses, which before kept two cows, amounted only to half an acre, in lieu of shackage, and the common of above 100 acres.

Improvement.—The method Mr. DURSGATE took to improve his waste was, after stubbing the whins, to plough and sow oats, which yielded a good crop; he then fallowed and worked it well for wheat, which crop was very bad. He then clayed it well, and sowed turnips,

which

ENCLOSING. 159

which were very fine, but from a tinge of black sand in the soil, stock did not prove on them. Then barley, which looked beautifully in the spring (1800), but fell off in the drought, and proved a bad crop, though much in bulk.

SHARNBOURN.—ENCLOSED ABOUT 30 YEARS.

Quantity.—Something under 1000 acres.
Corn.—Very much increased.

SHARRINGTON.

Quantity.—Common	- -	200 acres.
Half year land	-	120
Whole year ditto	-	570
		890

Rent.—Increased full one-third. The quality value of the common was 16s. an acre.

Common Rights.—37 ; and 3 acres of average value were assigned to each cottage-right house, to enable the cottager to keep a cow: their cattle before were starved, by the farmers superior stocking.

Corn.—Increase in the proportion of 150 additional acres cultivated; and the improvement of half year land converted to whole year.

Sheep.—Much the same.

POPULATION.

Baptisms from 1780 to 1789 - -	74
Burials - - -	43
Increase - - -	31
Baptisms from 1790 to 1799 - -	76
Burials - - -	36
Increase - - -	40

SHIPDHAM—NOT ENCLOSED.

Baptisms from 1780 to 1789	376
Burials	258
Increase	118
Baptisms from 1790 to 1800	477
Burials	258
Increase	219

HINGHAM—ENCLOSED.

Baptisms from 1780 to 1789	300
Burials	202
Increase	98
Baptisms from 1790 to 1799	358
Burials	228
Increase	130

SHROPHAM, 1798.

Quantity.—About 800 acres common: some half year land.

Rent.—Will be about 14s.: was before only 4s.

Corn.—Will be greatly increased.

Sheep.—Decreased.

Cows.—There will be as many as before.

Poor.—About 50 or 60 acres for fuel: very few kept cows; and will keep better than before: several little proprietors of two or three acres, and will now have double allotments added, and be better able to keep.

Rates.—About 4s. in the pound.

Tithe.—Land allotted for tithe.

SHOULDHAM AND GARBOISE.—ENCLOSED 1794.

Quantity.—Whole year lands - 820 acres.
Common fields, commons, and waste - 4750
                                    ——
                                    5570

Tithe.—Impropriations: remain subject.

Poor.—Two allotments at Shouldham, of 82, and 13 acres, for fuel: what they have is good, but not in the former plenty; this, however, is not their great complaint, but the deprivation of keeping live stock: they used to have cows, mares, geese, ducks, &c.; but now nothing; and their language is (I talked with several) that they are ruined. About 40 poor people kept cows at Shouldham; not all with what was esteemed a right; but if the commons were drove, the fine was small: those cottages that had rights and allotments, are now let merely as houses, and the allotments laid to the farms. The account a farmer gave me was, that many poor kept cows before; now, not more than one or two. It is sufficiently evident, therefore, that this enclosure classes with those which have been, *in this respect*, injurious to the poor. At Garboisethorpe, the poor before the enclosure kept about 20 cows; now none.

Course.—The former husbandry was:

1. Fallow,   3. Barley,
2. Wheat,   4. Oats, pease barley.

Now—1. Turnips,   3. Clover,
    2. Barley,   4 Wheat.

Corn.—The quantity raised, very considerably increased in both parishes.

Sheep.—About the same number kept as before at Shouldham: 3 or 400 fewer at Garboisethorpe.

Cows.—At least 50 fewer than before, at Shouldham.

Rent.

Rent.—The old rent at Shouldham was about 1300l.

Rates.—At Shouldham, 1s. 6d. to 2s. in the pound before; now 3s. 6d. on the new rental.

Soil.—At Shouldham, much good loam on marle; and in the vale an extraordinarily fine sandy soil of great depth.

SHOTTESHAM ST. MARY, AND SHOTTESHAM ALL SAINTS, 1781.

Quantity.—In all 3561 acres: about 314 acres of common enclosed by the act.

Rent.—It was good land; about 12s. an acre, quality price; now above 20s.

Course.—The greatest part (3-4ths) grass; the arable in—1. Turnips; 2. Barley; 3. Clover, wheat.

Corn.—Increased proportionably.

Cows.—Not diminished; cattle in general very much increased.

Sheep.—Increased.

Poor.—A common of 48 acres of good grass allotted to the poor, occupying under 5l. a year, and six acres of turf for fuel; also another common of six acres for the same purpose. To each cottage a right of keeping a cow, not merely common-right houses, but to poor inhabitants indiscriminately.

An account of the number of cows kept in 1792, by poor people:—In 1792, 11 cows; two occupiers of 7l. each, keep two, or 4; in all 15.

Note—before the enclosure, only four cows were kept by poor people in Shottesham.

By half a right since the enclosure, they can keep a *bud*; a colt not exceeding 12 months, or an ass, or three geese, a gander, and their followers. A whole right is a cow, a horse,

horse, or double the others. In 1800, increased and increasing: great competition.

Before the enclosure, cottages let from 40s. to 3l. Now Mr. FELLOWES has not increased the rent of any (all at 40s.), but the rent of others is increased 20s. a year more than before the enclosure. It has upon the whole been very beneficial to the poor; and all were pleased; their share of 60 acres in 300 was uncommonly large. No poor person here, if he can, and will pay, but may be supplied with milk. There are six additional rights, which produce as many guineas a year, laid out in improving the common, by keeping open the drains, carting on the earth from them, and mowing rushes.

Tithe.—Remains subject.

Rates.—About 1795, they were at 9d. in the pound; the year ending Lady-day, at rack rent 2s. in the pound. In 1789, 11d. in the pound; paid off 70l. for the purchase of the house, besides turnips. Now 4s. in the pound.

Mr. FELLOWES' father (and himself continued it) took care to have no children bound to weavers, which has been a principal cause of preventing poverty being an increasing inhabitant of these parishes. Another cause has been, Mr. FELLOWES having established a poor-house for all the Shotteshams, for the reception of very old and very young people; but which takes in none that can maintain themselves upon the sum which it would demand to keep them in their own cottages: this has also had a very good effect. Another plan that has been very successful, has been a steady determination in Mr. FELLOWES to get children, whether from the house or from cottages, to be put out for a year in farmers' services; he absolutely requires the attendance of all the farmers for that purpose, and on the day fixed, the children of a due age

age are brought out one by one, and offered to be let: if refused without a premium, 10s. is offered with her or him; if no offer, 15s. then 20s. and so on till some farmer agrees: as it does not suit Mr. FELLOWES to take them, he takes the burthen of any cripple, or other child that is particularly objectionable, and is himself at the expense of procuring them a situation without a premium.

Earnings in the house:

	£.	s.	d.
Lady-day 1798 to Lady-day 1799, by thirty people,	39	10	0

All very young or very old.

	£.	s.	d.
Lady-day 1788 to Lady-day 1789, by twenty-six people,	29	10	0

The old to the young, as three old to four young.

In 1794 each poor person in the house cost, exclusive of their earnings, 2s. 8d. per week.

In 1784 each cost 2s. per week.

Expenses.—The total.

	£.	s.	d.
Act and Law charges	218	10	0
Roads	105	0	0
Commissioners	69	0	0
Ditto expenses	18	13	4
Ditching	43	0	0
Surveyor	42	0	0
Sundries	104	15	$10\frac{1}{2}$
Total	£.600	19	$2\frac{1}{2}$

And yet this enclosure was opposed in Parliament.

POPULATION.

In October 1763, St. Mary and St. Martin, an accurate account:

St. Mary, including St. Botolph	190 souls.
St. Martin - -	68
All Saints - -	372
	630

In 1782,

St. Mary and St. Martin	314
All Saints - -	394
	708
Families—St. Mary and St. Martin -	56
All Saints - -	80
	136

Now they are certainly increased.

This great attention to getting the children of the poor into service and habits of industry, has had a very considerable effect in keeping down rates, which usually rise in proportion to negligence. The expense of these premiums is paid by the rate, and equally borne consequently by the whole parishes.

SHOTTESHAM ST. MARY.

Twenty years before enclosing:

Baptisms from 1760 to 1779 - -	155
Burials - - -	90
Increase - -	65

Twenty years since enclosing:

Baptisms from 1780 to 1799 - -	165
Burials - - -	62
Increase - - -	103

SHOTTESHAM ALL SAINTS.

Twenty years before enclosing:
Baptisms from 1760 to 1779   -   -   233
    Burials   -   -   -   187

Increase   -   -   46

Twenty years since enclosing:
Baptisms from 1780 to 1799   -   -   313
    Burials   -   -   -   213

Increase   -   -   -   100

SNETTISHAM.—ENCLOSED 1762.

Quantity.—Five thousand acres in the parish.

Soil.—Various; sand, sandy loam, &c. on marle and stone and gravel.

Rent.—About 4000l.; but in a valuation in 1801, has been at least doubled; since the enclosure perhaps much more.

Course.—The husbandry is six shift: 1. Turnips. 2. Barley. 3. and 4. Seeds for two years. 5. Pease. 6. Wheat.

Corn.—Very greatly increased.

Sheep.—Increased certainly.

Cows.—Increased; neat stock also much; many bullocks fattened in marshes open to the sea, before the enclosure.

Tithe.—Free of all tithes.

Rates.—On an average 2s. in the pound.

Expenses.—2200l.

Poor.—The poor had an allotment of 90 acres for cutting flag, which they make little use of; I did not see one stack on it: no right of common on it: there are ten little proprietors of commonable-right houses, who have

3, 4, 5, 6, or more acres of land, keep cows, and are in comfortable circumstances.

Population.—No register kept for many years previous to 1760; I can therefore only give the last twenty, as if no enclosure had taken place.

Baptisms from 1780 to 1789	219
Burials	218
Increase	1
Baptisms from 1790 to 1799	267
Burials	141
Increase	120

SPROWSTON.

Quantity.—Eight hundred acres in common; no half year land.

Rent.—Worth 6s. raised to 14s.

Corn.—Great increase.

Sheep.—Decreased.

Cows.—None; nor will there be any.

TACOLNESTON—ACT 1778.

Quantity.—Six or 700 acres: with Thorpe 1000.

Soil.—Pretty good mixt soil.

Rent.—Quality 15s. to 20s. but now let higher.

Corn.—All to corn, and fences well got up: a valuable improvement.

STIFFKEY AND MORSTON.—ENCLOSED 1793.

Quantity.—About 3400 acres, and 1200 of salt-marshes. Very little common. Value about 1400l.

Tithe.

Tithe.—Remains subject. Glebe of two parishes laid together to one parsonage. Livings consolidated.

Object.—The object of the enclosure was chiefly to lay the land together, and to extinguish rights of shackage, &c. The common did not contain above 30 acres, and therefore was a very small object.

Rent.—Lord TOWNSHEND advanced his farms 240l. a year, on 1900 acres, for the expense of 1150l. his share of the enclosure, besides the tenants doing the fences: at least 1000l. a year added to the whole rental. The marshes are valuable only as a sheep-walk at certain times, for the spring tides overflow them.

Poor.—The cottagers that kept cows, asses, or horses, had allotments of land.

Sheep.—More, and of a far superior quality.

Cows.—More kept now than before; for one of the farms, the pasture of which was at a distance before, is now become a dairy one.

Expenses.—12s. an acre, exclusive of fences, which were done by the owners or tenants.

Improvement.—It has been a very capital one in every respect, and the manners of the people much ameliorated—less wandering and idleness. The improvement is the more remarkable, as the usual object in Norfolk enclosures (commons) was almost wanting.

Corn.—Being half year land before, they could raise no turnips except by agreement, nor cultivate their land to the best advantage: they raise much more corn than before.

STOKESBY.—ENCLOSED 1722.

This was an enclosure of about 350 acres of marsh common, which is on the river leading to Yarmouth: I walked three miles from Acle, and crossing the ferry, made

made inquiries for some person who could give me information, but all were long since dead: I was assured that three or four acres were laid to each common-right house, besides an allotment in common of eighteen acres of marsh for the poor that had no cottages of their own, which remains so at present. All those allotments, however, for rights, have been sold long ago, as the whole parish is now the property of one person. By means of the common left, there are 17 or 18 cows now kept by the poor people, who buy hay or other winter food. Poor rates about 4s. in the pound. The parish register at Yarmouth.

SWAFHAM.—NOT ENCLOSED.

Ten years, from 1780 to 1789:

Baptisms	619
Burials	489
Increase	130

Ten years, from 1790 to 1799:

Baptisms	655
Burials	389
Increase	266

Rent.—The rental in Swafham in 1795.

Land at 16s. an acre	£. 3108
Besides above 2000 acres of common.	
Houses	1875
Stock in trade and mills	120
Rectory, tithe and glebe	387
Vicar	150
	£. 5640

Number

Number of inhabitants in Swafham in

1782	- -	1877
1792	- -	2031
Mean number from Jan. 1782 to Dec. 1791		1954
Average baptisms each year	- -	63
Burials ditto	- - -	45

Baptisms, one in 31.

Burials, one in 43½.

Died from 1784 to 1794—eleven years, 484, or 44 average:

Of which under 1 year	-	136
At 1 ditto	-	30
2 to 4	- -	23
5 to 9	- -	12
10 to 19	- -	26
20 to 29	- -	42
30 to 39	- -	25
40 to 49	- -	39
50 to 59	- -	33
60 to 69	- -	57
70 to 79	- -	42
80 to 89	- -	19
		484

From December 1798 to February 1799,

There died, 1	of	82
1	of	99
1	of	95
1	of	78
1	of	89
1	of	83
1	of	73
1	of	77

ENCLOSING.

In 1798, Males — — 1022
Females — — 1126
                    ——
                    2148

Males between 15 and 60 — 494
Waggons — — — 44
Carts — — — 89
Farm horses — — 227

Price of the best wheat and barley at Swafham market, on the market day after Christmas, as ascertained by a corn rent paid by the rector to the vicar.

	Wheat.	Barley.
1750	£.13. 0 0 per last.	£.6 0 0 per last.
1751	16 0 0	7 0 0
1752	15 15 0	6 15 0
1753	14 0 0	8 0 0
1754	12 0 0	6 0 0
1755	11 0 0	5 10 0
1756	21 0 0	9 10 0
1757	18 0 0	8 10 0
1758	11 0 0	5 5 0
1759	12 2 6	5 10 0
1760	12 0 0	5 15 0
1761	12 5 0	5 15 0
1762	16 0 0	10 0 0
1763	15 0 0	7 0 0
1764	21 0 0	9 0 0
1765	19 10 0	10 10 0
1766	15 0 0	7 0 0
1767	21 0 0	9 10 0
1768	16 10 0	7 0 0
1769	16 10 0	7 15 0
1770	19 0 0	9 10 0
1771	21 0 0	10 10 0
1772	24 10 0	12 10 0

	Wheat.	Barley.
1773	£.23 0 0 per last.	£.11 10 0 per last.
1774	25 0 0	11 15 0
1775	18 10 0	11 0 0
1776	18 0 0	7 15 0
1777	20 0 0	9 10 0
1778	20 0 0	9 10 0
1779	13 0 0	7 15 0
1780	25 0 0	8 0 0
1781	20 0 0	7 5 0
1782	26 0 0	14 0 0
1783	23 10 0	13 15 0
1784	24 0 0	10 10 0
1785	19 0 0	10 10 0
1786	17 10 0	10 5 0
1787	21 0 0	10 0 0
1788	22 10 0	9 10 0
1789	26 0 0	11 15 0
1790	23 10 0	11 5 0
1791	21 0 0	12 10 0
1792	19 0 0	13 0 0
1793	22 0 0	15 0 0
1794	25 10 0	14 10 0
1795	55 0 0	17 0 0
1796	28 0 0	11 6 0
1797	20 0 0	10 10 0
1798	21 10 0	12 10 0
1799	48 0 0	28 0 0
1800	66 0 0	38 0 0
1801	37 0 0	20 0 0
1802	26 10 0	11 10 0

Until 1792 the last contained 21 coombs; since that time, 20.

THE HEATH.

There is no clay under it, that has yet been discovered: a deep sand; and it is questioned whether it will answer enclosing: the town of Swafham feeds it, and cuts fuel; and has a sheep-walk over it. It contains about 2000 acres.

TERRINGTON, ST. JOHN'S, AND ST. CLEMENT'S.—ENCLOSED 1790.

Quantity.—Lands allotted, 868 acres, all a salt-marsh common, embanked from the sea. One hundred and eighteen commonable rights on it.

Improvement.—The tract was worth less than nothing: being injurious by the commoners' cattle being often swept away by the tides; when embanked, it was valued to the poor-rates at 25s. an acre, and is let from 20s. to 40s.; average, 30s.

Produce.—From yielding nothing, it is now all ploughed and cropped with wheat, oats, cole seed, and some beans. None of it laid to permanent grasss.

Expenses.	£.	s.	d.
The bank cost, and was then deserted, the sea breaking it	4535	0	0
But afterwards completed	8032	0	0
	£.12,567	0	0
Sluice - -	309	0	0
A partition bank -	254	0	0
Act, Solicitor, &c. -	759	0	0
Commissioners, &c. -	514	0	0
Survey - -	200	0	0
	14,603	0	0
Sundries - -	2367	0	0
Total -	£.16,970	0	0

Population.—Ten years before embankment:

Baptisms from 1780 to 1789	272
Burials	224
Increase	48

Ten years since the embankment:

Baptisms from 1790 to 1799	291
Burials	269
Increase	22

This account is surprizing, if we look to the balance; that near 900 acres of waste should be brought into operose tillage, and have no effect, or rather a bad one on population, seems extraordinary. It could not result, I should suppose, from the profligacy introduced by the bankers, though it might be something, for that cause could not, one would suppose, be more than temporary, even if the effect was certain. Has it been caused by removing so much further the effect of the tide in ameliorating the atmosphere of this low tract of country? However, the comparison must not be estimated altogether from the balance, as stated above, because men leave a parish, and are buried elsewhere; the christenings are a better rule, perhaps, and these have increased.

THORPE ABBOTS.

Here Mr. PITTs has 56 acres of common, and determining to improve it as quickly as possible, advertised for labourers, and had above 40 people at work: he ditched and hollow-drained the whole in six weeks; and has now clayed it 80 loads an acre. Quality price about 14s. an acre.

TITCHWELL.—ENCLOSED 1786.

Quantity.—A small common, and much half year land.

Soil.—A good loam.

Corn.—Greatly increased. The half year land was in very irregular management; now in six shifts.

Sheep.—More than doubled.

Cattle.—Neat cattle trebled. The whole parish could not muster more than 25 cows before the enclosure; now one farmer has wintered 100 beasts.

Tithe.—Remains subject.

Rent.—Three hundred acres of salt-marsh were embanked by the act, and raised from 1s. 6d. to at least 20s.

POPULATION.

Baptisms from 1780 to 1789	45
Burials	29
Increase	16
Baptisms from 1790 to 1799	39
Burials	25
Increase	14

THORNHAM.—ENCLOSED 1794.

Quantity.—Two thousand one hundred acres; of which a ling common of 300 acres.

Soil.—Sandy loam.

Corn.—Much more produced than before the enclosure.

Sheep.—Before enclosing there was but one flock in the parish, of about 300; now not less than 900.

Cows.—Rather lessened; but not amongst the poor.

Tithe.—Remains subject.

Expenses.—Twelve shillings and sixpence an acre.

Poor.

Poor.—The common-right houses had an allotment of three roods, in lieu of shackage; and the common marsh stinted: they had also 29 acres of the ling common allotted for fuel, on which each poor person, and the occupiers of a common-right house had a right to cut 60 whin faggots; so that they are in a much better situation than before.

Baptisms from 1780 to 1789	205
Burials	176
Increase	29
Baptisms from 1790 to 1799	172
Burials	121
Increase	51

THORNAGE.

Baptisms from 1780 to 1789	57
Burials	44
Increase	13
Baptisms from 1790 to 1799	52
Burials	31
Increase	21

WEST TOFFTS.

Three thousand and six acres.

Mr. GALWAY	2804 acres.
Lord PETRE	202 acres.
Rated at	£.590
Inhabitants	64
Cottages	13
Poor-rates, 1801	£.86 0 10

UPTON.

UPTON.

Quantity.—800 to 1000 acres of common, no half year land; but great advantage by laying properties together.

Rent.—By giving contiguity, land of 5s. made worth 20s. The common was under water; now drained by a mill, and worth 18s. to 20s. per acre.

Corn.—Will be increased.

Sheep.—None before; but there will be some.

Cows.—Very few; but there will be a great increase of grazing.

POPULATION.

Baptisms from 1780 to 1789	100
Burials	49
Increase	51
Baptisms from 1790 to 1799	79
Burials	44
Increase	35

WALSHAM ST. LAWRENCE.

Baptisms from 1780 to 1789	105
Burials	71
Increase	34
Baptisms from 1790 to 1799	118
Burials	41
Increase	77

WALSHAM ST. MARIES.

Baptisms from 1780 to 1789	81
Burials	55
Increase	26
Baptisms from 1790 to 1799	62
Burials	38
Increase	24

About 2000 acres in the two parishes. They are going to enclose the common, and half year lands.

WALLINGTON, ELLOW, AND NORTH COVE, 1797.

Quantity.—Four hundred and twenty acres of common: no open field.

Rent.—Before enclosing, the common was worth nothing: now 14s. an acre.

Corn.—Three hundred and twenty acres ploughed.

Sheep.—They kept very few: but now will keep many

Cows.—Much increased.

Tithe.—Remains subject.

Rates.—An Incorporated hundred.

WALPOLE.

In 1770, Walpole, St. Peter, and St. Andrew.

Pasture land	4130 acres.
Arable	2050
Common, or waste, about	2500
	8670

In the occupation of 103 persons, at the yearly rent of	£.4760
Assessed to the land-tax, at	2907

Farmers

ENCLOSING.

Farmers and their wives and children	275
Servants	160
Labourers, and wives and children	81
Poor maintained by the parish	22
Souls	538
Houses	120
Cows	200
Horses	310
Beasts, young and feeding	580
Sheep	10,000
Hogs	330
Rates in 1730—Church	2d.
Poor	6
Highways	0
1760—Church	4
Poor	8
Highways	0
1767—Church	2
Poor	10
Highways	3

Quantity.—One thousand three hundred acres of salt-marsh were embanked and enclosed in 1789.

WHEATING.—ENCLOSED 1780.

Quantity.—About 4500: common 400 acres.

Improvement.—The chief object was laying property in the former open field together: there were 50 allotments of the common to commonable rights, eight acres each.

Corn.—Not increased; by reason of a large portion lawned and planted by a nobleman who resides here.

Sheep.—Are better than before the enclosure; but not so many by some hundreds.

Cows.—Greatly lessened.

Poor.

Poor.—Much better off in every respect since the enclosure.

WINFARTHING.—ACT 1781.

Quantity.—About 600 acres of heath and common: no open arable.

Soil.—Like that of Banham, but hardly so good.

Rent.—Quality price, from 7s. to 20s.

Corn.—All turned to arable.

Sheep.—Six to 700 sheep were kept, and lessened to 300; but of a much better sort.

Cows.—The same as before.

Improvement.—All clayed, and hollow-drained where wet.

Tithe.—Remains subject.

Rates.—Two-and-sixpence to 3s.

Expenses.—Something under 2000l.

Poor.—Twenty acres allotted for fuel; many allotments to little people; and they are well content; and have kept them: they have much hemp, wheat, &c. and well managed.

An account of the number of baptisms and burials in the parish church of Winfarthing, for forty years last past:

Burials from 1762 to 1780	335
Baptisms	190
Increase	145
Baptisms from 1781 to 1799	342
Burials	189
Increase	173

WRENNINGHAM.—ALLOTTED IN 1779.

Quantity.—Two hundred and sixty acres.

Rent.

Rent.—Before 15s. an acre; now 20s. the whole parish.
Tithe.—Free: an allotment in land, 37 acres.
Poor.—Twenty acres for fuel allotted.
Corn.—A great increase.
Expense.—About 620l.

WOODBASTWICK, 1767.

Quantity.—About 300 acres of common.
Rent.—Worth above 14s. an acre at present.
Course.—1. Turnips; 2. Barley; 3. Clover; 4. Wheat.
Sheep.—Decreased.
Cows.—Lessened.

WOODRISING.—ENCLOSED ONLY PARTIALLY.

Baptisms from 1780 to 1789	27
Burials	16
Increase	11
Baptisms from 1790 to 1799	42
Burials	18
Increase	24

GENERAL OBSERVATION.

Mr. BURTON, of Langley, on all wet commons, recommends to under-ground drain before breaking up; and where level, to clay it also. To be began at Midsummer, then left to March, when it should be ploughed and oats dibbled: and if it was left a year or more for grass to get up, it might be the better. After the oats, if not clayed before, clay it: if it was clayed, scale in the stubble after harvest, and plough three or four times for oats, to be sown broad-cast. After that, fallow for turnips, and pursue the common husbandry, which would want no muck for a few years.

ENCLOSING.

In fifteen enclosures, in which Mr. BURTON has been Commissioner, there are 10,800 acres of common land; about half converted to arable, and proper for the five-shift husbandry. One thousand acres for wheat, producing 5000 combs, at 24s. 6000l.; 1000 barley, 8000 coombs, at 12s. 4800l.; 1000 oats, second crop after wheat, 8000 coombs, at 10s. 4000l.; 1000 acres turnips will feed 500 bullocks or cows, which will pay for the turnips only, 5l. a head, or 2500l.; and 1000 acres of grass, with the offal turnips, will feed 4000 sheep in winter, and fattened by a part of the new lay grass, to be off by June. The sheep, for the turnips and grass, will pay 3000l.: there remain 5000 acres of pasture, which will support 500 bullocks, bringing them forward for turnips, to 3l. per head, or 1500l.: and there may also be summer-kept 4000 sheep, which, with wool and profit on carcass, will pay 2000l. And, besides all this, 200 cows, at 6l. or 1200l.

Wheat	6000
Barley	4800
Oats	4000
Bullocks	2500
Sheep	3000
Bullocks	1500
Sheep	2000
Cows	1200
	25,000

There are 800 acres more unaccounted for: 5000 acres of arable will take 200 horses to till it, wanting each four acres arable and pasture; 400 of this for corn will give 10 coombs a horse, for 30 weeks, or 2000 coombs, at eight coombs an acre; 250 acres for the 2000 coombs: 150 acres remains for hay, which producing 200 tons, will,

will, with the barns, maintain them the winter; and the 400 acres of pasture will support the 200 horses, with mares and colts, in summer. The improved rent on the 10,000 acres, is for a lease of 14 years, 7725l. The tithe, for 14 years, about 1500l. a-year. Capital to stock, and improvement, about 10,000l. being additions to farms adjoining. The interest 500l. Labour for said lands, including harvest, 5000l. Seed corn, 2200l. Wear and tear; the blacksmith will be about 500l.; wheelwright, about 100l.; carpenters, 100l.; small seeds, 300l.; poor-rates, on an average, at 5s. 1031l. 5s.; capital for 5000 acres of pasture, 6000l.; interest, 300l. Contingencies, &c. may be estimated at 1000l.

	£.
Tithe	1500
Interest of capital, on the arable	500
Labour	5000
Seed corn	2200
Wear and tear	700
Small seed, 1000 acres, at 6s.	300
Rates	1031
Interest of capital, for pasture	300
Contingencies	1000
	13,531
Rent	7725
	20,256
Produce	25,000
Expenses	20,256
Profit	4,744

This calculation, which does Mr. BURTON credit, sets in a very clear light, the immense advantages which have resulted

resulted from the enclosures in which he has been employed.

In all the enclosures in which Mr. ALGUR has been concerned as a Commissioner, it has not been the practice to put poor men to the proof of the *legality* of their claims, but the mere practice, and if they have proved the practice even of cutting turf, it has been considered as a right of common, and allotted for accordingly.

In the several parliamentary enclosures of Snettisham, Ringstead, Heacham, Darsingham, Sharnborne, Fring, Sedgford, Thornham, Tichwell, and Hunston, which are all in Mr. STYLEMAN's vicinity, the effect has been, at least doubling the produce: many of the farm-houses were before in the villages, and the distance to the fields so great, that no improvements were undertaken. The houses are now on the farms, and the improvements very great. In respect to the rent, the first leases were at a low rate, from the allotted lands being in a most impoverished condition: but on being let a second time, where that has taken place, the rise has been considerable.

In enclosures, in which Mr. GOOCH, of Quidenham, has been a Commissioner, claying found of great advantage upon fresh land; it divides and mixes well with the decaying turf. Mr. GOOCH recommends one earth for oats, by all means dibbled; then a second crop harrowed: then turnips and barley, or oats laid with seeds; clover, trefoile, and a little ray, or better no ray. Leave it a year, and then clay, if a deep loose bottom, 100 loads; but if shallow, near brick-earth or clay, about 60. Leave it with the clay on it a year, and it incorporates immediately.

In the enclosures, particularly remarked by Mr. BRADFIELD, of Heacham, two methods have been pursued, but that intelligent farmer recommends on dry land, after grubbing the whins, to plough and dibble in oats; then to clay

or

or marle on the oat stubble, and summer-fallow for wheat; after that, to take turnips, and follow the four-shift husbandry, of 1. Turnips; 2. Barley; 3. Clover; 4. Wheat; which he thinks better for keeping land clean, than letting seeds lie two years.

FENCES.

In these several Norfolk enclosures, the fences consist of a ditch four feet wide and three deep, the quick laid into the bank, and a dead bush hedge made at the top.— Expense:

	s.	d.
Digging, banking, and planting - -	1	3
Bushes, a load 20s. does near 30 rods -	1	3
Quick - - - -	0	6
	3	0

The Banham and Carleton enclosures are well grown, and the fences excellent; but this much depends on good care and attention; for if the banks are made too steep, or neglected, they slip down, and gaps are the consequence.

Mr. REEVE, of Wighton, in forming new fences, gives a complete summer-fallow to the lines where the quicks are to be set, and dresses the land with a good compost: and instead of leaving the bank in a sharp angle at top, he flattens it, to retain the moisture.

Mr. HILL remarks, that in making new fences, a southerly aspect, in strong land, will be a fence four years sooner than a northerly one; but on light sandy land, the north best, for the sake of moisture; on mixed soils, he prefers the east and west.

Mr.

Mr. COKE has moveable gates and posts, if they may be so called, to place in rows of hurdles: they are a very useful contrivance; set down expeditiously, and moved with great ease.

CHAP. VII.

ARABLE LAND.

WE are now come to the grand object of Norfolk husbandry: in all the other branches of agriculture the county is not conspicuous for singularity of system, but in the management of arable land much indeed will be found interesting.

SECT. I.—OF TILLAGE.

It would be easy to expatiate on every branch of this subject largely, but as much, in that case, would be inserted in this Report, equally applicable to almost every county in the kingdom, the writer will confine himself to the observations he actually made in Norfolk.

Ploughing.—There is a great difference in ploughing in West Norfolk; on some farms I have remarked the furrow to be cut flat and clean, but on others *wrest baulked*, by tilting the plough to the left, which raising the share fin, makes that inequality, and is partly the occasion of my having found so many ploughs at work which would not go a single minute without holding.

In East Norfolk the ploughmen, to prevent the soil when moist from turning up in whole glossy furrows, which they term " scoring," tie a piece of strong rope-yarn round the plate or mould-board, which by these means is prevented from acting as a trowel upon the soil.

MARSHALL.—I found this the practice at present, but
was

was informed that it was not so generally wanted as formerly, which they attributed to better constructed *plats*, or mould boards.

Depth.—Mr. THURTELL, at Gorlstone, near Yarmouth, has a great opinion of deep ploughing; three or four inches are a common depth about him, but when land is clean he always ploughs five, and sometimes six or seven; he is careful, however, not to do this on foul land; he has no apprehension of breaking the pan, having many times gone depth enough for that without any inconvenience, and as to bringing up a dead soil, he has not seen any ill effects from it: his land is a good sandy loam, on a clay, marle or gravel.

Mr. EVERIT, of Caistor, is of a different opinion; he is not fond of deep ploughing; he thinks four inches deep enough; his ideas on this subject, however, seem to have been chiefly the result of a trial made by his father, who broke the *pan* by trench-ploughing a piece of land, which has ever since been full of charlock, &c. The difference of their soils will not account for this opposition of sentiment: Gorlstone is a very good sandy loam at 20s. or 25s. an acre, though certainly inferior in depth and goodness to the land at Caistor. In discourse with these two gentlemen on this subject, Mr. EVERIT remarked to Mr. THURTELL, that if he ploughed deeper than common, he ought to add manure proportionably to the quantity of soil stirred; an old idea of mine, and I remember well, combated by my friend ARBUTHNOT.

Mr. FRANCIS, of Martham, ploughs four or five inches deep.

Mr. CUBIT, of Honing, on a fine sandy loam, always as shallow as possible; and at East Ruston, where the soil is exceedingly good, the same: he thinks the smaller quantity of muck by this means answers.

They

They do not plough four inches deep at Scotto; Mr. Dyble remarked, that a piece was there ploughed five or six inches deep, and damaged for seven years: three inches enough: the soil much of it a fine sandy loam, manifesting no want of depth.

Mr. Palgrave, at Coltishal, applies deep ploughing in one case with singular judgment: he brings by water from Yarmouth, large quantities of sea ouze, or haven mud, this, on dry scalding gravels and sands, he trench-ploughs in without fear of burying, and finds, on experience, the effect very great, forming thus a cool bottom, so that the surface burns no more.

Mr. Johnson, of Thurning, thinks that it is common to plough too shallow; nor does he believe that any mischief results from depth. He has made a ditch one year, and thrown it down again the next, and the benefit was seen for seven years, without the soil being acted on by draining, or wanting it: nor is he nice to have his muck ploughed in shallow, having no fear of burying it. By ploughing a good pitch for turnips, they come slow to the hoe, but when they do get hold, thrive much faster than others.

Mr. Reeve, of Wighton, is an advocate for deep ploughing; he goes five inches deep; *if I did not I should get no turnips.*

Mr. Dursgate approves of deep ploughing; remarking, that he breaks up his *ollonds* deeper than most people.

Mr. Willis observing the marle on his land was sunk below the common path of the plough, turned it up again by going a deeper pitch, and found it to answer nearly as well as a new marling; and he suffered no inconvenience.

The two furrow work about Holt, &c. is to turn a furrow on lay or ollond; the plough then returns and throws it back with the untouched land that is under it,

into

into the former open furrow, and overlapping that, rests on the baulk left beside it.

Mr. MONEY HILL, in breaking up the strongest land he has, ploughs deeper than on the lighter, that is, four inches, and on light three and a half, and on that depth drills on flag; if twitch in the land, ploughs only three and a half: if beyond the usual depth, would hurt the crop and give weeds. Waterden, a thin and flinty soil. Mr. HILL's father lived for many years at Gateson, and was succeeded in the farm by Mr. PARKER. Mr. HILL's last crop yielded 400 lasts of corn, above 250 of which were barley. He generally ploughed four inches deep, and never more than four and a half. Mr. PARKER in the first year ploughed the second barley earth seven inches deep, sowing about eight score acres (the common quantity twelve score). He sold that year but twenty lasts: the seeds also were worse than usual: the wheat that followed, good; but in general, he had indifferent crops for fourteen or fifteen years.

In March he applies what he calls *one-furrow* work to a foul stubble, if he has such by chance; the land side horse (that on the left hand) always, after the first furrow, returns in it; it is left open; harrowed down with a heavy harrow; then the weeds gathered and burnt, and the next ploughing given across.

In June 1776, being at Wallington, adjoining Marshland, I found the high broad ridges begin, which thence spread over a great tract of country, nearly perhaps across the island; and many of the furrows were *then* twelve inches deep in water: but at present I was informed that much more attention is paid to taking water off.

Team, &c.—Mr. THURTELL, through the summer, ploughs with three horses two acres a day, one always resting: this, from finishing sowing spring corn to the end

end of turnip tillage. There is no doubt of their ploughing with ease an acre in four hours and a half.

In the clays of Marshland all are foot or swing ploughs; never more than two horses used: they do an acre a day, and in summer one and a half at two journeys.

At Hemsby, each pair of horses two acres a day, at two journeys.

Thirty years ago the common price of ploughing was 2s. 6d. an acre in every part of Norfolk, except Marshland: it is now 4s.; in some places 3s. 6d.

These notes of the quantity ploughed per diem, might be multiplied in every part of the county; the farmers in every district of it, get more land ploughed in a day by their men and horses, than on any similar soil in any other part of the kingdom: not altogether to be attributed to the merits of the plough, though it is certainly a good one; nor to any superior activity in the horses: the cause is more in the men, who have been accustomed to keep their horses and themselves to a quick step, instead of the slow one common in almost every other district.

Harrowing.—This operation is no where better performed than in Norfolk, where the farmers are very attentive to finish their tillage in a very neat manner. In no other county with which I am acquainted, have they the excellent practice of trotting the horses at this work, which gives a fineness and regularity in burying small seeds, not to be attained with a slow regular motion of the harrows. They harrow from 12 to 15 acres with a pair of horses once in a place per diem. The practice of walking the horses up hill, and trotting them *down*, in the same place, is an excellent one.

SECT.

SECT. II.—FALLOWING.

The grand fallow of Norfolk is the preparation for turnips, which will be mentioned under that article. The common summer-fallow takes place on strong, wet, and clayey soils; upon which, however, turnips are too generally ventured. One fault in the husbandry of the county, and of Norfolk farmers when they move into very different districts, is that of being wedded too closely to practices which derive their chief merit from a right application to very dry or sandy soils.

Mr. Overman, whose husbandry merits every attention, having taken a farm of Mr. Coke, at Michaelmas 1800, and the outgoing tenant possessing a right to sow some layers which were very full of spear-grass, &c. Mr. Overman gave him, to the surprize of his neighbours, 5l. 10s. per acre to desist; not that he might himself sow those fields, but for the sake of completely fallowing them. Some I found had undergone the operation, and were clean; others were in it, and almost green with couch. He destroys it by mere ploughing and harrowing, without any raking or burning; conceiving that by well-timed tillage, any land may be made clean; and that on these sandy soils, a July earth in a hot sun will effect it: but whenever or however done, his object, whether with much or little tillage, is sure to be answered; and as the successive cleanness of the land depends on its being once got perfectly free from weeds, his great expenses, he expects, will in the end prove the cheapest way of going to work. He gives four earths in all; the first before winter, only two inches deep; another in the spring; the third two, or two inches and a half deep, in July, in a hot time; the fourth after harvest.

1803.

1803.—He now tells me, that the land I saw full of couch, is at present as clean as a garden.

Mr. PITTS, of Thorpe Abbots, remarked to me the great consequence, for the destruction of weeds, of ploughing a summer-fallow just before, and also directly after harvest.

SECT. III.—COURSE OF CROPS.

IF I were to be called on to name one peculiar circumstance, which has done more honour to the husbandry of Norfolk than any other to be thought of, I should, without hesitation, instance this of the rotation of cropping.

I should not hazard, perhaps, too bold an assertion, were I to declare that, till the accession of his present MAJESTY, there were to be found few just ideas on this subject, in the works upon husbandry of any author preceding that period: if any thing tolerable occurs, it is mixed with so much that is erroneous, that credit cannot be given even for what is good. The fields of the rest of the kingdom presented a similar exhibition: right courses hardly any where, perhaps, no where, to be found. But in West Norfolk, the predominant principle which governed their husbandry at that period, as well as ever since, was the carefully avoiding two white corn crops in succession. Turnips were made the preparation for barley; and grasses, that for wheat, or other grain.

I have viewed various parts of the county, at different periods in the last thirty years, and have found these ideas steadily adhered to.

COURSES ON SAND, AND ON TURNIP LOAMS.

Some of the finest rye I have any where seen, was on Mr. BEVAN's farm, in 1802, after two successive years of sowing cole, eaten off with sheep. The rye put in on one earth; there are thirty acres of it, and fourteen of them on a black sand.

Mr. BRADFIELD, of Knattishall, tenant to Mr. BEVAN, pursues regularly this course of crops:

1. Turnips,
2. Barley,
3. Seeds,
4. Seeds,
5. Wheat,
6. Turnips,

But if seeds fail, changed to

1. Turnips,
2. Barley,
3. Vetches,
4. Turnips,
5. Barley,
6. Turnips;

by which means, in the sixth year, the variation ceases, and it comes, as in the other, to turnips again. The system, however, is open to two great objections: in the fourth year, he has no summer food for sheep; and, what is as bad, he doubles his quantity of turnips; he also loses wheat in the course. To have two successive years of vetches, appears to be a better system, and a much less interruption, or rather none at all. If the first vetches are to be fed, grass-seeds might be sown with them for the second year, and this would save the expense of seed vetches and tillage for that year.

In 1802, the barley crop is generally very fine; yet in the whole line from Holkham to Toffts and Thetford, I remarked many pieces which in colour were too yellow for Norfolk management in a good year. I have some suspicion that it has been caused by the very high prices of corn inducing some farmers to be too free with their land, and varying from the course of shifts, which, in steady times, they adhere to more exactly.

Mr. FOWEL, of Snetterton:
1. Turnips, drilled at eighteen inches,
2. Barley, ditto at nine,
3. Seeds,
4. Seeds,
5. Pease,

5. Pease, drilled at twelve inches; or wheat at nine, &c. and this is the rotation of the vicinity. Wheat, pease, oats, or rye, the fifth year; if rye, a bastard fallow for it: the second year, seeds.

In general about Hingham:

1. Turnips,	1. Turnips,
2. Barley,	2. Barley,
3. Clover,	3. Clover,
4. Wheat,	4. Pease,
	5. Wheat.

Nearly the same around Attleborough.
About Watton:

1. Turnips,	3. Clover, &c.
2. Barley,	4. Wheat.

Mr. BLOMFIELD, at Billingford, in one field near his farm-yard:

1. Winter tares, and then turnips,
2. Barley;

and the crops always good.

Mr. DRAKE gets better turnips after wheat, the stubble ploughed in, than after pease.

Mr. WRIGHT, of Stanhow, never takes barley or pease after wheat, though his soil is a good loamy sand: he thinks that no district where this is the practice deserves the reputation of having the true Norfolk husbandry.

Mr. DROZIER remarked, that upon the sandy land of Rudham, and that vicinity, the greatest improvement perhaps would be, to lay down for eight or ten years to repose, the land from turnips and corn, which would so freshen it as to render it productive perhaps in the stile of the first breaking up; but common grasses wear out, and will not pay the present rents after two years: they sow trefoil and ray.

Sir

Sir MORDAUNT MARTIN's course is a five-shift.

1. Turnips,
2. Barley,
3. Clover,
4. Wheat,
5. Potatoes, mangel wurzel, or vetches, &c.
6. Turnips,
7. Barley,
8. Trefoil and ray,
9. Pease,
10. Potatoes, mangel wurzel, vetches, &c.

Mr. OVERMAN, of Burnham, has found, from many observations, that pease do not succeed well if sown oftener than once in twelve years. Where he has known them return in six or eight years, they have never done well.

Mr. OVERMAN ploughed up a layer of four years, and drilled wheat on it. Then ploughed for winter tares. Ploughed the stubble once for a second crop of wheat, which I viewed: a very fine produce, and as clean as a garden. Three crops of great profit on only three ploughings, and yet the land kept perfectly clean. Not a little resulting from four years sheep feeding without folding from it.

His common course:

1. Turnips,
2. Barley,
3. 4. 5. Seeds three years,
6. Wheat,
7. Turnips,
8. Barley,
9. 10. Seeds two years,
11. Pease,
12. Wheat.

But with the variation of having part of the twelfth under pease on the three years layer, and also some tares. This course is partly founded on the experience of pease not doing well, if sown oftener than once in twelve years.

Mr. COKE:

1. Turnips,
2. Barley, drilled at 6¾ inc.
3. Seeds,
4. Seeds,
5. Wheat, drilled at 9 inc.
6. Turnips,
7. Barley,

7. Barley, drilled at 6¾ inc.   10. Pease, drilled at 9 inc.
8. Seeds,   or tares at 6,
9. Seeds,   11. Wheat, drilled at 9 inc.

Mr. PURDIS, of Eggmore, a very uncommon variation from the general husbandry:

1. Turnips,   4. Seeds,
2. Barley,   5. Tares,
3. Seeds,   6. Wheat.

Upon a large part of this fine farm the former rotation included a summer-fallow, which afforded (broken at whatever time) little food for live stock; tares now occupy the place, and support immense herds of cattle and sheep. What a noble spectacle is this farm! 300 acres of turnips, 300 of barley, 600 of seeds, 300 of tares, and 300 of wheat: 1800 acres arable, the crops luxuriant, and much the greater part of the farm very clean; all of it except the layers; on which, however, some thistles, so difficult to extirpate.

Mr. THURTELL, near Yarmouth, is in the four-shift, returning to turnips always after the wheat, for he thinks that nothing is so bad as taking a fifth crop.

At Caistor, in Fleg, the land excellent; they are in the five-shift of East Norfolk, that is,

1. Turnips,   4. Wheat,
2. Barley,   5. Barley;
3. Clover,

with two variations practised sometimes by Mr. EVERIT, at the Hall farm.

1. 2. Cole seed instead of turnips and barley, taking two years,
3. Wheat,
4. Barley; but not a great crop; and then turnips again.

The other is, to substitute pease instead of clover, followed

lowed by wheat, and then in the four-shift, to come again to turnips.

A remarkable circumstance in the rotation of crops here is, that spring corn will not succeed well after wheat which follows cole seed: they will give an excellent summer-fallow for this crop; spread 14 loads of fine dung per acre, and sowing wheat after the cole, get the finest crops, yet if barley or oats follow, the produce is seldom tolerable; oats better than barley, but neither good.

Some farmers at Hemesby, and among others Mr. FERRIER, on his own property:

1. Turnips,
2. Barley,
3. Seeds (clover once in 10 or 12 years),
4. Wheat,
5. Pease, or oats,
6. Wheat.

It may easily be supposed that the wheat of the fourth year is much better than that of the sixth. The course cannot be defended even on Hemsby land, and the wheat stubbles were some of them not so clean as they ought to be.

At Thrigby, Mr. BROWN, &c. is in the Fleg five-shift; barley after the wheat; with the variation, to avoid clover every other round, of sowing half the barley with other seeds, and dibbling pease on the other half.

At South Walsham, Mr. SYBLE, and others:

1. Turnips,
2. Barley,
3. Seeds, one or two years,

Unworthy of Norfolk in any case whatever.
{ 4. Wheat,
5. Barley, or oats,
6. Pease,
7. Wheat.

The variation of the seeds is to prevent clover coming two rounds together, as the land here, as elsewhere, is sick

sick of it. Upon a part, white clover, trefoil, and ray are substituted, and left two years; about one third of the wheat is on a two years layer. If the clover be a good crop, the wheat is better than after the other seeds. The barley after wheat (if that followed a two years lay), is better than after turnips. But Mr. SYBLE, if the land is foul after the first wheat, is sure then to take turnips. Sometimes pease on a two years lay, and then wheat; but Mr. SYBLE does not like pease, from their being so liable to failure. He is of opinion that the husbandry of Fleg and Blowfield wants variation, from having been kept too long in a regular course. One which has succeeded with him, is to sow barley after pease or vetches, in which way he has had great crops.

At Repps and Martham the common *Fleg five-shift* husbandry; that is, barley follows wheat; clover and other seeds alternate, and the wheat as good after one as the other.

At Ludham, the common five-shift.

At Catfield I found a variation; here the course is a six-shift husbandry:

1. Turnips,
2. Barley,
3. Clover, &c.
4. Clover, &c.
5. Wheat,
6. Barley.

Mr. CUBIT practises this in common with his neighbours; the seeds *riffled* the second year before harvest, that is, *rice-baulked, raftered,* half ploughed: some *scaled,* a clean earth, as thin as possible; the management Mr. THURTELL reprobated for his soil: and what is singular, they seem to do it with equal reference to dibbling and broad-casting.

At Honing the same as at Catfield. All around North Walsham the same; and thence to Preston, Cromer, and
Ayle-

Aylesham, in general the same husbandry. If seeds fail, some scale the stubble in, and keep it for pease.

Mr. MARGATESON, of North Walsham, assured me, that he considered the four-shift husbandry of turnips, bad; clover, wheat, as the best of all management, if the land will bear it; but clover has been sown so long in East Norfolk, that it is sure to fail in that rotation.

Mr. DYBLE, at Scotter, is in the six years course, which is also common through the country.

Mr. REPTON, at Oxnead, has been, from the year 1773, regularly in the six-shift husbandry, of

1. Turnips, 4. Seeds—ollond,
2. Barley, 5. Wheat,
3. Seeds—hay, 6. Barley;

which is common throughout the country. I took an account of several of his fields; and found but few variations; accidentally, wheat occurred on the first, instead of the second year's layer. Sometimes the barley omitted after the wheat. In a few instances, pease on the ollond, or hay and wheat, &c. after the pease: but the variations few; so as to shew clearly the established rule.

Mr. REEVE, of Heveringland, in the five-shift; the seeds two years.

Mr. BIRCHAM, at Hackford:

1. Turnips, 4. Clover, and other seeds
2. Barley, alternately,
3. Clover, and other seeds 5. Wheat,
   alternately, 6. Barley, oats, or pease.

But if land be out of condition, Mr. BIRCHAM's method to recover it, is to take,

1. Turnips, 3. Turnips,
2. Barley, 4. Barley.

The same six-shifts at Haydon.

Mr.

Mr. JOHNSON, of Thurning, thinks that the common course of,

1. Turnips,
2. Barley,
3. Seeds, two years
4. Seeds,
5. Wheat,
6. Barley;

which, is the usual system about him, would be improved by the following variation:

1. Turnips,
2. Barley,
3 Seeds,
4. Seeds,
5. Seeds,
6. Pease,
7. Wheat,
8. Barley;

in which the land would have rest for feeding, three years in eight, instead of two in six, as in the other.

Mr. ENGLAND, at Bingham:

1. Turnips,
2. Barley,
3. Seeds,
4. Seeds,
5. Wheat;

never adding barley after the wheat. Sometimes on tender land, not equal to wheat, drills pease on the ollond, and then, if the land be clean, takes barley, or even wheat, but not without rape cake.

Mr. REEVE, of Wighton:

1. Turnips,
2. Barley,
3. Trefoil, white clover, and ray,
4. Ditto,
5. Wheat, drilled,
6. Turnips,
7. Barley,
8. Clover,
9. Wheat.

Every idea of this most accurate farmer, merits much attention; and this course among the rest: whenever red clover is left a second year, it disappears, and the land is principally covered with ray grass: query—if it is not much better, when red clover is the object, never to leave it two years: this is an improvement in Mr. REEVE's intention, but it has not yet been his general practice.

Mr.

Mr. REEVE mucked a barley stubble for vetches, ploughed once for that crop; and then drilled wheat on one other ploughing. The stubble clean as a garden.

Mr. M. HILL:

1. Turnips,
2. Barley,
3. Seeds,
4. Seeds,
5. Wheat,
6. Turnips,
7. Barley,
8. Seeds,
9. Pease, or tares,
10. Wheat.

Mr. M. HILL has now (1802) on his farm a very fine field of wheat, drilled on the flag of a four years layer: he remarked it as an instance of confidence in his landlord (Mr. COKE), not to have broken it up sooner at the end of a lease.

The same farmer sows cole after winter vetches fed off; eats it off at Michaelmas; sows rye for spring feed; eats that off, and tills for turnips, getting four green crops to feed on the land in two years. Excellent husbandry.

Mr. HENRY BLYTHE, of Burnham, is in the five-shifts; the seeds for two years.

The following detail of all the courses pursued by Mr. DURSGATE, on his fine farm of Summerfield, is, in my opinion, a most satisfactory account of Norfolk husbandry, shewing the very considerable exertions made in this county for the great objects of keeping land clean and in heart: the particulars merit a close examination. These fields contain near 1050 acres.

No. 1.—Fring Break.

1797 Summer-fallow.
1798 Wheat, manured with half a ton of rape-cake per acre.
1799 Turnips, mucked, and fed with sheep.
1800 Barley.

1801 Clover and ray, mowed.
1802 Ditto, fed with sheep.

No. 2.—Upper-end of Cow-close.

1797 Half summer-fallowed. Half vetches, fed by sheep.
1798 Wheat, rape-cake on one-half of it, one-quarter of a ton; on the other half, one-third of a ton.
1799 Turnips, mucked, and fed with sheep.
1800 Barley.
1801 Trefoil and ray, half of it mowed.
1802 Ditto, fed with sheep.

No. 3.—Fox Close, new broken up fox-cover.

1797 Oats.
1798 Wheat.
1799 Turnips, drawn.
1800 Barley.
1801 Trefoil and ray, fed with horses.
1802 Ditto, fed with horses and sheep.

No. 4.—First Burnt Stock.

1797 Turnips, one-quarter of a ton of rape-cake per acre, fed with sheep.
1798 Barley.
1799 Turnips, mucked, and fed with sheep.
1800 Barley.
1801 Trefoil, fed with sheep.
1802 Ditto, ditto.

No. 5.—Church-Hill.

1797 Pease, barley, and vetches.
1798 Turnips, mucked, and one-quarter of a ton of rape-cake per acre; fed with sheep.
1799 Barley.
1800 Clover, mowed.

1801 Clover, fed with sheep.
1802 White pease.

No. 6.—Black Hurn.

1797 Ollond, two years, fed with sheep.
1798 Turnips, mucked, and fed with sheep.
1799 Barley.
1800 Trefoil, mowed.
1801 Ditto, fed with sheep.
1802 Vetches.

No. 7.—First part Cow-close.

1797 Wheat.
1798 Turnips, one-quarter of a ton of rape-cake per acre; fed with sheep.
1799 Barley.
1800 Trefoil, fed with sheep.
1801 Ditto, fed with horses.
1802 White pease.

No. 8.—Stack-yard Break.

1797 Turnips, mucked, and fed with sheep.
1798 Barley.
1799 Clover and ray, mown.
1800 Ditto, fed with horses and sheep.
1801 White pease.
1802 Wheat; one-third of a ton of cake per acre.

No. 9.—New-pit.

1797 Turnips, half mucked, and on the other half a quarter of a ton of cake per acre, fed with sheep.
1798 Barley.
1799 Trefoil, ray, and white clover, fed with sheep.
1800 Ditto, fed with sheep.
1801 White pease.
1802 Barley.

No. 10.—Hilly-piece.

1797 Turnips, one-quarter of a ton of cake per acre, and fed with sheep.
1798 Barley.
1799 White clover and ray, fed with sheep.
1800 Ditto, sheep fed.
1801 White pease.
1802 Barley.

No. 11.—Pound Close.

1797 Turnips, mucked, and sheep fed.
1798 Barley.
1799 Trefoil, white clover, and ray, sheep fed.
1800 Ditto, sheep fed.
1801 Vetches.
1802 Barley.

No. 12.—First 19 Acres.

1797 Turnips, mucked, and sheep fed.
1798 Barley.
1799 Clover and ray, mown.
1800 Ditto, sheep fed.
1801 White pease.
1802 Wheat; one-third of a ton of cake per acre.

No. 13.—First 14 Acres.

1797 Turnips, mucked, sheep fed.
1798 Barley.
1799 Trefoil, white Dutch, and ray, sheep fed.
1800 Ditto, sheep fed.
1801 Vetches, soiled.
1802 Wheat, part mucked, the rest one-third of a ton of cake per acre.

No. 14.—Second 14 Acres.

1797 Turnips, mucked, sheep fed.
1798 Barley.
1799 Trefoil, ray, and white clover, sheep fed.
1800 Ditto, sheep fed.
1801 White pease.
1802 Wheat; one-third of a ton of cake per acre.

No. 15—Lady Summersby Break.

1797 Barley, after mucked turnips, and sheep fed.
1798 Clover, ray, and white Dutch, mown.
1799 Ditto, sheep fed.
1800 Pease.
1801 Wheat; one-third of a ton of cake per acre.
1802 Turnips, mucked.

No. 16.—Saffron-row.

1797 Barley, after turnips, one-quarter of a ton of cake per acre, and fed with sheep.
1798 Trefoil, ray, and Dutch, mown.
1799 Ditto, sheep fed.
1800 Forty acres of it summer-fallow; 11 of it pease.
1801 Wheat; one-third of a ton of cake per acre.
1802 Turnips, mucked.

No. 17.—Thoroughfare.

1797 Barley, after turnips, mucked, and sheep fed.
1798 Clover, ray, and Dutch, mown.
1799 Ditto, sheep fed.
1800 Pease.
1801 Barley.
1802 Turnips, mucked.

No. 18.—Second Burnt Stock.

1797 Barley, after wheat.
1798 Trefoil, ray, and Dutch, sheep fed.
1799 Ditto, sheep fed.
1800

1800 Trefoil, and then bastard summer tilth.
1801 Wheat; half of it tathed; half of it manured with cake, one-third of a ton per acre.
1802 Turnips, one-quarter ton cake.

No. 19.—Eleven Acres.

1797 Trefoil, ray, and Dutch, sheep feed.
1798 Ditto, sheep fed.
1799 Ditto, ditto.
1800 Vetches, soiled.
1801 Barley,
1802 Turnips; one-quarter of a ton of cake per acre.

No. 20.—Brick-kiln Breck.

1797 Clover, ray, and Dutch, mown.
1798 Ditto, sheep fed.
1799 Ditto, 20 acres of it sheep fed; 33 ditto, bastard-fallow.
1800 Wheat, ploughed up 15 acres, and sowed Tartarian oats.
1801 Turnips, mucked; and part of it one-quarter of a ton of cake per acre; sheep fed.
1802 Barley.

No. 21.—Black Hurn, adjoining.

1797 Barley, after turnips, one-quarter of a ton of cake per acre; sheep fed.
1798 Trefoil, ray, and Dutch, sheep fed.
1799 Turnips; one-quarter of a ton of cake per acre; sheep fed.
1800 Wheat, drilled, one-quarter of a ton of cake.
1801 Turnips, mucked; sheep fed.
1802 Barley.

No. 22.—First part Long Snelling.

1797 Trefoil, ray, and Dutch, sheep fed.
1798 Ditto, sheep fed.

1799 Half vetches, soiled: half summer-fallow.
1800 Wheat; one-third of a ton of cake per acre.
1801 Turnips, mucked part; part caked, one-quarter of a ton per acre; one-half drawn; one-half sheep fed.
1802 Barley.

No. 23.—Seventeen Acres.

1797 Trefoil, ray, and Dutch, sheep fed.
1798 Ditto, sheep fed.
1799 Summer-fallow.
1800 Wheat; one-third of a ton of rape-cake per acre.
1801 Turnips; one-fourth of a ton of ditto, sheep fed.
1802 Barley.

No. 24.—Paddock.

1797 Trefoil, ray, and Dutch, sheep fed.
1798 Ditto, sheep fed.
1799 Summer-fallow.
1800 Wheat; one-quarter of a ton of cake per acre.
1801 Turnips, mucked, sheep fed.
1802 Barley.

No. 25.—Home Break.

1797 Trefoil, ray, and Dutch, one-half fed, one-half mown.
1798 Ditto, sheep fed.
1799 Wheat; forty acres of it manured with one-quarter of a ton of cake per acre.
1800 Turnips; forty acres, one-quarter of a ton of cake; twenty acres mucked.
1801 Barley.
1802 Clover; forty acres mown; twenty acres fed horses.

No. 26.—Ling Piece.

1797 Layer, second year sheep fed.
1798 Summer-fallow; clayed.

1799 Tartarian oats.
1800 Turnips; one-quarter of a ton of cake per acre.
1801 Barley.
1802 Trefoil, mown.

No. 27.—Second Nineteen Acres.

1797 Trefoil, ray, and Dutch, sheep fed.
1798 Summer-fallow.
1799 Wheat; one-quarter of a ton of cake per acre.
1800 Turnips; ditto, sheep fed.
1801 Barley.
1802 Clover, mowed.

No. 28.—Fourteen Acres adjoining.

1797 Clover, ray, and Dutch, mowed.
1798 Ditto, mown.
1799 Tartarian oats.
1800 Turnips; one-quarter of a ton of cake per acre; sheep fed.
1801 Barley.
1802 Trefoil, sheep fed.

No. 29.—Second Long Snelling.

1797 Layer, second year sheep fed.
1798 Summer-fallow.
1799 Wheat; one-quarter of a ton of cake per acre; on the other half, one-third of a ton.
1800 Turnips, mucked; sheep fed.
1801 Barley.
1802 Trefoil, mown.

A finer detail of courses, or of great exertions in excellent management, has not often been seen.

A singular management he has practised, has been that of breaking up a one year's layer, and sowing turnips, to the quantity of from twenty to fifty acres in a year, feeding them off with sheep, and then drilling wheat on one earth.

Part of one of Mr. DURSGATE's fields was summer-fallow and part pease; the whole then sown with wheat, which was better after the pease than after the fallow; the crop was damaged by the wire worm, against which he has found fallowing no security.

Mr. RISHTON, at Thornham, the old four-shift, of

1. Turnips,   3. Seeds,
2. Barley,   4. Wheat.

Clover, one round, and other seeds the next.

At Holm, on rich loam:

1. Turnips,   4. Clover, and tempered,
2. Barley,   5. Wheat,
3. Clover and ray-grass, 6. Barley.

Mr. STYLEMAN, at Snettisham:

1. Turnips,   4. Seeds,
2. Barley,   5. Pease,
3. Seeds,   6. Wheat,

Mr. GODDISON, at Houghton, and the farmers generally, are in the five-shift husbandry.

The same at Hillington.

Mr. BECK, at Riseing, the same. The seeds two years, and clover in alternate rounds. The fifth year some take wheat, some pease, and then wheat; but the land thus getting foul, Mr. BECK has not practised it of late years. When he has taken turnips after pease, he has fed them off in time for wheat, on one earth broad-cast on four-furrow work.

The old four-shift at Grimstone.

The same to the east of the Ouze around Downham, for some miles; sometimes the seeds are left two years, but in general only one: the course, however, is not by all adhered to, for some sow barley after wheat—some few beans. Mr. SAFFORY thinks that the grand point now in

Norfolk

Norfolk husbandry is a due change and variation of crops, as beans, carrots, &c.

At Watlington the same four-shift course: if clover fails, pease, and then wheat.

Mr. ROGERSON, at Narborough, on very poor sand, the five-shift, the seeds lying two years: but on the worst land, instead of wheat in the fifth year, rye, barley, vetches, pease, or Tartarian oats. He had no wheat this year (1802).

About Wymondham:
1. Turnips,
2. Barley,
3. Clover; or clover, trefoil, white clover and ray, one year.
4. Wheat,
5. Barley, but with exceptions. If the seeds fail, dibble pease sometimes, and take wheat after.

At Besthorpe:
1. Turnips,
2. Barley,
3. Clover, &c.
4. Wheat,
5. Barley;

but Mr. PRIEST leaves out this last crop.

Mr. TWIST, at Bretenham:
1. Turnips,
2. Barley,
3. Trefoil and ray grass,
4. Ditto,
5. Rye, on one or on three earths.

At Acle, on some of the finest wheat land in the county:
1. Turnips,
2. Barley,
3. Clover,
4. Wheat,
5. Pease,
6. Wheat.

At Halvergate also, very fine land in the same course. If clover follows once in four years, it is sure to fail.

Thirty years ago, a course I met with in Fleg hundred was:

1. Turnips,

1. Turnips,
2. Barley,
3. Clover and ray,
4. Ditto,
5. Buck-wheat, or pease,
6. Wheat.

It was remarked more than thirty years ago, near Norwich, that the crops of barley, after turnips fed late, were generally bad, which led to a practice that deserves noting: buck-wheat was substituted, which succeeded well, and was followed by wheat; an observation very applicable to Swedish turnips at present.

COURSES ON STRONGER LAND.

At Thelton, Mr. HAVERS, on his drier soils:
1. Turnips,
2. Barley,
3. Clover, once in eight years, the land being sick of it,
4. Wheat, dibbled.

But on heavy land:
1. Summer-fallow,
2. Barley,
3. Clover, once in eight years,
4. Wheat, dibbled,
5. Beans.

On either soil, in the intermediate course, trefoil, ray, and white suckling, substituted for clover; sometimes left two years, and then pease on the lighter land, and beans on the heavy—wheat following.

Mr. DRAKE, of Billingford, on heavy land:
1. Summer-fallow,
2. Barley,
3. Clover,
4. Wheat,
5. Beans or oats.

But as the land is sick of clover, he does not sow it oftener than once in eight or ten years; using white clover the intermediate round.

On his light land, the common four-shift husbandry.

Mr. PITTS, at Thorpe Abbots, on gravel:
1. Turnips,

1. Turnips,
2. Barley,
3. Seeds,
4. Wheat, barley, or pease.

But on heavy land:
1. Summer-fallow,
2. Barley,
3. Seeds,
4. Wheat.

Some put in pease or beans on the seeds, and then wheat.

Mr. MILDRED, on the Duke of NORFOLK's beautiful farm at Earsham, near Bungay; on his lighter land, the four-shift husbandry; but on the heavy:
1. Summer-fallow,
2. Barley,
3. Seeds,
4. Wheat, dibbled,
5. Beans or oats, the former dibbled.

His seeds for change are trefoil and white clover: he does not like ray grass, therefore sows as little as possible.

Mr. BURTON, of Langley, summer-fallows the strong land at Hempnal, for barley and wheat alternately; taking beans after either, and wheat after the beans.

On the sandy and gravelly loams of the hundred of Loddon, the four-shift husbandry; but as the land is sick of clover, they sow it but once in eight years: sow white clover and trefoil instead of it.

Mr. SALTER, near Dereham, on land so wet as to require much hollow-draining:
1. Turnips,
2. Barley,
3. Clover,
4. Wheat.

But it does not keep his land free from charlock.

1776, at Walpole:
1. Fallow,
2. Wheat,
3. Oats,
4. Beans,
5. Wheat.

Mr. FORBY's, at Fincham, on strong land:
1. Cabbages, dunged for, and worth, on an average, 5l. per acre;
2. Barley, $9\frac{1}{2}$ coombs;
3. Clover,

3. Clover, mown twice, produce three tons;
4. Wheat, dibbled, 8¾ coombs; has had ten round;
5. Oats, fifteen coombs.

About Harleston, on their good loams of 20s. or 25s. an acre, on a marle bottom, they pursue pretty much this rotation:

1. Fallow,
2. Wheat,
3. Beans,
4. Wheat.

With this variation:

1. Fallow,
2. Wheat,
3. Clover,
4. Beans,
5. Wheat.

The beans all dibbled, one row on a furrow; three bushels of Windsor ticks per acre: they used to manure for the wheat after them, but of late have got much into the practice of manuring for the beans, which has succeeded far better, not only for the beans, but with the wheat also. They most approve of ploughing the land for beans in the autumn, and leaving it in order, well water-gripped, for planting, after harrowing, the end of February or beginning of March, on this stale furrow. Crop, from eight to twelve coombs an acre. Dibbling is 6s. 6d. an acre, and hand-hoeing, twice, 10s.—5s. each time. They harrow, and roll in the clover on the wheat, in the spring.

Mr. SALTER, at Winborough, applied summer-fallowing, the first year of his taking his farm, much of which consists of various loams and sands, on a strong marley and clayey bottom, and abounding with springs; but after that, he has never fallowed, and never will.—His expression was, " *a man is a madman that summer-fallows.*" He is very regularly in the four-shift course of:

1. Turnips,
2. Barley,
3. Clover,
4. Wheat.

If

If clover fails, or on lands where he expects it to fail, winter tares or pease, instead of it. His wheat, on layers, all dibbled, and the turnips fed by sheep on the land; on the wettest soils, kept on them only by day, and laying on grass-land at night.

At East Bilney, Brisley, Gressenhall, Stanfield, Betdey, and Mileham, being adjoining parishes, the common course is:

  1. Turnips, 4. Wheat,
  2. Barley,   and some add,
  3. Clover, 5. Barley.

Also:

  1. Summer-fallow, 5. Barley,
  2. Wheat, 6. Clover,
  3. Oats,   7. Wheat.
  4. Turnips,

The Rev. Dixon Hoste, on some of the strongest and most tenacious land I have seen in Norfolk:

  1. Turnips,   6. Barley, drilled,
  2. Barley, drilled, 7. Tares, &c. as the land is
  3. Clover,   sick of clover,
  4. Wheat, drilled, 8. Wheat, drilled.
  5. Turnips,

Recurring thus but once in eight years, the clover stands.

At Goodwick, and the neighbouring heavy land parishes, the four-shift husbandry; the turnips on nearly flat lands! There are, however, some summer-fallows for wheat, in which case the course is:

  1. Fallow, 4. Barley,
  2. Wheat, 5. Clover,
  3. Turnips, 6. Wheat.

Mr. Porter, at Watlington, on strong land:

  1. Fallow, 3. Beans,
  2. Wheat, 4. Wheat.

One-fourth fallow. If fallow is had recourse to, how much better to introduce it thus:

1. Fallow,
2. Barley,
3. Beans,
4. Wheat,
5. Beans,
6. Wheat.

MARSHLAND.

At Wiginhall, St. Maries:
1. Summer fallow,
2. Wheat,
3. Beans,
4. Wheat.

This by good farmers; but some go on:
5. Oats,
6. Barley, or big.

On Governor BENTINCK's estate, in Terrington, by Mr. WILLIAM ARTON, a tenant:
1. Wheat,
2. Oats,
3. Wheat,
4. Potatoes,
5. Wheat.

Will any reader believe that this note could be made in the county of Norfolk?

Another field of the same farm:
1. Wheat,
2. Wheat,
3. Oats,
4. Potatoes.

Another:
1. Fallow,
2. Oats,
3. Wheat,
4. Wheat,
5. Spring wheat.

Bravo! for Marshland lads!

Other curious courses, from the same book:

1. Wheat,
2. Wheat,
3. Oats,
4. Potatoes,
5. Wheat.

1. Fallow,
2. Oats,
3. Wheat,
4. Wheat,
5. Wheat.

1. Fallow,

COURSE OF CROPS.

1. Fallow,
2. Oats,
3. Wheat,
4. Wheat,
5. Barley.

At Walpole:
1. Summer-fallow,
2. Wheat,
3. Oats,
4. Wheat.

Also:
1. Oats on grass,
2. Cole,
3. Oats,
4. Wheat,
5. Fallow.

Another:
1. Summer-fallow,
2. Wheat,
3. Oats,
4. Beans,
5. Wheat,
6. Barley or oats,
7. Clover,
8. Wheat.

Mr. SWAYNE:
1. Fallow,
2. Oats,
3. Wheat,
4. Beans,
5. Wheat.

Adjoining Marshland Smeth, to the North, &c. old land:
1. Summer-fallow,
2. Wheat,
3. Beans,
4. Oats,
   and some,
5. Wheat.

Also:
1. Cole,
2. Cole seeded,
3. Wheat,
4. Beans,
5. Oats or wheat.

Mr. SAFFORY's fen farms at Denver Welney, Fordham, Downham, west side:
1. Pare and burn for cole for sheep; the crop worth 30s. to 40s.
2. Oats, fifteen coombs,
3. Wheat, seven coombs,
4. Summerland cole for sheep, 25s.
5. Oats, fifteen coombs,
6. Laid

6. Laid to ray, one bush, white clover 8lb. red ditto, 8lb. for three years, fed in general with sheep and beasts, some mown 7½ load an acre.

Sometimes red clover only for one year, ploughed up, and wheat dibbled.

OBSERVATIONS.

One observation on the Norfolk courses occurs, which the practice of Mr. PURDIS, of Eggmore, I think, justifies:—he was long in the common system before he struck out a variation; and he was induced to it, partly from a conviction that the land generally wanted a change: the want of variety in the courses of the county, is the circumstance that I wish to allude to. For 60 or 70 years, the variations have, upon the whole, been very few: all have begun with turnips, followed by barley; then *seeds*, in which alone have occurred the chief variations, and those, by force of necessity, from failures. If there is a deficiency, I think it will be found in not having some substitutions of crops for so regular a routine. Mr. PURDIS introducing tares, appears a good idea: pease have been taken by some other farmers; but Mr. OVERMAN's curious observation, that they will not succeed, if taken oftener than once in 11 or 12 years, should be a caution. Chicory, followed by winter tares, amongst which some scattered plants rising, would be of no consequence, but rather an advantage, deserves attention. Upon very poor soils, this plant is essential to profit.

Another crop I shall take the liberty of naming for loams, is the bean. The notion, in Norfolk, that this is adapted only to strong soils, is very erroneous; it is more profitable on good sands, and pretty good sandy loams, than it is on clay; and would yield great crops on soils, wherein it is never found in Norfolk.

Carrots also deserve attention; for turnips have been repeated till the land is sick of them.

SECT.

SECT. IV.—TURNIPS.

It is proper to begin with the crop which, in Norfolk, is made the basis of all others.

1. Course,
2. Soil,
3. Tillage,
4. Manuring,
5. Sort,
6. Seed,
7. Steeping,
8. Hoeing,
9. Distempers,
10. Drilling,
11. Consumption,
12. Preservation,
13. For seed,
14. Is the land tired?
15. Swedish Turnip,
16. Importance of the culture.

1. *Course.*—At Massingham, on the first improvement above sixty years ago, it was common to take two crops running to clean the land, and it answered greatly: Mr. Car's barley, after the preparation, was greater than ever known in the common course: he had $6\frac{1}{2}$ quarters per acre.

Mr. Burton, of Langley, considers a wheat stubble as the best for turnips.

2. *Soil.*—Norfolk farmers are so wedded to turnips, that they sow them almost indiscriminately on all soils. Perhaps, the heaviest land I have yet seen in the county, is at Goodwick, on the farm of the Rev. Dixon Hoste: and I was petrified to see his turnips on such a soil, as well as his neighbours, on broad flat lands: it is true, he has hollow-drained well and carefully; but the very texture of the soil is adhesion itself, and greatly retentive of water; so that carting to remove the crop, is very hazardous; the consequence is, a barley crop inferior to the land; certainly, in many cases (even in this fine barley year, 1802), not more than the half of what would have followed beans or tares, well managed.

In

In discourse with Mr. JOHNSON, at Kempston, I found that he considered cultivation and turnips as synonimous: no farming without turnips:—*What, Mr.* JOHNSON, *on very wet, stiff, tenacious, poaching soils?*—" How are you to keep stock without them?" And, in Norfolk, they may be said to know nothing of the Northumberland culture, the only system that can make the crop advisable on such land.

The universality of this culture in Norfolk, whatever the soil, is singular; but the most extraordinary feature is, to see so many on the Marshland, clay the ridges almost flat; they are carted, or rather poached off, for cows and sheep.

The stronger, heavier soils of the southern parts of East Norfolk, will not bring turnips freely without marle, which acts by rendering the soil more friable. This is an observation of an ingenious writer; but, at present, marled or not marled, all is under turnips.

The universal system in Norfolk, whatever may be the soil, of sowing turnips, and cultivating them on flat, or nearly flat lands, must, without hesitation, be condemned: hollow-draining can never be praised too much; but there are districts, the soil of which is so tenacious, that no drains can make the husbandry admissible. Mr. FORBY's experiments on cabbages, very carefully made, and accurately reported, bear immediately on this point, were they necessary to establish it; but, in truth, few experiments are wanting to prove the point: for the many bad or inferior crops of barley I saw in 1802, a very great barley year, on such soils, would alone convince me that the turnip culture, in such cases, is mischievous. The difference between six or seven coombs of barley, and eleven or twelve, would buy lintseed cake for the consumption of straw; or pay the loss of fattening hogs for the

the same purpose. Straw must be converted into dung: these methods make better dung than turnips; nor are these the only resources.

3. *Tillage.*—About Watton not less than four earths given. The seed harrowed in—no drilling practised.

Mr. MONEY HILL, of Waterden, scarifies his turnip fallows in March, April or May, as it may happen; the scarifier attached to a frame on two wheels, made for that purpose, to save the carriage of the drill machine; but the second time of going over the land he fixes it to the frame of a roller, to which are added irons pierced for that purpose, the roller breaking any clods that contain the roots of the twitch grass, and freeing it to be taken up by the shares, or afterwards by the teeth of the horse-rake, the teeth of which are freed by working in the common manner of other couch-rakes, through a frame of wood.

A practice of Mr. THURTELL's on a pea stubble, which he has followed many times, is to scarify directly after harvest, and then throw it into four-furrow work for winter; in the spring takes off one bout from the ridge, then harrows well across, and leaves it for weeds to vegetate. Good.

Mr. BROWN, of Thrigby, ploughs five inches deep for turnips; his first earth taking up the stubbles deep; and the first stirring *scrapes the bottom.*

Mr. PARMENTER finds scuffling a practice of great utility on a foul turnip fallow.

Mr. BIRCHAM, at Hackford, scuffles his fallows much to his satisfaction; not to save ploughing, but additionally.

To these notes I could have added others; for the practice of adding the operations of the scuffler to those of the plough, is gaining ground in the county, but has not yet been so long established as to enable many to speak with much decision.

4. *Manuring.*

4. *Manuring.*—The practice of manuring for the turnip crop is universal in Norfolk. Before the culture had been for a long period general, good crops were sometimes gained without, but for many years past none are to be procured except on new land, without much attention to this necessary branch of the management. The more common method is, to cast the yard dung of the preceding winter into heaps, which are turned over and carted on to the turnip land before the last earth is given. There are variations in every step of the business, but these, where important, will be noted either in the present section, or in the chapter of manures.

Mr. THURTELL is of opinion, that on heavy land, autumn may be a very good time for carrying on muck for turnips, but on his light land he always scales it in before the seed earth, which is given deep enough to bury it properly. Mr. THURTELL mucks all his turnips from Yarmouth.

Mr. SYBLE, of South Walsham, informed me that the farmers in that vicinity were, not many years ago, in the practice of carrying out dung for turnips at Michaelmas, but they have left it off, as neither the turnips nor the barley proved such good crops as with other management.

Mr. LAYTON, of East Norfolk, and his neighbours, were in the practice of ploughing in the dung very shallow by the last earth but one, to harrow well for mixing with the soil, and then to plough and sow.

Mr. M. HILL applies twelve loads per acre of yard-dung or compost; sometimes turns it in by the seed earth, but when work is forward, scales it in, and leaves the land ready for the seed earth. He prefers, when it can be done, to lay the dung on in November, on the wheat stubble, and plough it in; this keeps stiff land open, and he thinks undoubtedly the best practice: the pulverization is such,

such, that an earth may be saved by it, and it ensures a greater certainty on cold wet land. This can only be done to the extent of what is made by six weeks feeding in October and November, after the horses are taken in; and this long dung is undoubtedly best for wet land, and as good as any other on all soils at that season. The turnips also come sooner to the hoe, and are less eaten by the fly. But if long muck be spread just before sowing turnips, the fly is increased, owing to the dung being not well buried.

Mr. COKE's turnips were putting in while I was at Holkham, and I found the business perfectly arranged for an equal and steady employment of all the teams and hands: three men filled the dung-carts, of which there were four; three drivers; ten horses in the carts; three men spread the dung, one throwing out roughly, and two with forks breaking the pieces and distributing equally. Five ploughs were at work; these followed by a light roller; then the drill plough; and after all, a pair of light harrows; thus 18 men and boys were employed, and 24 horses and mules. Six acres a day finished, upon which 60 loads of dung spread; the drive not a short one. The men out at four in the morning from their own houses, to the field at five, and finish at two; breakfast in the field: in the afternoon the ploughmen take care of the horses, and the carters in the hay. The ploughs work the same hours, but out an hour later, and home an hour later, as they turn in all the dung carried out: ten loads an acre.—For his application of oil-cake, see *Manuring*.

5. *Sort*.—Inquiring of Mr. BURTON, of Langley, if he cultivated Swedish turnips, he replied that the green round turnip stood the frost so well, that Swedes were not wanted for that object, and the produce much larger, and much more certain. The sort most general, is the large globe white.

6. *Seed*.—Mr. SALTER, on his various loams, some of which

which are wet and unkind for this crop, sows a quarter of a peck: I viewed all his fields, to the amount of about 80 acres, and the plants were in most of them very thick, but not too thick with harrowing: gives 8s. an acre for two hoeings: he had seventeen hoers at work.

At Hillington, from one to two pints an acre, but more on chalk.

7. *Steeping.*—Mr. SALTER, of Winborough, tried steeping the seed in lintseed oil, &c. and drying it with flour of brimstone, but the fly ate all.

GEORGE Earl of ORFORD for several years used Mr. WINTER's process for steeping turnip seed in train oil 12 hours, and was of opinion that it was beneficial against the fly: but if heavy rains fall after sowing, the effect lessens.

Mr. SHEPHERD dresses all his turnip seed with train oil and sulphur; three pints of the oil and one pound and a half of sulphur to a bushel of seed: dresses the seed with the oil by thorough mixing in a tub, and dries it with the sulphur, keeping it 12 hours before sowing: he has tried it repeatedly, and in comparisons, and is firmly persuaded of the benefit, from the superiority of the dressed seed in crops attacked by the fly, not only in experiments side by side, but also in saving crops when turnips have in general been destroyed.

It has been found that steeping old seed in water, and then drying in the sun, has brought it up sooner than sowing dry.

8. *Hoeing.*—Upon land which is exceedingly given to charlock and wallock *(Raphanus & Sinapis)*, Mr. DURSGATE has hoed by the day instead of paying by the acre, to have the greater security of thorough cleaning; or in other cases paying an extra price per acre.

The turnips in Norfolk are universally twice hoed; the operation is every where well done, except, I think, in one

one respect, that of being set out too thin, which with *dashing* hoers is a common evil.

9. *Distempers.*—Mr. COKE having heard that ducks had been used to clear turnips of the black canker, tried them on a field of 33 acres: he bought 400 ducks; on the 16th of July they were turned in, having water at one corner of the field, and in five days they cleared the whole completely, marching at last through the field on the hunt, eyeing the leaves on both sides with great care, to devour every one they could see.

The *anbury*, or external knots, each containing a small worm in the centre, depends on soil, and most of the soils in Norfolk are subject to it till they have been marled or clayed, which is an almost sure preventative.

10. *Drilling.*—The application of this mode of culture to the turnip crop has not yet made any considerable progress in the country; nor are the opinions of the farmers settled upon the question of its propriety.

Mr. FOWEL drills his turnips for bullocks at 18 inches, but for sheep at 12. I viewed several of his crops, and found them very regular; the drillings well joined, and very straight. Four strokes of the drill sow an 18 feet ridge, without a marker; the horse led by the lines of the furrows. He gives 3s. 6d. for hoeing the first time, and 2s. 6d. the second.

Mr. BLOOMFIELD, at Billingfold, has this year (1802) very promising turnips on a bad black gravel soil, which he enclosed and broke up from the heath. His culture is, to set out the ridges of two feet from the flat, with a double breast plough, and to lay the muck in the furrows: he then sows broad-cast, and splits down the ridges with the ground wrest of a double breast plough without its breasts, harrows across, and the turnips come up regularly on flat land in rows at two feet. The bailiff thinks that on this poor soil they should have had no turnips in the

NORFOLK.] common

common mode of cultivation; and that in this method 10 loads of muck are equal to 14 used in the common way.

Mr. COKE, at Holkham, sows none broad-cast; all are drilled.

Mr. REPTON, at Oxnead, drills turnips at one foot, and prefers the method.

Mr. ENGLAND, of Binham, tried the drill last year, but he found the plants too thick in the rows, and has observed the same in some other cases; they are then difficult to hoe.

Mr. REEVE, of Wighton, drilled them for two years at 11 inches and a half, but has left off the practice.

Mr. HENRY BLYTHE, of Burnham, though a very staunch friend to drilling corn, from 12 years experience, does not drill turnips, finding that they are not so easily hoed.

Mr. STYLEMAN. at Snettisham, drills them at 12 inches, and he thinks they hoe better than broad-cast crops, and that the produce is greater.

Mr. BLOOMFIELD, of Harpley, finding that his turnips were very apt to fail, like those of his neighbours, on a chalk soil, varied his husbandry; he spread the muck, and then sowed the turnip seed, and ploughed them in together, by two furrows meeting, but not lapping the one on the other, and the seed coming up along the centre of the flat ridge thus formed, before winter he ploughs between, to earth them up powerfully, for preservation against the frost. The success has been great, and much exceeding the common practice on that soil. Mr. GODDISON favoured me with this account—Mr. BLOOMFIELD not at home.

Mr. PRIEST, of Besthorpe, has this year turnips drilled at 18 inches, with COOKE's machine: I viewed the crop, and admired their regularity and size, for so unfavourable a year a year (1802); they were first hand, and then horse-hoed; the rows 18 inches asunder: hand-hoeing performed as easy as in broad-cast crops.

Mr. TWIST, of Bretenham, has seen drilled turnips, but did not like them well enough to adopt the practice, though a great driller of corn.

11. *Consumption.*—Mr. BEVAN has for some years pursued the common practice of drawing half his crop alternately by stitches, and carting them to his yard or to layers, for cattle; and feeding the other half on the land by sheep; he has long suspected that he lost by it: this year his barley is so inferior to what it ought to be, as to afford entire conviction of it; and he is determined never more to repeat it. The sands of Riddlesworth are not rich enough to bear this treatment.

Mr. DRAKE, of Billingford, carts off his turnips with quarter-carts, the horse and one wheel going in the furrow, and only one wheel on the land, and that on the crown of the ridge. The mischief thus done, he says, is less than in any other method he has seen. The soil strong and wet.

Upon good land Mr. COKE draws half and feeds half; but on the weaker soils feeds all.

It is common with many farmers in West Norfolk to draw out the largest roots for carting home to bullocks, and for feeding the smaller ones in the field by sheep. Carting damages many; but there is a great advantage in leaving the small ones, which resist the frost the best.

Mr. MITCHEL, of Houghton, having a great superfluity of turnips, in April, 1791 (a circumstance not uncommon in Norfolk), used a tool for cutting them into four quarters; it was a broad knife, crossed at right angles, with a handle about three feet long; women used it, and the expense was but a few shillings per acre.

I have

I have known crops carted, at a great expense, into ditches to rot.

Thirty years ago, three roods of turnips would fatten a beast of 45 stone, or six Norfolk wethers, in East Norfolk.

In 1770, I found the general method of consuming the crop, from Norwich to Yarmouth, to be drawing every other land for beasts, and eating the other half on the land by sheep.

At Thelton, the soil not being generally adapted to sheep, the crop is consumed by bullocks in the farm-yards (*par yards*).

Many drawn also at Billingford and Thorpe Abbots, as well as through all that country, sheep not being a common stock.

Mr. THURTELL, near Yarmouth, draws about one-third of every field for bullocks, kept loose in the yard; of tying up in stalls, his expression was, *it is done with!* ten beasts at liberty, make as much manure as thirty tied up: not that they may not fat something faster; the difference, however, is small, if there be sheds around the yard: if he fatted a beast on a bet, it should be tied up. The remaining two-thirds of the turnip crop are eaten by bullocks, and fat sheep in the field; and, in this consumption, he is in the Fleg system of drawing, and carting enough to spread a fair portion of the field cleared for the yard fatting; and the whole of the turnips consumed in the field, are pulled and thrown. This method is now common in Fleg, and the best farmers have an high opinion of it: the stock do better, and less offal is made, than where the roots are not drawn. Mr. EVERIT, of Caistor, is in the practice, and thinks, that to *pull and throw*, though in the same field, will make the turnips go further by one-fifth, and the stock doing better at the same time. One-horse carts the best for this work.

Mr.

TURNIPS.

Mr. FERRIER, at Hemsby, carts his whole crop to the *par* yard, the roots being first tailed in the field. At Hemsby, &c. in Fleg, 30 great cart-loads an acre; and single roots as much round as a middle sized man's body.

They have been sold at Ormsby, to the Yarmouth cow-keepers, at 7l. 7s. an acre. A price fixed by appraisement, at Michaelmas, to incoming tenants, often 4l. 15s. He has known 36 large loads an acre; and 24lb. a turnip, and quite brittle, no flockyness.

Mr. HORNARD, of Ludham, draws all his crop; he throws on his layers, on the wheat stubbles, and on wheat. He eats all his crop abroad, none in the yard or stalls; but the bullocks are brought home at night to hay or straw. The expense is something, but not heavy, as two horses, with one lad, will cart 30 or 40 acres in a season; the fields, however, within a furlong: some farmers do the same, though they have to cart a mile and a half.

At Catfield, and, in general, through all Happing, they draw all the turnips; and think that an acre thus managed, goes as far as five roods fed on the land: they are carted to bullocks in yards, or thrown on *ollonds*, or wheat stubbles: it is not uncommon to throw on wheat in February and March, and it is seldom hurt, if not done too late; but if in April, damage ensues; if a dry time succeeds, and the land not in good heart, it is generally injurious.

In South Erpingham the same practice; at Coltishal, all drawn, and thrown on wheat stubble, or eaten in the yards.

At Oxnead, all drawn for feeding on the *ollonds*, &c. till Christmas, or till too *jammy*, that is, too much trodden from moisture, then into the yards. But Mr. REPTON steams them, mixing turnips and their liquor with chaff, cut by horse work, and giving it with much success to young cattle, &c.

Mr.

Mr. REEVE, of Heveringland, draws the largest turnips, and throws them on *ollonds*, or wheat stubbles: he has thrown sometimes on wheat, and if the land be light, it answers very well.

Mr. BIRCHAM, of Reepham Hackford, reckons that 20 acres of turnips will muck in the consumption, 20 acres of *ollond*, especially if thrown for sheep; and leaving enough for eating on the land; so that if more was left, the barley would not be the better, but, from its bulk, perhaps worse. In drawing the large turnips, they go only in the furrows, picking up all that are broken by horses or wheels, using small carts with two horses; often drawing in this manner, more than half the crop.

Mr. JOHNSON, of Thurning, used to be very fond of *par* feeding bullocks, but has left it off for two or three years past, now feeding them on the land: he draws part, and feeds part, picking out the great turnips; by which means he improves 40 acres of land, by growing 20 of turnips; prefers throwing on wheat stubble to *jamm* it in for barley, which answers better than haulming.

In feeding turnips by sheep, Mr. JOHNSON remarks, that it is right to begin at the poorest end of a field, or where the worst crop is, as the flock, by falling back, will double dress it. Hay is beneficial to give to sheep while at turnips; but they will fatten without it.

Mr. JOHNSON remarks, that it is wrong to top turnips for fatting beasts; the top being beneficial, if freed from slime and rotten leaves.

Mr. REEVE, of Wighton, feeds on the land by sheep: of all practices he most condemns drawing turnips; it is a heavy expence, and all to do mischief: far better to buy oil-cake.

Mr. M. HILL wishes to feed on the land all that is possible; draws no more than for converting straw to dung.

Mr.

Mr. DURSGATE is such a steady friend to feeding turnips on the land by sheep, that he would not have a bullock on his farm, except for the purpose of treading his straw into muck. He would have no straw eaten. In drawing a crop for beasts, he takes all, and manures with rape-cake, to supply the loss to the barley.

At Thornham, &c. some farmers pick out the large turnips for carting to beasts; others draw alternate lands.

About one-quarter, or one-fifth of the crop drawn a Holm, for bullocks.

Mr. STYLEMAN, at Snettisham, feeds all his turnips on the land with sheep; and therefore sells most of his bullocks in autumn.

Much the greater part at Hillington, fed on the land by sheep.

12. *Preservation.*—The Rev. DIXON HOSTE practised a method, with this intention, that answered well; he took the coulters out of his ploughs, and then ploughed in the turnips; and they held good through a very bad March.

The Rev. Mr. MUNNING has published his method: it is drilling at eighteen inches, and two feet, and ploughing furrows between, to bury them as well as may be effected. This method has been practised with great success, by Mr. REPTON, at Oxnead, and other farmers.

13. *Seed.*—Much attention is paid by farmers who raise their own seed, to the choice of roots for that purpose, selecting such as are clean in the crown and neck; the footstalks of the leaves rising from a thin clean neck, and not from conglomerated protuberances, or coarse rough necks; and if, after some years attention, the turnips come too fine and delicate, they let the roots run to seed, without any transplantation, which corrects that tendency

14. *Is the land tired of turnips?*—Mr. THURTELL is confident that he has no land tired of turnips; nor has he ever

ever seen any: the oftener he sows them, the better the husbandry, and no declension, on that account, in the crops hitherto perceived.

Mr. EVERIT, of Caistor, in Fleg, has no idea of land being tired of turnips in that hundred: he thinks they will bear repeating better than any other crop.

Mr. SYBLE, of South Walsham, is of a contrary opinion, and thinks that land becomes, from repetition, tired of turnips, as well as of clover: and he grounds the idea on the great failure of the crops which have been experienced for seven or eight years past. From that time, to twenty years ago, he remembers them nearly a certain crop; but not so now; being often lost. This has caused him to vary his practice, and sow on a wheat instead of a barley stubble, which promises better; the crops so arranged as to throw one turnip year equally far from another.

Mr. FRANCIS, of Martham, concludes that land tires of turnips, from the circumstance that more seed is now necessary than used to be commonly sown; formerly, he never sowed more than one pint and a half; now, always three pints, and yet they are not so certain as they were then.

Mr. CUBIT, of Catfield, never observed that any land was tired of turnips; and he thinks the crops would be better if they followed wheat, than following barley succeeding wheat, though they would then come round once in five years.

Mr. CUBIT, of Honing, never heard of land being sick of turnips in the six year husbandry.

Mr. MARGATESON, of North Walsham, has some doubts on this point: he sows three pints of seed, because now more subject to the fly than formerly; and he has remarked, that when by accident there has intervened more years

years than usual between the turnip crops, they have been the better; and are always best on new land.

Mr. DYBLE, of Scotter, it may be supposed, admits no such fact as land being tired of turnips; for he positively asserted, that he never lost a crop in his life; but never gives any of the tillage while the land is the least wet; it cannot be too dry for turnips.

No land tired of turnips at Oxnead.

Upon this question Mr. BIRCHAM remarked, that 40 years ago, they could get almost as good turnips without dung, as they can now with it; but still muck will do the business well.

Mr. JOHNSON, of Thurning, remarks on this point, that he gets to the full as good turnips as his father did. He never sows less than three pints. However, one observation of his looks like the land being tired, for he remarks, that turnips, in his proposed course of eight years, will come better from the sowings being longer apart.

Mr. ENGLAND, of Binham, has no other idea of land being sick of turnips, than what results from the fact, that this crop was to be gained twenty years ago without dung, but not so at present.

Mr. REEVE, of Wighton, agrees in the idea of his neighbour, and makes a point of manuring all he sows.

Mr. M. HILL is clear that the land sickens of turnips, and that they are less in size every seven or ten years. His bailiff has been in the farming line fifty years; in discourse with him on the subject, he assured me, that there was no sort of comparison to be made between the crops at present and those formerly raised: he has known two turnips as much as a man could throw over the side of a cart. But he thinks the crops of corn better now; and he is also sure, that more sheep are kept now than in the former periods, notwithstanding the walks being broken up.

Mr.

Mr. HENRY BLYTHE, of Burnham, or his neighbours, cannot get any turnips without manuring.

Mr. DURSGATE never found any inconvenience from turnips being sown too often. At Sedgford he has often had them two years together, and the second better than the first. He does not admit, therefore, that land is sick of them.

Mr. COKE is clear, that at Holkham, if turnips are sown oftener than in common, they fail, as the land is sick of them.

Mr. FOWEL is clear that turnips have rotted much more for the last 10 years than they did 20 years ago, which he supposes to be caused by a change in the seasons. But he cannot by any means agree with those who assert that the turnips in Norfolk are inferior to what they formerly were, from long repetition; he is clear that they are just as good as ever.*

From these notes it appears, that opinions vary, and I wish the reader to have the ideas of the farmers, rather than any general notion of my own, formed from those opinions—such might be erroneously given. I make it a rule to let the county speak for itself on every point.

15. *Swedish Turnips.*—Mr. WALKER, of Harpley, in Norfolk, has cultivated them for some years, with great success: generally has from 20 to 30 acres; feeds them off with sheep and bullocks, and can depend on them, when common turnips are all rotten. His crop in 1800, notwithstanding the drought, was very fine.

Mr. OVERMAN, of Burnham, had in 1800 a field of this plant, and among them a new-comer, the root somewhat resembling them, but the leaf much more like a

---

* In the neighbouring county, the Duke of GRAFTON has made the same observation. Turnips were cultivated at Fuston as early as at Rainham, yet His Grace is clear, that for the last 40 or 50 years they have not declined at all.

common

common turnip; and it buries itself in the ground more than any other, but the size inferior. It afterwards proved good for nothing.

Mr. COKE, of Holkham, has 30 acres this year—has cultivated them several years, with the greatest success, and esteems them as a very valuable acquisition.

Mr. BEVAN sowed them in 1792, at the same time with common turnips, and the crop was so inconsiderable, as to prove the time quite improper. 1802.—He has sowed this plant since repeatedly, and at the right season in May and June; but the fly has always eaten it, so that he has never had a crop.

Mr. FOWEL, of Snetterton, has sowed Swedish turnips for seven years successively, but has never had a crop: the fly ate all.

Mr. M. HILL, this year (1802) sowed Swedish turnips twice, and both sowings were taken off by the fly.

Now (1803) sowing Swedes and tankard turnip together, to draw the latter for autumn use.

Mr. SYBLE, of South Walsham, had last year a crop of these roots, which came to a good size; as heavy nearly as common turnips, but they were so hard that no stock which he tried liked them. They were white fleshed. Swedes, in East Norfolk, in the opinion of Mr. PALGRAVE, *rather* coming in: Mr. BARTLET GURNEY has them at North Reps, and a few at Coltishal.

Mr. BIRCHAM, of Hackford, has tried them, but did not succeed; he believes they were not sown early enough.

I saw a piece on the farm of Mr. REEVE, of Wighton.

Mr. H. BLYTHE, of Burnham, had a field of them last year, and found them of great use in the spring: this year he has ten acres.

Mr. WILLIS, of Choseley, has a few acres.

Sir MORDAUNT MARTIN, of Burnham, a promising crop.

Last year Mr. DURSGATE had very fine Swedish turnips, sowed in May: he approves much of them.

In Mr. MACKIE's nursery, at Norwich, they are much infested with insects: a species of aphys.

Turnip Cabbages.—More than 30 years ago, Mr. HOWMAN, of Bracon Ash, cultivating this plant, observed that those which were left in the seed bed, came to much the larger size. The same remark has been often made on the great common cabbage. In consequence of that observation, I then recommended the practice of sowing cabbages where they are to remain; but know not that it was adopted by any person except the late Mr. BAKEWELL.

16. *Importance of the culture.*—The general feature of the wet districts of the kingdom, is that of cleaning land which has become foul by the culture of white corn, by means of a summer-fallow; but in sandy and other dry countries, and more especially in Norfolk, the same object is attained by the turnip husbandry; and the great additional advantage secured of supporting great flocks, and herds of sheep and cattle. The system is, at present, pretty well known in most parts of the kingdom; but no where practised on such a scale and so universally as in Norfolk. The difference between a barren fallow and an ameliorating crop, which admits so much tillage and successive hoeing, is generally known and admitted; and to expatiate on the importance of cattle and sheep in manuring, would be idle; but (confining ourselves to Norfolk) it merits inquiry, whether the practice, so common, of cultivating this crop on nearly all the soils in the county, whether sandy and dry, or heavy and wet,

be

be so really advantageous, as to justify its recommendation generally to the kingdom at large.

From all I have at various times seen in Norfolk of this custom, and I have often viewed the conduct of the farmers in winter, on strong, heavy, wet lands, I must freely confess that it is carried too far. The reader should keep in his mind one material circumstance, that the tillage of the county is very generally performed on flat, or nearly flat lands, *stitches*, ridges, or by whatever name they are called: high and arched lands are unknown, and the Northumberland culture of drilling on narrow ridges, no where practised. Hollow-draining is the only dependence, and excellent as that husbandry is, it will not prevent much poaching, either by eating on the land, or carting the crop off.

Every degree of treading, poaching, or kneading in the spring, or when the spring is approaching, is on these soils pernicious: drying winds must follow before the ploughs can get to work, and then the furrow cuts whole, and what is called *livery*, soon becomes hard and tenacious, so that a very favourable succession of moderate showers and fair weather must ensue, or the tillage will be either disturbing hard clods, or poaching in the mire. On such soils, and in such seasons, to give the turnip land only one furrow when nearly dry, and dibble in beans, would be far preferable, than against circumstances to determine for barley: and in conversation with several good farmers on sowing turnips on *really* wet lands, I have heard them admit that *it is bad husbandry: a dead fallow would be better; but we are tempted against our judgments.*

When we speak therefore of what only deserves the title of true Norfolk husbandry, we ought always to confine the remarks to sand or sound loams.

It

It is not merely a question between turnips and fallow, but beans should be more generally adopted: they are coming in, but move slowly.

SECT. V.—BARLEY.

NORFOLK is the greatest barley county in the kingdom, this grain forming the chief dependence of most of the farmers, in all except the very wettest parts of the district.

The notes taken may be thus arranged:

1. Course,
2. Tillage,
3. Time of sowing,
4. Sort,
5. Seed,
6. Depth,
7. Drilling,
8. Dibbling,
9. Produce,
10. Awns,
11. Malt.

1. *Course.*—Mr. HAVER's bailiff assured me, that he gets as good barley on a fallow without muck, as he does after turnips on the same strong land that was well dunged: a good crop ten coombs; rarely less than eight in any management. The husbandry is well conducted. After harvest the fallow is laid on to ten furrow ridges, so that in spring they have only to plough and sow: to scarify and sow would be better, the horses going only in the furrows.

Mr. PITTS, of Thorpe Abbots, also gets much better barley on a fallow without manure, than after turnips well manured for, if the land be heavy.

Mr. CUBIT, at Catfield, and his neighbours, get more barley after wheat than they do after turnips drawn; but the

the cleanest and the best coloured is after turnips; and they find that the barley after wheat, in a six year's shift (the seeds lying two years), is much better than in the five shift, in which the seeds remain but one year.

Mr. CUBIT, of Honing, also gets the best crops of barley after wheat, but attributes it to the pulling and throwing turnips on the wheat stubbles; but he remarks, that the throwing business seldom answers, except in very dry weather.

Mr. REPTON, at Oxnead, generally gets better barley after wheat than after turnips, but the latter all drawn.

Mr. JOHNSON, at Thurning, gets as good barley after wheat on four earths, as after turnips.

Mr. EVERIT, of Caistor, in Fleg, observes, that if turnips are fed off by sheep early, that is, by December, then the barley is much better than what is gained after wheat; but if the turnips are fed in the spring, in March for instance, the barley after wheat beats it.

Mr. FRANCIS, of Martham, upon the whole, has rather better barley after turnips than after wheat, but the latter the greatest bulk: the turnips carted away.

Mr. PARMENTER, of Aylesham, has compared the barley after turnips fed on the land, and carted off; and the superiority of the former is very great, greater than he should have imagined; of course he has many doubts on the common practice of carting off for bullocks.

Mr. STYLEMAN has observed a manifest superiority in his barley, from giving the sheep hay while feeding the turnip crop.

2. *Tillage.*—About Watton all is put in on two earths and an half; the first clean, the second *two-furrow work*, or ribbling, and the third clean burying the seed. No one-horse ploughs. But if the turnips are very late on the ground, then only one earth, and the seed harrowed in.

Through

Through East Bilney and the adjoining parishes, they give two clean earths, and an half ploughing; some three clean ploughings: if the turnips are late, but one, and harrow in the seed.

Mr. SALTER, of Winborough, always gives three clean earths for his barley, and will not admit any idea of lessening this tillage: finer crops than his farm exhibits (1802) were never seen: I guessed them at 15 or 16 coombs an acre ; and he has had above 20. He puts all in with one-horse ploughs.

About Hingham they plough the turnip land twice clean and one half ploughing: no such thing as ever sowing on a stale furrow. Some plough in the seed, and a few harrow it in.

Mr. BEVAN, upon trying the drill husbandry, some years ago, and being well satisfied with it, adopted the one-horse ploughs for that purpose, and they answered very well, doing an acre and a half a day ; after which he used the scuffler, one man and three horses, doing seven or eight acres a day; being induced by the good crops which his tenant, Mr. BRADFIELD, gained in that manner ; and he prefers scuffling, as being cheaper. The crop comes up equally, and he conceives that if the drill is superior, it must arise (not from the seed being deposited at an equal depth, because the same object is attained in other methods, but), from the seed being crowded together as in cluster sowing, which has, in certain experiments, proved highly beneficial. Of course this applies to the present Norfolk practice of neither horse nor hand hoeing being applied in the drills.

At Thorpe Abbots, they generally give three clean earths to their turnip land for barley.

At Hemsby some farmers give four. There are some also that put it in on one, but the crop is not so good.

BARLEY.

At Repps and Martham, three earths for barley, whether after turnips or wheat. Mr. FRANCIS is very careful not to touch the land till it is dry in March, and thinks that all winter ploughing is mischievous. He has tried putting in barley on one earth, but never knew it answer.

At Ludham, three earths to turnip-land, for barley; but Mr. HORNARD, this year, put some in on two, and never had a better crop. He ploughs it in with one-horse ploughs, and now has a double plough for the same work, which answers better.

At Catfield, three or four earths to turnip-land, for barley. Mr. MARGATESON, at North Walsham, three and four on wheat stubbles, sometimes five. He, in common with the practice of the country, ploughs in the seed after turnips, and harrows in that sown on wheat stubbles. Here and there are one-horse ploughs, but not common.

Mr. PARMENTER, of Aylesham, puts in with one-horse ploughs.

Mr. DYBLE, of Scotter, sometimes three earths, but oftener four, and has given five after turnips, and also after wheat.

At Oxnead, Mr. REPTON, four or five earths for turnip-land barley.

Mr. BIRCHAM, three earths.

Mr. ENGLAND, of Binham, if his turnip-land is quite clean, rarely more than two earths to prepare for the drill, but three if wanted.

Mr. REEVE, of Wighton, this year (1802) scaled some of his turnip-land in January, and left it till the beginning of March, then harrowed, and gave a stirring earth; rolled it with a light roller, and left it till the end of March, or the first week in April, and after the first shower, harrowed and drilled directly; the soil stiff and stubborn

stubborn (what a Norfolk sand-farmer assigns these expressions to); the crop very great.

Mr. COKE, at Holkham, gives three clean earths, invariably.

Mr. MONEY HILL, three earths—part with one-horse ploughs, the rest drilled.

Mr. HENRY BLYTHE, of Burnham, drills his barley on two earths, after turnips; sometimes on three.

It is common about Thornham to plough for barley thrice, if broad-cast; twice for the drill; Mr. RISHTON thrice.

Mr. STYLEMAN, at Snettisham, if his turnips are late on the land, ploughs but once for barley, rolls and drills, and gets as good crops as any: but on the land fed early, ploughs thrice. It depends on the season, which shall prove the best crop.

Mr. PORTER, of Watlington, ploughs his turnip-land once, for barley drilling, and gets as good crops as with more tillage.

3. *Time of sowing.*—GEORGE Earl of ORFORD, tried at Houghton, some experiments on sowing barley much earlier than common, which were interesting. The soil, sand. In 1785—23 acres, after turnips fed off: sown Feb. 7; sharp frosts, with and without snow, followed, and the seed laid five weeks before it appeared. Produce, five quarters one bushel and one peck per acre.

In 1786—14 acres, sown the 8th of February; produce, four quarters seven bushels and one peck.

In 1787—30 acres, sown the 6th, &c. of February; produce, five quarters one peck.

Mr. COKE is a friend to early sowing; he would wish always to begin by the 20th of March, and finish by the 15th of April, and never be a moment later than the 20th.

Mr.

Mr. OVERMAN thinks the best season for sowing barley, is from the 1st to the 20th of April: it will vary with seasons; but about that period the genial warmth of the earth will, on an average of seasons, take place. He has not made observations on the foliation of trees with this view.

Mr. ENGLAND, of Binham, as early as possible in April.

Captain BEACHER, at Hillingdon, the earlier the better: this year (1802) some the last week in February; and though it was cut by the frosts, and looked for a time badly, yet he never had a better crop.

When the buds of the oak are breaking, a few days before the expansion of the leaves, no time should be lost in getting in the seed-barley: a rule in East Norfolk.—From mid April to mid May, their general time.—*Mr. Marshall.*

4. *Sort.*—Mr. BEVAN has cultivated Egyptian barley two or three years; it bears sowing a month later than the common sort, and produces two coombs per acre more, but of a coarser sample. He has this year 20 acres of it, the crop good. Seed, two bushels and a half per acre.

Naked barley was tried by Mr. OVERMAN, of Burnham, on comparison with common, and produced scarcely half the crop.

5. *Seed.*—Mr. SALTER, on heavy land, four bushels broad-cast.

At Thelton, three to four bushels; six pecks and a half, drilled.

Mr. THURTELL, near Yarmouth, ten pecks, drilled at six inches: if he thinks hoeing may be wanted, then at nine inches.

Mr. EVERIT, of Caistor, in Fleg, three bushels to three and a half.

Mr. SYBLE drilled three bushels; four sown. Mr. FRANCIS, of Martham, four bushels; turnip-land requires rather more than on wheat stubbles.

Mr. HORNARD, at Ludham, sows four bushels.

Mr. CUBIT, of Catfield, &c. three bushels and a half.

Mr. DYBLE, of Scotter, sows three bushels and a half; drills two.

Mr. PALGRAVE, at Coltishall, two bushels drilled, four broad-cast.

Mr. REPTON, at Oxnead, drills two bushels and a half an acre; not at all agreeing to any thin seeding land in the drill husbandry.

Mr. REEVE, of Heveringland, drills at six inches, 10 to 11 pecks.

Mr. BIRCHAM, at Hackford, three bushels broad-cast.

Mr. JOHNSON, at Thurning, drills four bushels, the rows at four inches and a half.

Mr. ENGLAND, of Binham, finds that 11 pecks are the best quantity drilled.

Mr. REEVE, of Wighton, 10 pecks, rows six inches and three-quarters.

Mr. OVERMAN drills on light land three bushels. Mr. COKE, at Holkham, the same.

Mr. HENRY BLYTHE, of Burnham, drills nine to ten pecks.

Mr. STYLEMAN, of Snettisham, drills six pecks.

Captain BEACHER drills two bushels and a half, at six inches three-quarters: broad-cast on *marmy* land, four bushels.

Mr. BECK, of Castle Riseing, drills ten pecks.

Near Downham, broad-cast, two bushels to four.

At Watlington, three bushels. About Wymondham, three to four bushels.

Mr.

Mr. Twist, at Bretenham, drills two bushels: sows three pecks to three bushels broad-cast.

6. *Depth.*—Mr. Heath, of Hemlington, had some barley this year drilled three inches deep, and the crop suffered. Mr Syble, of South Walsham, had part of a crop drilled at one inch and a half, and part at two inches; and the former was evidently the best. Mr. Hornard, of Ludham, ploughs it in broad-cast, two to three inches. Mr. Dyble, of Scotter, thinks it cannot be too shallow, if buried.

Mr. Palgrave, at Coltishal, one inch to one and a half, drilled.

Mr. Repton, at Oxnead, drills two inches deep.

Mr. Bircham, at Hackford, Mr. Johnson, at Thurning, and Mr. England, at Binham, two inches and a half.

Mr. Reeve, of Wighton, not more than two inches: in most cases not so much; but if the weather be dry, deeper.

Mr. Overman drills his barley one inch deep. Mr. Coke, at one inch three-quarters.

Mr. M. Hill, two inches.

Mr. Rishton, at Thornham, one inch and a half deep.

Mr. Styleman, of Snettisham, thinks this a point of great consequence; drills two inches and a half deep.

Mr. Porter, at Watlington, two inches.

Mr. Priest has found that in both drilling and broad-casting, the shallower barley is put in, if it be but buried, the better.

7. *Drilling.*—Mr. Fowel, of Snetterton, puts in his barley at nine inches on one stale furrow, and thinks it a great improvement, as well as saving. He used to follow the common method of ploughing thrice, but prefers his present method greatly. He pens his turnips for sheep,

across

across the ridges; then works the land in the same direction with Cook's scarifier; and ploughs as soon as the close is finished, *with* the ridges, leaving it till seed time: harrows, and drills, at nine inches, on this stale furrow. The soil is a good sand, but light. He sows the grass-seeds after the harrows, and before the drilling. The success answers every expectation: his crops, which I viewed, were fine. He has twenty acres, half drilled at nine inches, and half at six inches three-quarters, for comparison. This marked variation from Norfolk management came from Suffolk: he has a relation near Ipswich, where spring ploughing is going out of fashion amongst the best farmers.

For barley, several farmers have remarked, that the great saving by the use of the drill, is forwarding business: the common practice has been to give three earths; but by means of the drill, one is saved; this a material object, as the farmer gets sooner to his turnip-fallows. About half a bushel of seed is also saved. As to crop, the *bulk* is reckoned less than from broad cast sowing, but the corn as much, and of rather a better quality. They have not observed any difference in ripening. Seeds are sown broad-cast, and harrowed in; no horse-hoeing in this case. This management depends on the land being clean. When barley is sown broad-cast on two earths, it is harrowed in, and does not come so regular, not being put in at an equal depth.

I have rarely seen the drill so superior to the broad-cast, as in a large field of Mr. Bevan's: in 1802, the crop drilled was not only considerably superior to the broad-cast, but vastly freer from weeds; especially poppies, which had damaged the broad-cast much; as neither had any hoeing or weeding, this effect is remarkable, and what I cannot account for, nor could Mr. Bevan. The whole after turnips; one-half fed, one-half carted, alternately.

Mr.

Mr. HAVERS, at Thelton, drills barley, and has this year a very fine crop, in that method.

Mr. PITTS, of Thorpe Abbots, drills barley on the flag, one earth on white clover, and trefoil layers; by this means he gets it in much earlier than common, which, on a burning gravel, he finds of great consequence; otherwise the crops are apt to go off in July, however well they might look in May and June.

Mr. THURTELL, near Yarmouth, has for two years drilled much: he drills on the second earth on his turnip land; saving the third, usually given for broad-cast barley: the rows at six inches.

Mr. HEATH, of Hemlington, gets very fine crops, by drilling on two ploughings.

Mr. SYBLE, of South Walsham, approves much of the husbandry, and intends practising it more, but will give the three earths, which he thinks necessary. He has this a year a very fine crop drilled, which I viewed; the rows at seven inches.

Mr. PETRE, at Westwick, drills at nine inches, with COOK's drill, and thus gets his best crops: eight to ten coombs.

Mr. DYBLE, of Scotter, this year (1802) drilled some barley on three earths, saving one, and these were his best crops, by two coombs an acre; the rows at six and nine inches; but he prefers seven.

Mr. PALGRAVE, at Coltishal, gets as much barley from two bushels drilled, as from four broad-cast.

Mr. REPTON has, from much experience, an high opinion of drilling barley. He began in April, 1790, when he drilled 46 acres, with 30 coombs 3 bushels of seed; saving 15 coombs 1 bushel, which, at 11s. 6d. was 8l. 6s. 3d. Twice horse-hoeing the nine inch rows, 2l.

5s. Net

5s. Net saving by drill, 6l. 1s. 3d. Produce, 427 coombs, but five acres were destroyed by the wyer worm: per acre, nine coombs one bushel and a half: and if the five acres be deducted, 10 coombs one bushel and a half per acre. Sowed broad-cast, 38 acres, with 38 coombs of seed: produce, 321½ coombs; or eight coombs two bushels per acre: from that time he went on drilling, being convinced of its superiority, and has now no sown barley. THOMAS FOX, his bailiff, remarked this year (1802), that in a field where was both drilled and broad-cast, that the straw of the latter was *faint*, and the ears short; but that the drilled straw was stiff, and the ears long. *Would you drill, if you had a farm of your own?*— His answer to me was, *I really think I should.* He approves much of dibbling wheat.

Mr. REEVE, of Heveringland, drills his barley at six inches, and finds the crop better than broad-cast.

Colonel BULLER, at Haydon, drills all his barley, and finds that it beats the broad-cast *out and out*.

Mr. JOHNSON, at Thurning, drills all on three clean earths, the rows four inches and a half, with ASHBY's drill, which he thinks a very good one, though it is not easy with it to make so straight work as with COOK's: answers much better than broad-cast: the straw stiffer, and the crop larger.

Mr. ENGLAND, of Binham, drills all.

Mr. REEVE, of Wighton, all; the rows at six inches three-quarters; but thinks six, if the land is good, would be better.

Mr. M. HILL drills nearly all his barley at six inches, three bushels of seed an acre; when he sows broad-cast, three and one-quarter: does not hoe. Sows clover, &c. the day after. In 1802, he had two coombs an acre more from

from covering the seed with one-horse ploughs, five furrows to a yard, than from the drill.

Mr. COKE drills at six inches three-quarters, and gets immense crops: finer barley cannot be seen, than I viewed on his farm in 1800, 1802, and 1803.

Mr. DURSGATE drills all his barley at six inches three-quarters.

Mr. RISHTON, at Thornham, drilled all; some at six inches three-quarters, and some at nine inches.

Mr. STYLEMAN, at Snettisham, drills all his barley on his large farm of about 2000 acres; the rows at six inches three-quarters, and nine inches; and he thinks, upon the whole, that his best crops are at nine inches. Drilling much superior to broad-cast: he has this year part of a field drilled, and part broad-cast; the former, the best crop, and even the labourers confess it. The broad-cast has a weak faint straw, on comparison with the drilled barley; and Mr. STYLEMAN attributes this superiority to the uniform depth of the drilled seed.

Mr. BECK, of Castle Riseing, drills at six inches and a half, except on his very sandiest soil, on which he ploughs in with one-horse ploughs, to bury the seed deeper than the drill.

Mr. EDW. SCOTT, of Grimstone, on a farm of only 200 acres, drilled his barley in 1801, and it turned out much to his satisfaction.

Mr. PORTER, of Watlington, drills all his barley; and hoes all at 2s. 6d. to 4s. an acre, covering the clover seed: the operation does much good to the crop: he had this year 13 or 14 coombs an acre: 17 loads from four acres: from another four acres 16 loads: from five acres reaped 10 loads, each from five to six score sheaves.

Mr. PORTER, of Tottenhill, this year drilled 190 acres

of barley; some at six inches, some at nine, and thinks nine (especially on a wheat stubble) the best.

Mr. ROGERSON, of Narborough, was amongst the earliest drillers in Norfolk, and on a very large scale, especially for barley; but this year (1802) I found he had put in all his crop with one-horse ploughs, preferring this method, after long experience: he never had a better crop.

Mr. PRIEST, of Besthorpe, drills his barley at nine inches, and horse-hoes it if he has not sown grass-seeds; and he has observed an evident benefit from the operation. His threshers admit that drilled corn is fuller bodied than broad-cast. At Shropham he tried seven, eight, and nine inches distance of rows, and nine proved the best.

Mr. TWIST, of Bretenham, drills all his barley, and has much better than when he sowed broad-cast.

8. *Dibbling.*—Mr. DRAKE, of Billingfold, dibbles barley on his lighter land on one earth; one row on a furrow, and then sows a cast and harrows; and this he thinks pays better than wheat on land much subject to poppy, in which he has had wheat that cost from 20s. to 30s. an acre weeding and yet a bad crop; but of barley never gets less than nine coombs an acre, and the land clean.

Mr. REPTON, at Oxnead, has tried dibbling barley, but gave it up, as it would not answer.

9. *Produce.*—In a tolerable season the poor sands in the south west part of the county, of the rent of 5s. will produce five or six coombs an acre; and in a good, that is in a wet season, six to eight: the better soils in the same district, intermixed with the preceding, give from six to ten, which is not an uncommon crop in a wet year. The richer lands, from Quidenham, by the line of separation on the map, and from Swafham, by Castleacre, to Holkham, are very fine barley lands, and yield great crops, not, however,

however, without a mixture of inferior soils of several varieties. It is in West Norfolk, as in many other districts, the best land is where the soils change; between chalk and clay, between sand and clay, &c. there is usually a breadth of mellow loam of good quality. The crops of barley are generally good on tolerable land. Dividing West Norfolk into two districts, one of 5s. and the other of 10s.; the first rising in tracts to 7s. 6d. and falling to 3s. 6d.: the other rising to 15s. and falling to 8s.; and I should average the barley of the former at six coombs, and of the latter at ten. I have considered this point under many corrections, and do not think it far from the truth: and I am of opinion that if the land was managed in an inferior stile, the poorer districts would not produce four: nor the richer more than seven.

The greatness of Mr. SALTER's barley (1802) may be easily conceived from this circumstance: he set 18 men to mow 18 acres, they worked all day, 22 men all the next day, and 18 men till nine o'clock the third day. There were about 120 loads of it, forming a stack 28 yards long.

They have a whimsical term about Holkham to denote a good crop; they call it *hat* barley: if a man throws his hat into a crop it rests on the surface if good; but falls to the ground if bad. " All, Sir, is *hat* barley since the drill came."

About Watton seven coombs on an average: I saw much in a good season that produced ten and twelve.

Average at Langley, on fine loamy and gravelly sands, ten coombs an acre.

Caistor, and average of Fleg, eleven coombs; sixteen have been known.

Hemsby, five quarters; but some light land and open field: sixteen have been known.

Thrigby

Thrigby and Fleg, in general nine to ten coombs.
Martham, nine coombs; and of late more.
In 1793 a great crop in East Norfolk.

The same in 1800; Mr. FRANCIS, of Martham, had that year:

			Coombs.
54	acres of wheat, which yielded	460
63	barley,	823
14	oats,	187
6	pease,	63
137	$11\frac{1}{5}$ coombs per acre,	1534

The Wheat $8\frac{1}{2}$ coombs.
Barley 13
Oats $13\frac{1}{3}$
Pease $10\frac{1}{2}$

which, combined with the high price of the year, and *here* not a bad harvest, gave a good farming account.

Catfield and Happing, in general nine to ten coombs.
Happsborough and Walcot, fourteen coombs.
Honing, nine to eleven coombs.
North Walsham, eight to nine coombs.
Scotter, eight coombs.
Around Coltishal, ten coombs.

Mr. REPTON, at Oxnead, favoured me, from his books, kept with uncommon accuracy and care, with an account of his crops from 1773:

	Acres.	Produce per acre.	
		Coombs.	Bushels.
1773	113	7	1
1774	84	7	1
1775	74	7	3
1776	99	8	0
1777	73	7	3

BARLEY.

	Acres.	Produce per acre.	
		Coombs.	Bushels.
1778	77	8	1
1779	73	8	1
1780	93	8	2
1781	99	10	0
1782	116	7	1
1783	98	7	3
1784	81	9	0
1785	90	7	3
1786	91	8	3
1787	107	9	2
1788	100	9	3
1789	94	9	0
1790	85	8	2
1791	90	9	3
1792	71	8	3
1793	90	9	0
1794	79	8	0
1795	90	9	3
1796	90	7	2
1797	84	9	0
1798	90	8	3
1799	91	8	3
1800	94	9	$3\frac{1}{2}$

1801 Book not made up.

Average of the whole, $8\frac{1}{2}$ coombs.

On the lighter soils at Heveringland, seven or eight coombs; on the better lands, nine or ten.

At Causton, the soil a good barley land, and the produce averages ten coombs.

Mr. BIRCHAM, &c. at Hackford, ten coombs.

At Haydon, eight coombs.

At Thurning, &c. seven to eight coombs.

At Burnham Westgate, eight coombs. Mr. M. HILL varies, from difference of soil, eight to ten and a half; most on the lightest land.

Mr. COKE, at Holkham, nine coombs.

At Holm, eight to ten coombs, and some more.

At Snettisham, eight coombs.

At Houghton, eight coombs.

At Hillingdon, from five to twelve coombs.

The vicinity of Downham, eight coombs.

At Watlington, ten coombs.

About Wymondham, nine coombs.

Harvest.—" When wet in the swath, it is not turned in East Norfolk, but *lifted*; the heads or ears raised from the ground with a fork or rake, admitting air underneath the swaths."—*Mr. Marshall.*

It is the present practice of a large part of the county, especially where large farms prevail, to stack the corn, particularly barley, in the fields where it grew: evidently to save time in the harvest period.

On a great Norfolk farm I found the work in carting a very heavy crop of drilled barley, the bulk of which I guessed at four waggon loads an acre, going on in this manner: Four putting in; four pitching to two waggons; four loading; six women raking; two boys leading the horses; six men driving away; fourteen at the stack, eight of which, in two parties, were forking in holes and from a scaffold, as the stack was high; ten waggons and thirty horses. From eleven o'clock in the morning to night they cleared, by estimation, twelve or thirteen acres. The drive about half a mile. The expense is heavy. Each man (1800) had in money and beer about 7l. for the harvest; if we call this five weeks, it is near 5s. a day per man; hawkey, &c. &c. will make it up 5s.

32 Men,

BARLEY.

	£.	s.	d.
32 Men, at 5s.	8	0	0
6 Women,	0	6	0
2 Boys,	0	2	6
30 Horses, at 1s. 6d.	2	5	0
	10	13	6
Wear and tear 10 waggons,	0	5	0
13 Acres,	£.10	18	6
Per acre,	0	16	9

I conceive that with one-horse carriages this might have been done much cheaper.

I found Mr. BURTON's fields, at Langley, clearing under the following arrangement:

2 Pitchers,	2 Drivers,
2 Loaders,	4 Waggons,
2 Rakers,	10 Horses,
4 Unloading,	

and cleared 12 acres of barley a day; 36 loads; from one-quarter to three-fourths of a mile distant, and some even a mile.

Mr. EVERIT, of Caistor, in Fleg, assigns ten acres per man: Mr. FRANCIS, of Martham, the same.

Mr. BIRCHAM's arrangement on his farm of 720 acres arable, for a gang:

2 Pitchers,	3 Waggons,
2 Loaders,	5 At the stack
2 Rakers, boys,	7 Horses.
2 Drivers,	

Mr. JOHNSON, at Thurning, 14 to 16 acres per man.
Mr. DURSGATE, 14 acres per man.
Mr. STYLEMAN, 17 acres per man.
At Houghton, 20 acres per man.
Mr. PRIEST, at Besthorpe, 10 acres per man.

10. Awns.

10. *Awns.*—To free barley from the awn, in years or crops when it is very tough and adhesive, Mr. BEVAN has a horse rode by a boy repeatedly over the floor when six to eight or nine inches deep in barley, and it is found effectually and cheaply to free it.

11. *Malt.*—In 1800 Mr. GILPIN, of Heacham, a considerable maltster, bought some beautiful barley that had not received a drop of rain, and trying a small parcel of it, found it malted badly: he tried a most uncommon experiment, and founded upon an idea very contrary to all common ones on the subject: he kiln-dried it by a gentle heat, watering it lightly with a watering-pot twice or thrice, six hours intervening; dried it: after which operation it malted well, every grain sprouting, and no malt could be finer. Hence observes the very intelligent gentleman* from whom I had this account, it is evident that a good shower of rain in harvest, or a sweat in the stack, is beneficial to the maltster.

By the same gentleman it was remarked to me, that malt keeps better with the *comb* in it, screening when wanted.

The best trial is to swim it in water; all that swims is good malt; what sinks, is barley rather than malt.

SECT. VI.—CLOVER. SEEDS.

I PURSUE the general rotation in treating of these crops. *Seeds*, as they are usually called, are universally sown with barley that succeeds turnips.

Thirty years ago they had for some time found their clover crop failing, from its recurring too often; this

---

* MAXEY ALLEN, Esq. of Lynn.

caused

caused the variation of substituting trefoil for one round, and the clover being sown but once in nine years, the evil was removed.

I found the same account every where in the South of the county, that the land (whatever the soil) was what they call sick of clover. Formerly it was sown every fourth or fifth year; but now if it returns so often it fails, for acres together: they therefore sow clover in one round, and then substitute white clover and trefoil, adding a little ray-grass, but as little as they can help. Whether the wheat is as good after these seeds as after clover, is rather an unsettled point. In discourse with Mr. BURTON, of Langley, a most intelligent observer, upon this question, he said, that he himself got as good wheat after white Dutch as after red clover, but that he believed the true change for the soil would be to sow no seeds at all; and he shewed me a large field of red clover, part of which was very regular and good, and part inferior: the former was in a course where no seeds had been sown, and the latter where Dutch and trefoil were introduced: a strong confirmation of his remark.

Mr. FOWEL, of Snetterton, six pecks of ray, six pounds of clover or trefoil, four of white clover, for two years. I recommended him to try chicory, as well as to substitute cocksfoot for ray.

The land around Hingham is tired of producing this crop, and causes the variation of sowing ray, trefoil, and white clover: but Mr. HEATH has sown no ray of late years, for he has found it injure the land, so that he never observes such good wheat after it as after other grasses.

About Watton, if clover alone, ten or twelve pound; but if on land which has been found apt to fail, eight pound of clover; six or eight of trefoil, and half a peck of ray: but Mr. ROBINSON has a bad opinion of ray, especially

cially on heavy land. When cows are to eat clover, they reckon a small mixture of ray beneficial, to prevent hoving. Clover is very apt to die in the winter; they have repeated it so often that the land is *sick*; this has occasioned the substitution of trefoil, ray, and white clover.

Upon the various soils near Dereham, towards Bilney, and the adjoining parishes, ray is much sown, yet does badly. The opinion in favour of it lessens gradually; they admit that after Midsummer it is good for nothing.

All over Earsham hundred the land is sick of clover, so that it will not stand if sown oftener than once in eight or ten years. White Dutch, trefoil, &c. are substituted, but the wheat is not equally good after them.

Over the hundred of Loddon the same remark is applicable; they use as little ray-grass as may be.

Mr. THURTELL, near Yarmouth, ventures clover not oftener than once in eight years; substitutes white Dutch, trefoil, and ray-grass; but he thinks that land tires of these seeds as well as of clover, and therefore on a portion of his land omits all, and sows pease. Mows clover twice —seeds once.

Mr. EVERIT, of Caistor, sows clover but once in eight or ten years, either substituting white Dutch and trefoil, or *baulking* it of seeds entirely: he mows clover twice for hay, and the wheat is the better for such mowing; remarking at the same time, that when he has soiled a part of a field of tares and left the rest for wheat, that crop has been better after the latter than after the former: a sure proof, he observed, of the benefit of shading the ground.

Mr. FERRIER, of Hemsby, finds no difference in his wheat, whether it follows clover or other seeds.

Mr. SYBLE, of South Walsham, thinks that nothing prepares for wheat so well as a good crop of clover.

Mr. FRANCIS, of Martham, has found, in common
with

with his neighbours, that the land is sick of clover; when he *baulks* it by substituting pease, then the four-shift husbandry instead of their usual Fleg five-shift, turnips coming after the wheat. Having no marsh or meadows, he always keeps a layer two years, which is convenient also for throwing turnips: puts in wheat equally on one earth.

Mr. HORNARD, of Ludham, is not ready to admit that the land of this country is so tired of clover as many are ready to assert; and he instances the case of a field which to his knowledge has been 28 years in the common course, sown once every five years; there is nothing particular in the soil, and the last crop is as good as any former one.

In Happing hundred I find the approach of a change in practice: they admit at Catfield, that if clover recurs too often the land will not yield it, but their method is not an alternate substitution of other seeds, or baulking the land for a round, but to take a six course shift instead of a five, and mixing white clover and trefoil and ray, by which two precautions they succeed well. The clover is often mown twice for want of hay, fed the second year; it is of great importance, for there is no natural grass in the whole hundred, or next to none; all arable and commons. Such as sow their land in five-shifts, cultivate vetches for soiling their teams.

Mr. WISEMAN, of Happsborough, relies chiefly on white clover, preferring it to any other: sows nineteen pound an acre, though twelve the common quantity; it makes the best of hay, and he gets two waggon loads an acre.

Mr. CUBIT, at Honing, ten pound of red clover, three or four pound of white, and on the stronger soils a quarter to half a peck of ray-grass, for two years; the first year he he mows once, and feeds the second growth; the second year feeds all: on strong land he ploughs after the first feeding of the second year, and sows buck-wheat to plough under for wheat; but on light land he *riffles* the lay, or else scales it thin; after harvest harrows well, ploughs clean, and dibbles wheat. Red clover never fails with him, if sown only once in six years.

Mr. MARGATESON, of North Walsham, very rarely misses of clover in the six-shift husbandry; if it does fail the loss is great, for he thinks trefoil very uncertain for hay; when it does happen, the clover of the next course is sure to be good. The first growth of the first year is always mown; the second fed, as well as the second year. There will in the second year be as much grass upon one acre mown only once the first year, as on three acres that were mown twice: this is a remarkable observation, and goes directly to the point of mowing or feeding new lays of permanent grass.

Mr. DYBLE, of Scotter, does not find that land sickens of clover in the six-shift husbandry; sows clover-suckling and ray for two years.

Mr. PALGRAVE, at Coltishal, has sown his seeds on the barley, after it was up, but it did not succeed for want of rain. He sows only red and white then, and after harvest, the ray, on the first rain coming; by this means it is backwarder when the clover is mown: in the common management, the ray is in seed when the clover is in blossom, which damages the hay, besides the land being loaded with perhaps a sack an acre of the seed, which hurts the clovers of the second year. On the whole, however, he *hates ray-grass*, and has for three years omitted it; but candidly owns that his success has not been good.

Above thirty years ago, clover from Norwich to Yarmouth, was very generally mown twice; and as general an opinion,

opinion, founded on much experience, that the wheat following was better than after feeding.

Mr. LAYTON, of East Norfolk, was clear from numerous observations, that clover mown twice for hay, gave better wheat than clover fed, by the difference of a small coat of dung: the soil sandy. He attributed the effect to covering the soil from the sun.

"In East Norfolk it is universally made into large cocks, as soon as it is weathered enough to prevent its damaging in these cocks; in which it frequently stands a week, or perhaps a fortnight. The leaf and the heads are thus saved before it become too crisp; but heavy rains do it injury in this state."—*Marshall.*

The *trifolium procumbens*, called red suckling, cultivated about Norwich for the profit of the seed, as it yields a large quantity: said not to have any merit comparable to clover or to trefoil.

Mr. REPTON, at Oxnead, twelve pound of clover, three pound of white suckling, and half a peck of ray-grass; clover will not stand the second year, but the white and the ray succeed.

About Aylesham the land is sick of clover.

Mr. REEVES, of Heveringland, changes his rounds from clover to trefoil and suckling; when it fails, takes pease.

Mr. BIRCHAM, of Hackford, finds that clover fails if sown oftener than once in twelve years: he was in the four-shift course, but the land grew quite sick of clover: now in one round are substituted one peck of trefoil, two pecks of ray, and three or four pound of white Dutch. Whether these, or clover and ray, they are all for two years.

Mr. JOHNSON, of Thurning, finds that clover has long worn out the second year, but now half of it is lost even the first. His substitutes are white suckling, black nonsuch

such (trefoil), and ray-grass. He does not like trefoil, but ray-grass much: it makes the best of hay, and if kept fed very close, it is good all the summer, carrying much stock, which secures corn. He conceives that clover itself is not so good as ray, as he remarks that cattle feed down the borders when turned to clover much sooner and closer than they do when in ray-grass fields. After drilling his barley he harrows, rolls, and sows one bushel of ray, seven pound of clover, four pound of white. Mows the first growth, feeds the second, and the second year.

Mr. ENGLAND, of Binham, red clover, one round, and trefoil and white the other; ray-grass, one peck with either: if any thing like a failure, which thus rarely happens, takes it up the first year, instead of leaving it two.

Mr. REEVE, of Wighton, alternate rounds; clover, one course, and trefoil, white and ray, the other; but both for two years, if they will stand: intends the red clover only for one year. He has a great opinion of ray-grass, thinking it the best grass they sow, and gets as good corn after it, as after any other. He finds that trefoil stands a drought better than either red or white clover.

Mr. M. HILL, 9lb. of clover; 3lb. of trefoil; 6lb. of white Dutch; and from one-quarter of a peck to two pecks of ray: but his land is very sick of clover. He leaves the layer two years, breaking up part by a bastard fallow, and leaving part for one earth on the flag. In the consumption, prefers mowing one year and feeding the second. He has made observations on the comparison of mowing and feeding, and thinks the difference not striking in the wheat; but has remarked, that seeds broken up the *first* year after mowing, have yielded as good wheat as the second year's seed, though manured. This is curious; but, quere, whether the clover did not predominate in one case and the ray in the other?

In

In some cases, when his seeds fail, he lets his two years layer remain three, breaking up the failing land for other crops: and he remarks, that the fields which are thus left three, and, even in some few instances, four, instead of two years, he breaks up with a much better prospect of success, than if they had been laid only the usual time.

Time of sowing.—Mr. Purdis, of Eggmore, was recommended by a friend, whose management he had seen and approved, to sow his seeds at twice: half of each sort (white and red clover, and ray), at the time of sowing barley; and the other half before the rollers in going over the young crop: and this practice he intends to pursue in future. He thinks it will give them a better chance of succeeding. He has 600 acres of seeds: he sows the great quantity of 14lb. an acre of white clover, 8lb. of red, and one bushel of ray-grass. The last he esteems much in spring; and when an observation was made against it, said, that in April and May he had 3000 sheep that found the excellence of it.

In 1784, registering the husbandry of the spirited cultivator of Holkham, it is remarked, that " those who have been conversant in the husbandry of old improved countries, know that a common complaint is the failure of red clover. It has been sown so repeatedly, that the land is said to be surfeited with it. In the same district it comes to nothing on the old improved lands, yet yields immense crops on any accidental spot, where never, or rarely sown before." The observation is so common, that no doubt can remain of the fact; however, it may be attributed to certain methods in management pursued in this county. Pease and tares had been tried as substitutes, but they are tillage crops, and what these thin soils, harassed with the plough, want, is rest. Mr. Coke turned his views to a different and better quarter, to other artificial grasses,

which

which would answer the same purpose as clover and ray-grass. I had recommended to him, on a former occasion*, trefoil, white clover, cow-grass, rib-grass, and burnet. Mr. COKE applied them with no inconsiderable sagacity to the present purpose, and that the experiment might not be delusive, tried them spiritedly upon 30 acres in the middle of a large piece, laid with clover and ray-grass. The quantities of seed he has found will vary according to circumstances; but in general,

Of cow-grass	8 to 10 lb.
White clover	5 to 8 lb.
Rib	5 to 8 lb.
Burnet	5 to 12 lb.
Trefoil	5 to 8 lb.

according to the price, and also the intended duration of the lay. The success of the first trial induced him to lay down a yet larger space the second year. And the third (with the barley of the last spring), no fewer than 221 acres: this is, in truth, doing justice to a new husbandry. Mr. COKE has found that those seeds fill the land completely with plants, which are abiding two, and even three years; and how much longer they may last, is more than he can pronounce, as their appearance is yet as good as ever. I rode over all the pieces, and never saw a finer or more regular plant than they exhibited. And he has, on several occasions, remarked, that sheep give a preference to these grasses, whenever sown in the same field with clover and ray-grass."

In regard to the continuance of these trials, some of the pastures now remain, and are as fine as the soil will yield: thick, clean, and sweet.

* It was on finding a second crop of turnips in succession on the lawn facing the south front, as a preparation for grass; naming those plants, which were accordingly sown, and have ever since succeeded well; and now form a very fine turf.

Mr.

Mr. Coke considers hard stocking the seeds as the best preparation for wheat, and the safest means of saving the expense of oil-cake.—Sound doctrine, adds Mr. M. Hill.

Mr. H. Blythe, of Burnham, finds his land sick of red clover, and therefore sows it in alternate rounds with ray, and also with the trefoil and white clover of the other round. Rests two years, feeding most of the layers through both.

Mr. Wright, of Stanhow, does not like white clover; he thinks it a bitter food, and that sheep do not eat it kindly; so that while much food seems to be on the ground, stock do badly. This is an uncommon opinion, but I remember Mr. Bakewell starting the same idea.

Mr. Dursgate finding his land sick of clover, sows it alternately, a round with, and a round without, substituting white clover and trefoil; ray-grass with both: no doing without that; all other seeds should come but once in ten or twelve years.

Mr. Rishton, at Thornham, clover one round in the four-shift husbandry, and other seeds alternately: finds little difference in the wheat.

Mr. Styleman, at Snettisham, sows seeds for a two years lay; clover, 12lb. one round, and in the next, trefoil, 6lb. white clover, 4lb. and half a bushel of ray-grass; in which system clover stands.

At Hillingdon the land is sick of clover, and therefore the seeds are varied alternately by trefoil and white Dutch. 8lb. of red clover, 4lb. of trefoil, or white, and one-quarter or half a bushel of ray: variation, 8lb. trefoil, 4lb. of white, with the ray If it is good enough to come to the scythe, it is all mown the first growth.

At Grimstone it is no wonder the land is sick of clover, for they are in the four-shift husbandry: they vary it with trefoil and white Dutch alternately.

All

All round Downham the land, in some measure, sick of clover; and in that case, their variations are vetches, and by some, potatoes. Also sowing white clover, trefoil, and ray grass, leaving it two or three years, and breaking up for pease.

At Besthorpe, tired of clover; change it for trefoil, if it stands, mow the first growth for hay, and the second often for seed.

In Marshland I saw many very fine crops of pure red clover, and the malady, of the land being sick of it, is unknown. Mr. DENNIS, of Wigenhall, St. Maries, sows it on his wheat in the spring, eating off the crop and harrowing well, before and after sowing the clover, and is sure to succeed: mows it twice for hay, and the best of all their wheats succeed—gets six coombs an acre.

From the preceding notes it appears, that one of the greatest difficulties which have for some years been found in the Norfolk husbandry, has been the failure of clover. I have often heard this, as a general fact, denied by men whose practice ought to have taught them better: in the common management there can be no doubt of the fact, and it well deserves the serious consideration of the farmers of this respectable county, whether there may not be devised some methods, beyond those already tried, to correct the evil.

An observation I made, during nine years that I was in the constant habit of viewing the farm of Mr. AR-BUTHNOT, in Surrey, may here merit some attention. When he began to farm, the land was sick of clover, insomuch, that it was almost sure to fail, from having been, perhaps for a century, sown every four or five years. My friend adopted the course of—1. beans; 2. wheat; 3. clover, in which it occurred once in three years, and

the

the farmers predicted an absolute failure: I viewed three courses, and better crops, of pure red clover, were never gained. He began with ploughing treble the depth of that to which the land had been usually stirred, and he manured very amply for every crop of beans, partly with night soil, from London. In what degree the success arose from depth of tillage, and in what degree from a variation in manuring, cannot be ascertained; but the experiment proved that these agents were equal to the cure of the malady.

Some farmers in Norfolk, as appears in the Notes upon Tillage, have moved out of the common sphere, and ventured to plough deeper than their predecessors; nor have they found any inconvenience in so doing. It merits consideration, whether this practice will not prove in some measure a remedy to the failure of clover. As to manuring, and especially in great variations, the means are generally limited, and a change in this respect, however desirable, is rarely in their power.

The only effective remedy hitherto practised, is that of omitting clover altogether, for one or two rounds, which points out the great importance of introducing as many new artificial grasses as possible.

RAY-GRASS.

Mr. SALTER, of Winborough, on various loams called heavy, and some are so, sows but little.—*We can't sow too little; perhaps none would be better.*

Mr. M. HILL thinks that the common prejudice against ray-grass arises from a mistaken practice; approves the use, but not the abuse of it. Whenever it is sown for feeding, he particularly recommends the bare feeding in the spring; if suffered to grow more than two inches long, it will imperceptibly rise and run to bent, and then only it is injurious.

Walking over Mr. MONEY's farm at Rainham, with Mr.

Mr. HILL, I remarked, that ray had been sown with clover, on a soil perfectly adapted to the latter plant; and condemning the practice, Mr. HILL agreed, and made an observation that deserves noting: ray is sown with the clover, and if from a bad delivery of the seed, from the wind driving the lightest further, or from any other cause, some of the land misses its plant of ray; the clover in such spots is much more vigorous; a sure proof of the exhausting quality of the ray.

Mr. COKE, on his fine farm at Holkham, used to sow ray-grass with sainfoin, but has left off the practice, and in general sows as little as possible of the grass, being convinced that it exhausts; his corn is not so good after it, as where it has not been sown.

Mr. OVERMAN had half a field sown with ray, trefoil and white clover; the other half with the two latter plants only. When the whole was broken up for wheat, the crop was much the better where no ray had been sown: the difference so great as to be visible at the distance of half a mile.

Mr. HAVER's bailiff observed to me, that the less ray-grass is sown the better, as he never found it kind for wheat: pease do better on it.

BURNET

Was introduced at Stoke 35 years ago, as I then registered, with great success; but it never made any progress, though it yielded luxuriant food for many horses in February.

The reputation of this plant made so much noise in the world, that Mr. COKE formed an experiment at Holkham, to examine carefully its merits, and, with the spirit that characterizes his husbandry, sowed 40 acres, mixing a small quantity of white clover and rib-grass with it.— The result was as decisive as can be imagined; the field has been fully and incessantly stocked with sheep, and was constantly pared as close to the ground, as a favourite spot in a pasture is by horses.

Mr. Bevan has found burnet to be the most wholesome food for sheep, in a wet spring, and a certain remedy for the flux.

1802. He continues of the same opinion, and is never without 20 acres of it.

COCKSFOOT.

Sir Mordaunt Martin, in 1788, observing, by an experiment, that this grass grew four inches in less than three days, determined to attend more particularly to it: he remarked, that when sheep were let out of a fold, they ran over every thing, to get at a baulk that was full of it, and there ate it in preference to other grasses. In some parts of Norfolk it is called cow's grass, from their being very fond of it. He began to cultivate it in 1794. It grows at Midsummer, in a drought, when every thing else is burnt up. He sows it with nonsuch, instead of ray-grass, and finds it much more profitable.

Mr. Overman, observing the eagerness with which sheep, when let into a field at Burnham-market that had some cocksfoot grass in it, ran over ray-grass, and every thing else, to get a bite of this plant, thought it worth cultivating, and sowed about an acre, on the dry gravelly part of his farm, just above the marsh. This spot was the only one, in a large field, that did not burn in the severe drought of 1800, and convinced him of the excellence of the grass.

This gentleman, shewing me a beautiful crop of drilled wheat, which could scarcely be estimated at less than four quarters and a half per acre, pointed out a part of the field, superior, if any thing, to the rest; and said it was an experiment on the cocksfoot grass: he had found it an excellent plant for sheep, but having examined the roots, perceived them to be so strong, that he had some suspicion they might

might exhaust the land, and therefore sowed this piece for a trial: the result has satisfied him that all apprehension of the kind was ill founded, and he intends substituting it for ray-grass.

Observation.—I have cultivated this grass on a large scale for many years, and have found it of great use. It is a most valuable plant when kept close fed.

CHICORY.

Through all the southern district of West Norfolk, in which are great tracts of poor sand, the layers quite contemptible, I was petrified to see spontaneous plants of luxuriant chicory, pointing out what nature is ready to perform, were she assisted by adapting the plants to the soil. Here are thousands of acres which would, without other expense than that of a few shillings per acre in seed, be doubled and trebled in value; and were such layer fed by sheep, without folding from it, the succeeding crops of corn would be as superior to the present products, as the number of sheep kept would exceed the present stock. If such lands are so open, that folding is really necessary, let it be where the flock is fed on the layer, and this would open the farmers' eyes to the vast importance of changing their grasses. The common mellilot is another plant luxuriantly indigenous on the same poor soils; yields seed plentifully, is much affected by sheep, and would work great improvements, though not equal to chicory: but nobody makes the trial of either, though I have incessantly, for twelve years, been urging the farming world, in the *Annals of Agriculture*, to open their eyes to the value of these and other native plants, far exceeding that exhausting one of ray-grass.

Crossed a large field of turnips, of Mr. TWIST, at Bretenham; a miserably poor crop, with spontaneous plants of chicory, seeded, three feet high, and, had it been cut at the proper ages, would have out-weighed any of the

the turnips it grew amongst, though, I suppose, hoed off when the turnips were set out.

Mr. BEVAN sowed an acre of poor sand, worth not more than 2s. 6d. rent, with chicory, in 1793, and the next year it produced 7l. 10s. in seed.

In company with the Member for the County, at Swafham, on the 19th of June, we could get nothing but salt butter. I hope the farmers there have found out that chicory is not worthy of attention.

Observation.—I have taken several opportunities of recommending this grass in Norfolk. On large tracts of poor land in that county, I am confident it would increase the produce ten-fold, and it well merits trial on every soil in it. The objection, which has been founded on its not being easily extirpated, is of no importance, for tares should be sown after it on some soils, and turnips on others, in which system its destruction is unquestioned.

SECT. VII.—WHEAT.

THIS is in general the crop for which *seeds* of various sorts are the preparation. The notes may be thus arranged:

1. Course;
2. Tillage;
3. Sort of Wheat;
4. Time of sowing;
5. Quantity of seed;
6. Steeping;
7. Dibbling;
8. Drilling;
9. Depth;
10. Sown with turnips;
11. Feeding;
12. Hoeing
13. Tathing, or feeding turnips on wheat;
14. Mildew;
15. Smut;
16. Roots of wheat;
17. Reaping;
18. Stubbles;
19. Produce;
20. Profit, compared with that of oats;
21. Price.

Course.

Course.—The Rev. Mr. HOSTE, on the strong soils of Goodwick, which require hollow-draining, ploughs up his second years' clover lay after the first crop, a clean earth; harrows well, then cross-ploughs clean, and harrowing again, gives a third earth, and fresh harrowing for drilling the seed; throwing the surface into *stitches* or *lands*, just of the proper breadth for one movement of the drill plough, the horse treading only in the furrows. Two clean earths and much harrowing, therefore, are incurred for the sake of drilling: and Mr. HOSTE is clear, that on his stiff land it cannot be drilled well on a single earth, as in the Holkham district.

Mr. OVERMAN, of Burnham, shewed me a field of very fine wheat, perfectly clean, drilled on a six years layer; he remarked, that it was laid down positively free from spear-grass, and when that is the case, you will find it clean after six years, as well as after two.

Mr. FRANCIS, of Martham, when he keeps a layer two years, puts in wheat equally on one earth: he has set it in the same manner on a three years layer, the crop very fine, and got great oats after it.

Tillage.—One of the most remarkable circumstances of the Norfolk husbandry, and the most difficult to account for, is the system, very common, of ploughing a lay intended for wheat, three or four times, beginning in June or July. In Suffolk, and in other well cultivated counties where the soil is good, no preparation for wheat is better than clover ploughed once at sowing time. In this country, on our goods lands, we never think of giving more tillage, and get as fine crops as can be seen: now, the necessity of tearing a loose soil in pieces, the fault of which is too great looseness, while no such necessity exists on much stiffer soils, appears to be quite a paradox. I made particular inquiries of Mr. COKE on this point, and found that

that he had tried it experimentally several times, on clover and ray-grass lays two and three years old, and that the superiority was regular and great in favour of the tillage I have described. Such being the case, all we can do is to reason upon it, and reconcile it to principles; for agriculture can never become a science while irreconcileable facts exist.

There appear to be three ways of accounting for it:—first, by supposing the land, from laying so long, abounds with the red or wyer worm, and that the ploughing and harrowing given to prepare the land, destroy or check them so much as to lessen greatly their depredations: the second, that the ray-grass, if ploughed but once in this loose soil, where tillage is given very shallow, is not destroyed by one ploughing, but rather encouraged to vegetate stronger, as if horse-hoed; and by its growth checks the progress of the wheat: lastly, we may conjecture, that, on these light soils which have been surfeited with clover, the plant of that grass is thin and weak, and nearly worn out in two or three years: the consequence of which is, the ray-grass and weeds only remaining, which every one knows afford little manure in their roots to the wheat, which red clover does very amply. In this case, more good may result from destroying them by tillage, than can arise from turning them under at one ploughing. But, in richer soils, when unmixed with ray-grass, it is exceedingly great, and proves a most ample manuring. The effect we are considering, may be partly owing to all these reasons. With such consideration, is the husbandry good? I shall not hesitate in declaring against it: any circumstance that drives a man to give such summer tillage, in order to loosen a soil, the principal fault of which is too great friability; and the characteristic of its improvement, giving it tenacity by clay, and the treading and kneading of bullocks and sheep, in the winter feeding of turnips; any such system is repugnant to that just theory which is the child, and ought to be the parent, of practice: it results from, and ought to regulate it. To see wheat thriving admirably on stiff soils with one ploughing, and to be told that four are necessary on light ones, is a contradiction to common sense. And yet the fact certainly is so, while an old ray-grass layer is alone in question. How avoid the evil? By dibbling pease upon the lay on one ploughing, and taking the wheat after the pease. I have little doubt of this avoiding the mischief. Another mode is, to follow Mr. COKE's husbandry of grasses, to the exclusion of ray, in which method, I am of opinion (barring the red worm), that one earth would give better wheat than more. And his own course of crops, already inserted, proves that this method will do. For, however useful, and even necessary in some cases, ray-grass may be, never let it be forgotten, that it is comparatively an exhausting plant, and not the best preparation for corn.—Either of these modes is consistent with the nature and character of the soil, the great feature of which, is the want of tenacity; but the practice now pursued, coincides with no other idea but the want of friability.—*Note, in* 1784.

Nothing is more injurious, in Mr. THURTELL's opinion, than to give any previous tillage, commonly called a bastard fallow, to a layer for wheat: this husbandry, once so common, he says, is quite done with.

It is, however, almost universal in the northern part of Happing hundred, to North Walsham, and also about Aylesham.

Mr. CUBIT, of Honing, set wheat on the flag for many years, but found it better to rise baulk or riffle, and the crops cleaner. He sets about half his crop.

Mr. MARGATESON, of North Walsham, breaks up the

the ollonds of the second year, by riffling before harvest; he harrows down, and ploughs as fleet as he can; after that, a full pitch for dibbling; he is clear that this is the best way for the wheat, but as certainly a loss in the barley crop; for the previous tillage has brought the flag into such a state, that the wheat exhausts far more of the benefit than when set on a whole furrow.

In general, they riffle the second year's lay before harvest at Westwick, &c. for dibbling; but some on one furrow; and many this year; the dry weather having prevented some breaking the lays.

Mr. DYBLE, of Scotter, has made the comparison of riffling a layer for dibbling wheat, and leaving a part unbroken for dibbling on whole furrow, and found the former best by a coomb or six bushels an acre; and the barley following, is as good, if the flag was not *too much* broken for the wheat; this, however, partly admits the evil.

Mr. PALGRAVE, at Coltishal, gets the best wheat by dibbling his layers on one earth; if it could be drilled, would be equal, if not superior, but it will not drill well; he has trench-ploughed for it with two ploughs, but the soil was thin, and it did not do well. It is remarkable, however, that among his tools I espied a skim-coulter plough of DUCKET's, laid by and never used: it is on the large construction, for four horses; but the hint of the skim surely invaluable. More wheat is put in on whole furrow, than on broken layers; all of which are of two years.

Mr. REPTON, at Oxnead, *riffles* his ollonds before harvest, then, however, crosses and harrows, and works well, as it should be loose for *dabbing*.

Mr. BIRCHAM, at Reepham, sometimes riffles layers for wheat, but only on the stronger lands: on light lands,

lands, dibbles in on whole furrow: sometimes he dibbles on riffled land; but intends, in future, to put all such in by the drill.

Mr. JOHNSON, of Thurning, puts in his wheat generally by dibbling on a whole furrow; tempering was the practice, but it is left off by the best farmers.

Mr. ENGLAND, of Binham, drills all his wheat on the flag, and prefers it to tempering; and this not only here on a soil comparatively light, but also at Hindringham, where it is more stiff. He ploughs his ollonds directly after harvest, to give time for the weather to pulverize the surface; but if there be not time for this, effects it by harrowing and rolling, &c. Directly after the plough, he goes with a heavy two or three horse roller to press the flag firm; and weights COOK's machine, which so does well in every case he wants it for: Mr. ENGLAND is clearly against the practice of tempering light land, which often brings poppies, that would not otherwise appear. Tempering is now done only by old fashioned farmers, and for the sake of four-furrowed work.

Mr. REEVE, of Wighton, whose farm is in such order, that much attention should be paid to his practice and opinions, always drills on whole furrow, if the land be clean: he had this year a remarkable experiment on this point; trying part of a field on one earth, and part of it tempered, and the former was the better crop, by at least two coombs an acre: and whenever he has made the comparison, he has always found the result the same. There is no difficulty, he observes, in drilling on whole furrow; immediately after ploughing, he rolls down the flag with a heavy roller; then leaves it two or three weeks, the longer the better, as a stale furrow much exceeds a fresh one; then harrows twice lengthways and twice across, after which it is in due order for the machine.

Mr.

Mr. M. Hill's system just the same as Mr. Reeve's; he admits all the preceding practice.

Mr. Henry Blythe, of Burnham, drills all on the flag, or one furrow, before Michaelmas, but never after; then on tempered land. He finds no difficulty in drilling on whole furrow; he rolls after the plough, and harrows, then drills, and covers with the harrow.

Mr. Dursgate, of Summerfield, whose great success in husbandry gives much weight to the opinions he draws from his experience, is not partial to putting in wheat upon ollond, except it be done very early; before Michaelmas it does well, but should never be ventured after; all then sown or drilled, should be on tempered land.

Mr. Styleman, of Snettisham, who puts in his wheat on pea stubbles, ploughs the layer in February; rolls the end of that month, or the beginning of March; and harrowing to a tilth, drills the pease; hand-hoes them; ploughs the stubble once, between three and four inches deep, scarifies, harrows, and drills wheat.

Mr. Goddison, at Houghton, tempers about half his layers, and keeps the rest for sheep food.

At Hillingdon the layers are sown for two years, but if they fail, they are broken up for wheat at one year; sometimes for pease.

Mr. Dennis, at Wigenhall, in the clays of Marshland, sows his wheat on ten-furrow ridges, thinking six-furrow work a loss of land.

Sort.—There is a notion about Riddlesworth, that red wheat will not do on black sand; white succeeding much better; on this account I found the distinction made on Mr. Bevan's farm.

Mr. Salter, of Winborough, sows red wheat only; white sorts do not succeed so well on the heavier soils. It goes by the name of the *old red.*

Mr.

Mr. M. HILL prefers the red chaff, red wheat, to the white; less likely to grow in harvest; but white better sample and price.

Time of sowing.—Mr. ROBINSON, of Watton, an intelligent attentive farmer, of 30 years experience, is clear that on their soils, which are not light, the sooner the wheat is sown, the better the crop.

Mr. MASON, of Necton, is a great friend to early sowing: he has put wheat in even in harvest time; old seed, kept in the straw and threshed just before sowing; and his success has proved the efficacy of early sowing. He has had it green in August, and not the worse; but in general, he reckons September the prime season: he sows old wheat to chuse, and does not steep, nor has he been troubled with the smut.

At East Bilney, and the adjoining parishes, they think they never sow wheat on heavy land too soon : but on light soils, and all given to red weed, a month after Michaelmas; nor is such land able to carry through forward crops.

Mr. SALTER, of Winborough, begins dibbling immediately after harvest, using old wheat.

Mr. WHITING, of Fring, is a great friend to early sowing of wheat. In 1799, he drilled on the 27th of August, and this harvest got the crop up before any rain fell; while nine-tenths of the wheat through the whole neighbourhood, is growing in the *shocks*, or in the ear, as it stands. He would wish always to have his seed in the ground within the month of September. The crop abovementioned was one of the very best he has ever had.

Mr. OVERMAN wishes to have all his wheat in on lays by Old Michaelmas.

Mr. M. HILL prefers from the 1st to the 20th of October, and never wishes to be later on the flag ; but on the tempered land, no objection to a fortnight later.

WHEAT.

In 1782, Mr. COKE made an experiment, to ascertain whether the time of sowing wheat in Norfolk (November and to Christmas) was not too late, by sowing a large field the last week in September, which was a month earlier than any near it. The wheat flourished away very finely through the winter; but the farmers predicted that it would not *prove* well in the spring. The fact turned out so, for when much poorer winter-looking crops began to rally, and spread upon the land in April, this went off, and gave at harvest a very light produce.

In general, the farmers in East Norfolk begin the middle of October, and continue till December; sometimes to Christmas: but for dibbling, at Michaelmas.

Mr. BURTON, of Langley, begins wheat-dibbling a week before Michaelmas, and continues till three weeks after: early sown generally the best, and it saves half a bushel of seed.

Mr. MARGATESON, of North Walsham, reckons the best time to begin is at Michaelmas; and to finish in two or three weeks; some are earlier, but they are apt to suffer.

Mr. ENGLAND, of Binham, as early as he can in October.

Mr. DURSGATE, of Summerfield, puts in no wheat on ollond after Michaelmas: all from that time on tempered land: seasons may prevent it, but he would wish to have all his ollond wheat in by that time. But on turnip land, it may be put in to Christmas.

Mr. MARTIN, of Tottenhill, drilled wheat in February, and the crop good.

Quantity of seed.—About Watton, dibbled six or seven pecks. Mr. ROBINSON, if he sows before Michaelmas, two bushels; afterwards, two and a half.

Mr. SALTER, at Winborough, near Dereham, four bushels

bushels broad-cast, dibbles ten pecks to three bushels. At Wissen, ten pecks dibbled, three bushels broad-cast.

At East Bilney, and the adjoining parishes, two and a half to three.

Mr. HAVERS, &c. at Thelton, dibbles six or seven pecks, early; but eight later.

Mr. BURTON, of Langley, seven.

Mr. THURTELL, near Yarmouth, seven to eight, dibbled.

At Caistor, in Fleg, six or seven pecks, dibbled.

Mr. FERRIER, at Hemsby, six pecks.

Mr. BROWN, at Thrigby, begins with six pecks, and finishes with eight, dibbled.

Mr. SYBLE dibbled two bushels early, but more late.

Mr. FRANCIS, at Martham, dibbles two bushels: the quantity sown is ten to twelve pecks.

Mr. CUBIT, at Catfield, &c. dibbles from six pecks to two bushels, according to time, early or late.

Mr. MARGATESON, of North Walsham, two rows on a flag, and three or four kernels in a hole, which is about two bushels an acre. He was once very attentive to the droppers—they put in six pecks, and he never had a better crop: but it is good to allow for carelessness. He sows very near three bushels.

Mr. DYBLE, of Scotter, two to two bushels and a quarter, whether dibbled or under furrow.

Mr. REEVES, of Heveringland, drills, at six inches, seven to eight pecks.

Mr. BIRCHAM, at Hackford, on summer-land, one bushel and a half; in dibbling, five pecks.

Mr. JOHNSON, of Thurning, dibbles three bushels.—Many here have lost by too thin a plant, half the wheats being under-seeded.

Mr. ENGLAND, of Binham, seven pecks, early; eight late.

late. In a favourable year, has had a great crop from six.

Mr. REEVE, of Wighton, on whole furrow, two bushels; on tempered, seven pecks.

Mr. M. HILL, at first eleven pecks; later, a peck more.

Mr. HENRY BLYTHE, of Burnham, drills seven to eight pecks an acre; the common quantity, broad-cast, ten to twelve.

Mr. DURSGATE begins with six or seven pecks, and never more than eight.

Mr. STYLEMAN, of Snettisham, six pecks, drilled.

Mr. GODDISON dibbles and sows two bushels to ten pecks.

Captain BEACHER, at Hillingdon, nine or ten pecks, drilled at nine inches.

Mr. BECK, of Castle Riseing, drills from six to eight pecks.

Mr. DENNIS, of Wigenhall, in the clays of Marshland, five to six pecks, broad-cast.

At Watlington, two bushels.

About Wymondham, dibble in six to eight pecks; broad-cast, three bushels.

Mr. OVERMAN, seven pecks, drilled.

Steeping.—Mr. ROBINSON, of Watton, for many years has had no other smut on his farm than what has been caused by accidentally sowing a head-land, or finishing a corner of a field with dry seed; but if steeped, the prevention infallible. His method is, to steep it in a brine made with common salt, of strength to bear an egg, for twelve hours, and then to dry with lime.

Mr. DOVER, of Hockham, had great plenty of pheasants, but lost them all, without knowing to what cause to attribute their disappearance: he found out, however, that

that it was entirely occasioned by his using arsenic in steeping his wheat-seed. Mr. ALGUR confirmed it, by observing, that he once found a covey of partridges dead or dying, from the same cause.

Mr. SALTER, of Winborough, dresses with salt and lime, without steeping, and never has the smut: it is only to be concluded that he has always sown clean seed.

Mr. M. HILL slakes the lime with salt, dissolved in a small quantity of water; dips the wheat in a skep, in plain water only, lays it on the floor, and incorporates it with the salt and lime: dries with lime. Inquiring of him whether he had ever tried this method with very black wheat, as I conceived in such a case it would fail, he said that he had not.

Mr. OVERMAN stirs his seed well in pump water, then lays it in a heap to drain, and adds half a pound of salt to every bushel, stirs it well together, and dries with lime: this he finds sufficient against the smut. Whence I conclude that his seed is always free from that distemper, or assuredly he would find the process to fail, for he does not leave it any time limed.

" The salt is dissolved in a very small quantity of water; with this salt the lime is slaked, and with this saline preparation, in its hottest state, the wheat is candied, having previously been moistened for the purpose with pure water." This was the practice in East Norfolk, reported by Mr. MARSHALL. It is not very general at present, but pursued by many.

Dibbling.—Mr. DENTON, of Brandon, sets all his wheat, and nearly all his oats, and his neighbours very generally do the same—one row on an eight-inch flag. I observed, however, that many of their rows were nine inches, from the men, I suppose, carrying a wider furrow than directed. He prefers this practice to drilling, which

has

has been tried, and is yet done by some. In his farm in Burnt-fen, he sets every thing. The drill roller has been tried there, but it would not do : the horses drawing by too great a purchase, tread in too much, and the roller *drives* the furrows.

At Oxborough, and its vicinity, great tracts are dibbled with wheat, oats, pease, &c. and found to answer much better than drilling, which has been tried. They put in but one row on a flag of wheat, six pecks an acre, and hoe well : the result very beneficial.

Mr. SAFFORY, of Downham, dibbling a field adjoining to one of Mr. CREASY's, which was drilling at nine inches, borrowed the drill for two lands, for comparison. Those lands and Mr. CREASY's field were mildewed, but the dibbled crop escaped.

About Old Buckenham, much wheat set, and generally one row on a flag, which they plough as narrow as they can, and put in six pecks of seed : this practice they find better than two rows. There is drilling, but Mr. ALGUR, &c. prefer setting greatly. Many oats also dibbled.

" Dibbling pease, practised time immemorial near Attleborough; but that of wheat, introduced by a labourer, JAMES STONE, of Deepham, about the year 1760, did not become very common till about the year 1770."— *Marshall.*

Twelve years ago, and how much sooner I am not well acquainted, they had, both in East and West Norfolk, discovered that this practice was not to be pursued under the notion (very common soon after its introduction), of saving seed : the usual quantity was risen to two bushels and a quarter and two bushels and a half per acre, and this both between Norwich and Yarmouth, and also around Houghton. Mr. LAYTON remarked, that *setting*
has

has failed in many instances, in proportion to the saving of seed.

Mr. ROBINSON dibbles all his wheat on layers, and is clear that he gains a coomb an acre more than he did broad-cast, with his land at the same time cleaner.

Sir THOMAS BEEVOR, with almost all the vicinity, dibbles every thing: drilling is known, and some few practise it; but the other method answers much better.

Mr. FELLOWES, of Shottesham, dibbles all, but neither wheat nor oats late; only while the season suits: for wheat late in the season, when the land is wet and cold, he thinks the water lodges in the holes and perishes the grain. There are drills; but not one acre so put in in twenty.

Mr. SALTER, of Winborough, dibbles all his wheat on layers, or on whatever land is proper for the practice, and what deserves particular attention, all is done by women, with only one confidential man for superintending them: he gives 10s. 6d. an acre for two rows on a flag. He thinks that women dibble better than men, from being more obedient and manageable. This practice deserves universal imitation: his women also reap wheat.

At East Bilney, Brisley, Gressenhall, Stanfield, Beteley and Mileham, much wheat is dibbled; also some oats and barley.

Most of the wheat about Hingham is dibbled, as well as both pease and oats. They pay 9s. and even 10s. an acre for it. Generally two rows on a flag, but on land much subject to poppy only one, for the benefit of hand-hoeing.

Mr. BURTON thinks, that on reduced land the best of all is to dibble one row and put in the same quantity of seed: it beats the drill: he drill rolls at five inches; no red weed; only nine-inch furrows, and good room for a hoe.

Not

Not so much dibbling as before. A new drill is in use; four inches for barley; it has cups and pipes, but not Cook's; price 30 guineas.

At Thelton and the vicinity, some farmers so much approve this method of putting in wheat, as to practice it even on summer fallows; they pass a heavy roller over the land, which prevents the moulds filling the holes of the dibbles.

In Loddon hundred there is much wheat dibbled, one row on a flag.

Mr. THURTELL dibbles, as general in his neighbourhood, two rows on a flag; if hand-hoeing be necessary, one row; but in that case some few drill at nine inches. Mr. THURTELL would prefer dibbling one row. He has tried putting the seed in by spike rolling; but he thought it made the wheat root fallen, which he attributed to the seed being too shallow. Nothing in his experience so good as dibbling; but drilling does well.

At Caistor, &c. in Fleg, three-fourths of the wheats are dibbled two rows on a flag. One bushel of seed would do, in Mr. EVERIT's opinion, but as it depends on the droppers, they put in from six to seven pecks. He made a comparative experiment in dibbling: the common way is, so to spread the two rows on a flag, that they are apt to be too near the seams. He made the dibblers keep their hands as close together as they could work them, setting the two rows very near each other in the centre of the flag. The result proved the excellence of the practice, for the crop was beautiful.

Mr. FERRIER, of Hemsby, two rows on a flag, and as close as he can get it done.

Mr. BROWN, of Thrigby, thinks that there are five times as many acres of wheat and pease dibbled in Fleg as are sown broad-cast. He remarked, that their lands, though

though they contain so much sand, will not do well if wheat is put in in the wet; pretty dry weather does much better.

Mr. SYBLE, of South Walsham, this year (1802) dibbled 60 acres; Mr. HEATH, of Hemlington, the whole of his crop on a farm of 500 acres. Mr. FRANCIS, at Martham, all for some years past.

There is a great deal dibbled all through Tunsted and Happing hundreds; and also through all North and South Erpingham: it rather declines about North Walsham, for want of good droppers.

Mr. PETRE dibbles about one-fifth of his crop, amounting to 100 acres; some drilled, and the rest sown; the best ears from the dibbled, but the drilled good.

About Norwich they have dibbled one row on a flag; but not putting in the same quantity of seed. They generally prefer two rows, and nine pecks an acre: Mr. CROW never saw a good crop of wheat in his life that was thin: this remark has thorough good sense in it, and he further notices, that if there is any mildew stirring, it is sure to attack such a crop severely.

Drilling has been tried about Blowfield, Acle, &c. but dibbling preferred by many as superior. They put in two rows on a flag, and find it the best practice of all.

Mr. DYBLE, of Scotter, dibbles much of his wheat, and has compared it with broad-cast in the same field, which it exceeded sufficiently to give him full satisfaction.

Dibbling common about Aylesham, but does not increase.

It increases about Reepham.

The cleanest crops at Haydon are those which are dibbled on a whole furrow. Ten shillings to ten shillings and sixpence an acre.

Mr.

WHEAT.   287

Mr. JOHNSON, of Thurning, thinks dibbling so excellent a practice, that equal crops are not to be gained in any other way; but three grains should always be put in every hole; for on various examinations he has found, that a single kernel in a hole has almost always produced a faint ear, scarcely ever a good one.

Mr. RISHTON dibbled his wheat when put in on a whole furrow, but would not have done it, only his men not skilful enough with the drill for that work.

Much dibbled and well done at Holm.

Mr. GODDISON, at Houghton, dibbles from 20 to 30 acres every year. But of all ways which he has tried of putting in wheat upon tempered land, he is inclined to prefer that of spreading the muck, then sowing broad-cast, and ploughing both in together into six-furrow ridges. He drills none.

Very little dibbling at Castle Riseing: Mr. BECK was in the practice, but left it off for drilling.

Dibbling well known all over Marshland.

Mr. PORTER, at Watlington, dibbles much, and thus gets his best crops, rather better than by the drill: but as he does not drill ridge work, his best land may be dibbled.

Dibbling very general about Wymondham, at 9s. to 10s. 6d. an acre.

Mr. PRIEST, of Besthorpe, dibbles his layers, if they do not plough well for the drill. He this year compared one row on a flag with two, and the former had the best ears, and was the stouter crop. He scarified the single rows, but being clean did not hand-hoe.

The greatest part of Mr. TWIST's farm at Bretenham is rye land; but he has some wheat land; he had 40 acres dibbled in autumn 1801; and he dibbles much rye. He has 50 acres of drilled wheat this year, and the best crop he had.

Leaving

Leaving Holt in the way to Holkham, came to four-furrowed work wheat stubbles.

Drilling.—More than thirty years ago Mr. FELIOWES, at Shottesham, drilled wheat at eighteen inches, the rows equi-distant, which produced equally with the common crops of the country.

Mr. DALTON, of Swafham, has drilled largely at Bilney; but his success for the two last years has been so bad that he leaves it off, convinced that the broad-cast answers better. If he drills early, the poppy gets greatly a-head; if late, the frosts turn the drilled wheat out of the ground; by ploughing the seed in he avoids the latter evil.

Mr. REPTON, at Oxnead, dibbles his wheat, as he cannot drill it on one furrow.

Mr. REEVES, of Heveringland, drills at six inches.

Many drill about Reepham; the practice answers best on light land; and Mr. BIRCHAM is of opinion that the layers should be broke previously for it.

Mr. JOHNSON, of Thurning, drilled forty acres of wheat, two years ago, on a whole fallow: the crop very good.

Mr. ENGLAND, of Binham, drills all at nine inches.

Mr. REEVE, at Wighton, at nine inches: he thinks dibbling a great improvement on the broad-cast husbandry, but that drilling exceeds it; and he never saw greater burthens of wheat than what has been produced by drills at nine inches. He had this year four good waggon loads an acre, from land so managed. I viewed his stubbles, and found them beautiful spectacles of masterly management; I rode cross and cross a field of nine acres, and do not think nine weeds were to be found in them: and all his wheat stubbles, on examining the intervals, I found in a perfectly friable and pulverized state.

Mr.

Mr. M. HILL drills all his wheat with COOK's machine, at nine inches; scarifies twice in March or April; no hand-hoeing, but all weeded. Quantity of seed three bushels, but the first put in before Michaelmas less.

Mr. COKE, at Holkham, drills all his wheat at nine inches: hand hoes twice, and uses COOK's fixed harrow once.

Mr. HENRY BLYTHE, of Burnham Westgate, drills all his wheat. Hearing much against the practice, he made an experiment for his own conviction, by sowing five acres broad-cast on four or six-furrow ridges, in an 80 acred field, drilled with COOK's machine, and at harvest he reaped one acre of the broad-cast, and threshed the produce immediately, which was seven coombs three bushels; it was a good wheat year, and on his best land: he then took half a land adjoining of the drilled, which measured three roods and 34 perches, and threshing it also immediately, the produce per acre was eight coombs three bushels, three pecks and a half, or per acre five bushels more than the broad-cast; besides which, the saving of seed was about three pecks. This convinced him that the drill method was the best, contrary to a prevailing opinion at that time.

Mr. DURSGATE drills all; some at nine inches, and some at six 3-fourths; hoeing twice at 4s. and weeding. I viewed his stubbles, and found them very clean.

Mr. RISHTON, at Thornham, drilled all at nine inches, and was well convinced of the superiority to broad-cast.

Mr. STYLEMAN, at Snettisham, drills all at nine or twelve inches; hand-hoeing once or twice, as wanted.

Mr. GODDISON, at Houghton, puts in some wheat with a drill harrow, which marking channels, the seed is sown broad-cast. I did not see the tool.

Mr. BEACHER, of Hillingdon, drills at nine inches, and hand-hoes.

Mr. BECK, of Castle Riseing, drills at nine inches, and hand-hoes twice, at 1s. 9d. or 2s. each time.

Mr. PORTER, of Watlington, drills his wheat on layers, if on the flat, but not on ridges: he has 100 acres of wheat, and half of it drilled: dibbles much, a practice he has a very high opinion of, but thinks he gets nearly as much by the drill.

Mr. PRIEST, of Besthorpe, dills on strongish land at nine inches, and intends trying at twelve this year; but if the furrows on layers do not whelm well, but stand on edge, he then dibbles, as the drill does not work well. He scarifies in March, and if necessary hand-hoes, if not, horse-hoes twice, if he has the opportunity. Success much depends on the ridges being well formed for a bout of the machine, and they ought to be rolled and harrowed, and left some time for the air to pulverize, before drilling.

" *Scarning-School Farm*, *Aug.* 10, 1802

" DEAR SIR,

" In compliance with your desire and my promise, I will endeavour to describe, as accurately as I can, the method I used in drilling the field of wheat which you saw, when you did me the honour of calling upon me at Scarning.

" The field consists of about twelve acres, and is a mixed soil: last year it grew clover and ray-grass after barley, and as soon as the first crop of clover was reaped, and the second crop fed off with sheep and cows, I broke it up in order to temper it for wheat. It was ploughed twice and scuffled twice, which, with many harrowings,
brought

brought it, by the beginning of October, into a high state of pulverization: in this state I began my operation by rolling such a quantity as might be sufficient for two days work. The surface then being as flat as possible, I set out the work thus: in the middle of the field (which was fixed upon because no side of it was straight) I set up two sticks, in order to draw a line as straight as possible to direct the drill. My drill is a small barrow with two hoppers or boxes, one on each side the wheel, and is pushed forwards by a man. In the line thus formed by the sticks the barrow was directed, depositing the seed from the hopper on the right side of the wheel upon the flat work. Immediately after the barrow, at the distance of about ten yards, followed a plough to cover up this line of seed, by turning the mould of a fleet furrow upon it: when the barrow, followed by the plough, had reached the end, the man with the barrow turned towards the left hand, and at the distance of one yard from the line of seed already deposited, dropped from the barrow another line of seed parallel to the former: now a second plough followed him as before, whilst the first plough, which had covered the first line of seed, was *backing* its own furrow. The barrow-man, arrived at the end of the second line of seed, turned to the left as he had done before, and dropped a third line of seed one yard from the first and parallel to it, and was followed by the first plough, whilst the second was *backing* its own furrow. In this manner I worked, my barrow depositing lines of seed, at one yard distance from each other, and my two ploughs alternately covering the lines of seed, and backing each its own furrow, till I had completed my morning's work, at the end of which you will observe, that except the work of the barrow, I had merely set out the tops of the four-furrow work of about two acres of land, and there remained the balks to be split;

split. This was the operation of the afternoon: for whilst my double barrow was directed upon a balk, depositing seed in the furrows from the two boxes on each side of the wheel, a double breasted plough, drawn by two horses, split the balk, covered the wheat so deposited, and completely made up the four-furrow work which had been set out in the morning. The next day I repeated the work precisely the same as the day before, by setting out fresh work from a line formed by two sticks, as at first set up across the field, in a direction parallel to the first line drawn, and at such a number of yards from the last line of seed dropped, as I thought would afford work for the day. Thus was the whole of the twelve acres laid into four-furrow work, with three rows of wheat upon every stitch, at the distance of nine inches between the rows, and eighteen inches for the furrows, with no more than five pecks of seed-corn per acre, and performed by three men, two ploughs, and four horses, in a morning, and two men, one plough, and two horses in an afternoon; and the whole two acres were finished in a day. I rolled it afterwards to please the eye, level the work, reduce the depth to which the seed was deposited, and afford mould in the furrows to support the wheat on the sides of them. As soon as the wheat came up I cleaned the furrows by a plough with expanding wings, drawn by one horse. In the spring I contrived to fix upon this plough two scarifiers, and taking off the expanding wings, I used it to hoe the furrows, and at the same time scarify two rows of wheat, one on each side of the furrow: afterwards I put on the expanding wings, and substituting hoes for scarifiers, I by one operation of this plough hoed the furrows and two rows of wheat, and at the same time moulded them up: this operation was performed twice.

"Thus, Sir, have I given you as clear an account as
I am

I am able, of the manner in which I drilled (if I may be allowed the expression) the wheat you saw.

"I am, Sir,
   "Your obedient Servant,
      "ST. JOHN PRIEST."

From the preceding notes, it appears that drilling wheat has made a remarkable progress in Norfolk. In the north western district, amongst the great and intelligent farmers who have rendered their county famous, it is become so established a practice, that it no longer admits any question of its utility, on a soil like theirs. In some smaller cases, dibbling is preferred; nor has it been sufficiently ascertained which of these methods will give the greater crop. The inquiry, however, is not of consequence in North-west Norfolk, for they have no population equal to dibbling becoming general: and a circumstance which tends much to impede this husbandry, is the imperfect manner in which it is performed, for the sake of making great earnings; this has, in many instances, given a preference to drilling.

Depth.—Mr. THURTELL, near Yarmouth, has found two inches, which are commonly half way through the flag, to be the best depth in dibbling.

Mr. EVERIT, of Caistor, in Fleg, thinks that the deeper it is dibbled the better; two inches, to chuse.

Mr. SYBLE, of South Walsham, finds two inches, in dibbling, the best depth.

Mr. MARGATESON, of North Walsham, prefers one inch; and three or four kernels in a hole.

Mr. DYBLE, of Scotter, remarked, that there is never any fear of the seed being dibbled too deep, even if through the flag: they plough only three or four inches.

Mr. PALGRAVE, at Coltishal, dibbles two inches.

Mr. REPTON, at Oxnead, dibbles one inch.

Mr. BIRCHAM, at Hackford, two inches and a half.

Mr. JOHNSON, of Thurning, two inches; rather more than half through the flag is best.

Mr. ENGLAND, of Binham, two inches and a half, on a flag.

Mr. RISHTON, at Thornham, one inch and a half.

Mr. STYLEMAN thinks the depth a matter of great importance; he drills two inches and a half deep, by weighting COOK's machine, and using two horses; if only one, he should be of extraordinary strength.

Mr. PORTER, at Watlington, two inches.

Mr. PRIEST, of Besthorpe, thinks, that if wheat be not buried from one inch and a half to two inches, it is apt to be root-fallen.

Mr. M. HILL one inch and a half; but by no means more.

Mr. OVERMAN drills his wheat one inch and a half deep on sandy soils, rendered light by cultivation; but on layers once ploughed, one inch only.

With turnips.—Mr. WALKER, of Harpley, some years ago, introduced and practised a husbandry in which he was entirely original. I viewed his farm, while these experiments were going on, for two successive years. In order to give a greater degree of stiffness to his sandy soil, he thought of putting in wheat without any ploughing at all, immediately before sowing. He began with six acres of turnips, hoeing in the wheat seed at the second hoeing of the turnips: these were eaten on the land by bullocks and sheep. The wheat proved good, and answered expectation. The next year he did the same on 35 acres; this also succeeded; but the best wheat was where the turnips were eaten in the driest weather. The following year he extended the culture to 70 acres, which also succeeded

succeeded to his satisfaction. The year following, he had 100 acres. Upon the whole, the culture produced not better than the common crops, but equal. The most adverse circumstance is a wet season for eating the turnips, but at the worst, it can amount to no more than the loss of the seed. The course in which he practised this management, was,

  1. and 2. Ollond, 4. Turnips,
  3. Oats,   5. Wheat;

his seeds being sown alone on the wheat stubble ploughed once; he has tried sowing both in autumn and in spring; both succeeded well, but the autumnal rather the best: upon his farm of 600 acres, the saving in horses has been unquestionably five: the three earths for barley, in a busy time, were entirely saved.

Feeding.—Mr. SALTER, of Winborough, whose crop I found extraordinarily great, this year fed all his wheat twice; and he says, that if he had not done it, it would all have been *laid.*

Mr. COKE—Do you feed your wheats?—Never, Never, Never!

Mr. M. HILL— Never!

Hoeing.—Not one farmer in twenty hoes any wheat about Aylesham.

Mr. ENGLAND, of Binham, scarifies early in April; and hand-hoes twice at 2s. each time per acre.

Mr. REEVE, of Wighton, hand-hoes once or twice, as wanted; at 2s. each time: but if the land be clean, once at 2s. 6d. with the use of the fixed harrow, answers every purpose. His stubbles beautifully clean.

Mr. HENRY BLYTHE, of Burnham, hoes his drilled wheat twice, for 3s. 6d. an acre, and weeds at 6d.

Drilled wheat in all the north-west angle of Norfolk, hand-hoed: but some do it not at Hillingdon.

On the clays of Marshland, where there are other signs of bad management, they are much pestered with red weed, May weed, clivers, &c. they are forced to weed much; Mr. DENNIS, of Wigenhall, has paid a guinea an acre for it: he hoes all his dibbled wheat.

Tathing.—This is a singular husbandry, which I did not meet with till I entered Fleg, from Yarmouth. It consists in carting turnips on to wheat in February and March; they call it to *pull and throw* on wheat, eating them on that crop by sheep and bullocks, if sheep are kept; but if not, by bullocks alone. Mr. EVERIT, of Caistor, assured me, that the wheats thus treated, are the heaviest crops they gain; it makes the straw as stiff as reed. It is not practised as a preventive of the crop being root-fallen—for that is little known here; when finished, the field is harrowed, and if necessary, hoed.

Mr. FERRIER, at Hemsby, and his neighbours, are in the same practice; and sometimes do it too late; with cows and bullocks.

Mr. BROWN, of Thrigby, confines this practice to land that is light, and subject to red weed; never later than March, and then wishes for some showers directly after: he finds that this practice does not make the wheat too rank, but the contrary, by stiffening the straw. The practice pretty general in Fleg; and some, but not so much, in Blowfield hundred.

Mr. SYBLE does it like his neighbours, *here and there*, on a piece given to poppy. The same rule at Repps and Martham; but some doubt whether it answers: Mr. FRANCIS has been *caught* sometimes by drought, and he has seen others. He remarked that the open fields give as good wheat as the enclosures, yet never have turnips thrown on them. He thinks it better to throw turnips on wheat stubble, treading all well down for barley.

Mr.

Mr. HORNARD, of Ludham, on his light soil is in this practice: he had two pieces under the same circumstances, one tathed and the other not, and the former was the best crop; on an average of seasons, he thinks it certainly does good. He was the first here that practised it 28 years ago; but had seen it near Norwich: he has continued it ever since; begins in February, and continues through March; he has no sheep, but if he had, would do it with bullocks also.

On his light land, Mr. CUBIT, of Honing, has carted muck on to his wheat in winter, throwing turnips after for sheep: he has done it for bullocks, but sheep answer the purpose much better.

About North Walsham, the practice of throwing turnips on wheat, is known, and that is all: Mr. MARGATESON does not approve of it with bullocks; he has done it, and mischief was the result; but with sheep only, it is good husbandry with those who keep a sufficient number.

Mr. PETRE, of Westwick, keeps 17 score sheep, and is in this husbandry: his bailiff, CROWE, informed me, that he practises it on light land with much success.

Mr. DYBLE, of Scotter, has known it done now and then, but it is no general practice here; nor does he approve of it.

Mr. REPTON, at Oxnead, has tried it, but the crop was damaged by the *jamming*, and he left it off.

Mr. JOHNSON of Thurning, never does it; but once saw it at Edgfield, and it was thought to answer.

Mildew.—Mr. FRANCIS, of Martham, remarked, that if wheat there be thin or backward, it is very liable to the mildew.

Mr. MARGATESON, of North Walsham, could never observe that thinness was the cause of mildew; but that the

the berberry bush will occasion it, he has ascertained by observations that could not deceive him.

The light soils of Saxethorpe, are very subject to the mildew, and about Reepham, fine barley land the same. Mr. PARMENTER, of Aylesham, a considerable miller, has, however, had some of his finest wheats from Saxethorpe.

The Rev. Dr. BAKER, at Causton, observes, that the wheat in that parish is so subject to the mildew, that he has found on various occasions, oats to be a more profitable crop; he has known from 16 to 18 coombs of oats, where more than seven of wheat could not be expected, and that, probably, much damaged by this distemper. The soil is very fine barley land.

The mildew is so mischievous at Hillingdon, that some farmers do not sow wheat at all; Captain BEACHER, Steward to Sir M. FOLKES, has not sown any for three years past: as it is not here the custom to cut early, upon account of this distemper, no wonder it should make such havock.

" This distemper has been accurately traced from a point across a field to a berberry bush in a hedge. Several similar instances; and Mr. MARSHALL produced the distemper by planting a small bush in the middle of a large piece of wheat; all clean, except a stripe where the berberry mildewed the crop."

The parish of Elsing is well known for the wheat that grows in it being very liable to the mildew, arising (as every person in it knows) from the number of berberry-bushes which abounded in the hedges, till much attention was given to extirpate them; and still, arising from the same cause, on the lands of those who are careless in this business. At Harpley-dam, a hedge of Mr. JOHN BLOMFIELD's had been cleared, and the wheat consequently safe from this cause; but this year (1802), observing a streak

streak of mildew in his wheat, he examined the hedge, and found a stub, missed by the workmen when they were set to eradicate the plant, which had thrown out only two shoots, not more than two or three feet high, and from that point the streak of mildew took its course. I had this from his own mouth.

Smut.—Very little smut is known in Fleg. Mr. BROWN adds salt to sea-water, to make it swim an egg; skims off the *cosh* (husks) carefully, as they find that harbours smut if there be any; he only wets the wheat by dipping in the basket; dries with lime slaked with sea water.

Many farmers about Houghton are troubled with the smut, but Mr. STANTON, of Darsingham, sows only old wheat, and never having been known to have any smut from such seeding, others are getting into the same practice.

Captain BEACHER, at Hillingdon, always sows old wheat, and never has any smut: no brining or liming, as that injures old seed. He has sown old and new seed in the same field, and had the smut in the latter, but not an ear in the former.

Red Worm—Often makes great havock in the vicinity of Watton and Hingham, as well as the cock-chaffer grub in grass-lands; but they have encouraged rooks every where, with some effect: and sea-gulls fortunately resort very much to lay their eggs on an island in Scoulton Meer, and rear great numbers of young, undisturbed, as they are known to feed only on worms and grubs, no seeds having been found in their crops. They come the end of March, and migrate in July.

Roots.—Mr. THURTELL has traced the fibres of the roots of wheat, five feet deep, on the side of a marle pit: also the root of a turnip, drawn by hand, two feet and a half in length, in a light soil.

Reaping.

Reaping.—Mr. SYBLE, of South Walsham, cuts very low, and consequently does not haulm the stubbles; which he thinks a very inferior practice, if a farm is kept clean.

Mr. PARMENTER, miller, at Aylesham, a considerable farmer also, and a very intelligent sensible man, remarked to me, that the farmers let their wheat stand too long before cutting. They were apt to have a notion, that when millers gave this opinion, it was speaking for their own interest: but he cuts his own wheat before it is ripe, and would do so on the largest scale, if he was not a miller: the quality is far superior, and the crop just as good.

" Mowing has been practised at Hainford; the crop very clean, and dead ripe. The gatherers followed the scythe, and the waggon the gatherers. Secured, at a trifling expence, without any risque from weather."—*Marshall.*

Mr. M. HILL prefers cutting green, and never began harvest but he wished he had began three days sooner.

Stubbles.—Mr. BURTON, of Langley, always haulms his wheat stubbles, for littering the yards; and it is the common management in Loddon hundred: by this means he finds he can support his farm in heart, without buying dung. At first coming, he bought barrack muck, at 5s. a load, laying on eight an acre: three years of this brought his farm into such a state, that he discontinued it.

Produce.—About Watton, three quarters, on an average. The finest wheat I saw in thirty miles, was a small broad-cast field at Tofts, of Mr. PAYNE GALWAY's; it promised to yield five quarters an acre. Tofts is in a very poor district, but I have often remarked, that if a good piece of land is found in such, it is usually uncommonly good.

At Langley, &c. average seven coombs. Caistor, and average of Fleg hundred, seven to eight coombs: fourteen have been known, and were reaped by J. HUNTINGDON,

DON, Esq. at Somerton, last year; he had 32 coombs on two acres and a quarter.

Hemsby, eight coombs: fourteen have been known.
Thrigby, seven coombs.
South Walsham and Blowfield, seven coombs.
Martham, seven coombs.
Catfield and the vicinity, seven to eight coombs.
Happsborough, Walcot, and Barton, nine coombs.
Honing, seven. North Walsham, six to seven.
Westwick, six.
Scotter, six. Mr. DYBLE, seven. As the average barley is but eight, the wheat seems high for a sandy loam, a good barley land; but drawing the turnips, and throwing them on the ollonds for wheat, explain it.

Around Coltishall, six coombs.

Mr. REPTON, at Oxnead, favoured me with an accurate account of all his crops for some years back.

	Acres.	Produce per acre.	
		Coombs.	Bushels.
1773	77	3	3
1774	50	5	2
1775	46	6	2
1776	56	5	2
1777	30	6	3
1778	71	5	1
1779	50	6	0
1780	61	5	0
1781	53	9	0
1782	46	7	2
1783	47	6	3
1784	50	5	3
1785	58	7	0
1786	59	8	2
1787	42	8	0

WHEAT.

	Acres.	Produce per acre. Coombs. Bushels.	
1788	56	7	1
1789	47	7	0
1790	58	7	2
1791	58	8	2
1792	57	7	0
1793	54	9	1
1794	48	5	$1\frac{1}{2}$
1795	59	5	$1\frac{1}{2}$
1796	52	8	0
1797	53	7	1
1798	53	9	2
1799	57	5	0
1800	55	7	3
1801	Books not made up.		
Average		6	$3\frac{1}{2}$

On the poorer soils at Heveringland, not much wheat is sown; the best of it, five coombs: on their better soils, six to seven.

Mr. BIRCHAM, at Hackford, Reepham, &c. seven coombs.

At Haydon, five coombs. At Thurning, better than five.

Mr. COKE, at Holkham, six coombs.

Mr. M. HILL, average of last twenty-one years, six coombs.

At Burnham Westgate, five coombs.

At Holm, some very rich land, eight to twelve; sometimes more.

At Snettisham, six coombs.

At Houghton, five coombs.

At Hillingdon, not four coombs, on an average—much damaged by the mildew.

About Downham and the vicinity, seven coombs.

At Watlington, seven coombs.

WHEAT.

At Wigenhall, St. Mary's, the average, five coombs per acre; sometimes not more than ten to twelve bushels. If this is true, it stamps their husbandry sufficiently.

At Walpole, five coombs the small acre; more than three roods. By another account, five and a half.

Near the Smeeth, seven to ten coombs.

About Wymondham, five.

General average of the county, six coombs.

Profit, compared with Oats. — Mr. FELLOWES, of Shottesham, remarks, that oats pay better than wheat, in many cases even which are supposed to be particularly favourable to wheat, as in clover lays. He has found it more profitable to plough a clover lay late in autumn; but before much wet comes, and to leave it till the spring, and then, as early as may be, harrow in oats or barley, on that stale furrow, and the crops have been so great, as to pay better than wheat; but oats better than barley. This seems to be, among others, one proof that wheat has been too cheap.

Price. — In discourse with Mr. PARMENTER, a miller of Aylesham, on what ought to be the price of wheat, reference being had to the farmer and to the poor, he gave it as his opinion, that 30s. a coomb, on the average of samples, would be that fair price for all parties. He remarked, that the very high prices we have seen, were not advantageous to millers, by reason of the great capital demanded, and from a want of proportion in the price of bran. When wheat was 4l. a coomb, bran was not higher than 1s. a bushel. Within four years he has bought a great deal of wheat, as he shewed me by his books, at 21s. and 22s. bran at 6d. and 7d.

SECT. VIII.—RYE.

Upon the light sands of the South-western district, this is a common crop, and follows the seeds generally upon a bastard fallow. The quantity of seed usually two bushels, and the produce more uncertain than with wheat, for this plant is very liable to be damaged by frosts. The culture ought to lessen every where, except perhaps in mountainous districts, since wheat has been found to produce, on poor sands, as many bushels per acre as rye.

Mr. Bevan's rye, in 1802, was sown in February, by which means it escaped the severe frosts which did so much mischief to this crop in June. I have rarely seen finer crops.

Colonel Cony had once six quarters per acre, at Runcton.

There is a practice in the South-west district, which has merit—that of ploughing up the rye stubbles in harvest, and sowing one or two pecks of seed additional per acre, with intention of burying the scattered grains, and thus having a crop of spring food for sheep. The tillage has its use for the following turnips, consequently the expence merely consists in the small portion of seed added.

SECT. IX.—OATS.

This is by no means a favourite crop in Norfolk, where, if the land is in order, they greatly prefer barley; and if it is not in order, they are too good farmers to sow any corn.

Mr. Hill, at Waterden, in common with the geneality

rality of his neighbours, buys all the oats he consumes; he can grow as much barley per acre as of oats, and is at the same time a more certain crop.

Upon the good barley sand, for some miles around Holkham, the farmers think oats a so much more exhausting crop than barley, that they buy all their oats at Wells; carrying wheat, barley and pease thither, the waggons bring oats back.

Mr. DENTON, of Brandon, dibbles oats with great success; puts in more than a coomb per acre of seed, observing, that he did not approve of adopting any method as a way of saving seed.

Admiring a very fine oat-stubble at Mr. FRANCIS's, at Martham, and inquiring the preparation, he informed me that wheat preceded, the stubble of which he scaled before Christmas, ploughed it across thin, and then a full pitch for the seed oats. The wheat was on a two year's layer.

" The surface of a piece of ground, sown several days with oats, but which were not yet up, was run, by heavy rains, into a batter, and baked to a crust: the owner ploughed the ground, notwithstanding they had begun to vegetate, under a fleet furrow. The success was beyond expectation."—*Mr. Marshall.*

Seed.—Mr. FELLOWES, of Shottesham, has found, from many observations in a long and attentive experience, that the more oat-seed he sows, the better is the crop. This he finds to take place as far as seven bushels an acre, and even to eight.

On strong land at Wissen, five bushels broad-cast. Mr. SALTER, six.

About Watton, four bushels; some five.

About Bilney, four and five.

At Thelton, four to five bushels.

Mr. EVERIT, of Caistor, in Fleg, five to six bushels.

Mr. SYBLE, of South Walsham, five bushels.

Mr. CUBIT, &c. at Catfield, five to six bushels.

Mr. REEVES, of Heveringland, drills at six inches, three to four bushels.

Mr. REEVE, of Wighton, three bushels, drilled: of Tartarian four bushels.

Near Downham, four bushels.

At Watlington, six bushels. Mr. PORTER there harrows in, because no drill cups will deliver enough.

About Wymondham, four bushels.

Produce.—At Thelton, ten coombs an acre common, and even more: some twelve to fourteen coombs.

On the fine loamy and gravelly sands of Langley, &c. average 11 coombs.

Caistor, &c. in Fleg, and the average of the hundreds, fourteen coombs: twenty-four have been known.

Hemsby, twelve coombs: twenty have been known.

Thrigby and Fleg, twelve coombs.

South Walsham and Blowfield, fourteen coombs.

Martham, eleven coombs.

Catfield and Happing, twelve to fifteen coombs.

Honing, eleven to fifteen coombs.

North Walsham, ten to twelve coombs.

Coltishal, twelve coombs: but few sown.

Mr. REPTON, of Oxnead, favoured me with an accurate account of his crops for many years:

	Acres.	Produce per acre.	
		Coombs.	Bushels.
1773	10	14	0
1774	14	15	0
1775	20	11	0
1776	9	15	0
1777	26	13	2
1778	12	11	0
1779	14	11	0
1780	15	10	3

OATS.

	Acres.	Produce per acre.	
		Coombs.	Bushels.
1781	14	12	1
1782	21	12	2
1783	10	10	0
1784	13	8	0
1785	14	6	3
1786	23	9	0
1787	17	10	0
1788	0	0	0
1789	10	5	1
1790	9	8	0
1791	8	13	0
1792	23	14	0
1793	4	14	0
1794	18	12	3
1795	13	7	3
1796	15	11	0
1797	11	12	0
1798	11	12	0
1799	10	9	2
1800	10	8	2
1801	Books not made up.		
Average	-	11	2

On the light soils at Heveringland, eight to ten coombs: on the better land, twelve coombs.

Mr. BIRCHAM, at Hackford, fifteen coombs: has had twenty-six coombs one bushel, on old land that had had vetches only.

At Thurning, eight to ten coombs.

Mr. REEVE, of Wighton, has had twenty-five coombs an acre of the Tartarian.

At Snettisham, ten coombs.

The vicinity of Downham, ten to sixteen coombs.

At Watlington, fourteen coombs.

About

About Watton, average ten coombs.

At Wigenhall, St. Mary's, six or seven coombs.

At Walpole, twelve coombs. Another account, twelve to fourteen coombs; on fallow, sixteen coombs.

Near the Smeeth, thirteen or fourteen coombs.

SECT. X.—PEASE.

Mr. Overman, carrying me into a crop of broad-cast pease in his neighbourhood, desired me to examine the strongest tufts to be found, to shew that the poppies, so far from being destroyed, were erect and ready to force themselves through when the pease fall, though overtopped at present. He remarked, that the common observation, that pease are apt to foul land if weak, and clean it, if stout, was erroneous: if red-weed or spear-grass are in the land, no crop will destroy them; and if they are not in the land, the pease cannot generate them.

Mr. Overman, from various observations, is of opinion, that if pease are repeated oftener than once in eleven or twelve years, they are very apt to fail.

Mr. Syble, of South Walsham, has found that pease are a very uncertain crop: this is known to all farmers; but he has remarked further, that they will not bear repeating. If sown often on the same land, they are almost sure to fail.

Mr. Everit, of Caistor, has this year (1802) a vast crop of pease, which I viewed with pleasure; he lays them at twelve coombs an acre at least: they followed wheat; and his remark was, that they had *shaded the ground so completely*, that he had no doubt of very fine wheat after them again.

Mr. Fowel, of Snetterton, ploughs such of his layers

as

as he intends for pease, in December or January, and rolls and harrows before drilling on this stale furrow. His mode of drilling this crop, is to have two rows at five inches, and intervals of fifteen inches for horse-hoeing. I viewed his crops, they were very well *hung*, that is, loaded with pods, but too many poppies for a driller. Horse-hoeing, in this system, he finds far more effective than the hand-hoeing he gave to equi-distant rows at twelve inches. For this drilling, he made a new beam to Cook's machine.

Mr. Reeves, of Heveringland, drills his pease at nine inches; and highly approves the method.

Mr. Johnson, at Thurning, drills all at nine inches.

Mr. Styleman, at Snettisham, drills all; some at nine and some at twelve inches: hand-hoeing, once or twice, as wanted.

Mr. Rogerson, of Narborough, drills all that he does not dibble.

Mr. M. Hill drills at nine inches on the flag, *scarifies* twice, and weeds; and prefers that practice to all others.

Pease are very generally dibbled at Thelton; crops, eight to ten coombs per acre.

Wherever found, thence to Yarmouth, and my route through the Flegs, Blowfield, &c. &c. to North Walsham, dibbling very general: about the latter town, few, but what there are, dibbled: they do not hoe them. The pea stubble reckoned *kind* for wheat.

Few pease are dibbled on ollonds at Scotter; but some are; and Mr. Dyble remarked, that pease cannot well be dibbled too deep.

Mr. Repton, at Oxnead, dibbles with great success: he had this year (1802) seventy loads from twelve acres.

The maple grey is a great favourite at present in Fleg hundred. Mr. Tuthill, of Southwood, selected them from a sample that came into the country, and cultivating them

them carefully, they have established themselves very generally; they produce ten or twelve coombs an acre, on land that would not give five or six of the old sorts; the straw is not so long as some other sorts, but long enough for shade, which Mr. BROWN, of Thrigby, holds to be a great point in a pea crop.

Mr. DURSGATE finds the pearl pea to ripen a fortnight sooner than other sorts: but this year the frost damaged his crop.

Seed.—About Watton, two bushels.

About Dercham and to Bilney, three bushels.

At Thelton, grey pease dibbled, two bushels.

Mr. SYBLE, of South Walsham, dibbles three bushels.

Mr. CUBIT, &c. at Catfield, four bushels, whether dibbled or broad-cast: a thick crop they consider as a great object in preparing for other corn.

Mr. REPTON, at Oxnead, dibbles, four bushels, two rows on a flag.

Mr. REEVES, of Heveringland, drills at nine inches, ten pecks.

Mr. REEVE, of Wighton, four bushels of the large pea, drilled.

Mr. HENRY BLYTHE, of Burnham, drills, ten pecks.

Mr. BECK, of Castle Riseing, drills, ten to twelve pecks, at nine inches, and hand-hoes.

About Downham, two bushels and a half to three bushels.

At Watlington, four bushels.

Produce.—At Langley, &c. when the year is favourable, the average produce is eight or nine coombs.

At Caistor, and average of Fleg, eleven coombs: fifteen coombs have been known.

At Hemsby, Mr. FERRIER has known fourteen coombs.

At Catfield, if good, ten coombs.
At Thurning, six coombs.
At Snettisham, when no failure, eight coombs.
About Downham, good, seven coombs.
At Watlington, if good, ten coombs.

SECT. XI.—BEANS.

On the stronger land about Watton, the failure of common red clover, from long repetition, has had one very good effect, that of inducing some farmers to plant beans. They plough the barley stubble once, and dibble in one row of horse-beans on every furrow, keep them clean by hand-hoeing, and sowing wheat after; get excellent crops, oftentimes better than after any other preparation.

I viewed carefully a small field of wheat at Scarning, belonging to Mr. Nelson, of Dereham, half of which followed beans and half potatoes: the superiority of the former considerable. The Rev. Mr. Munnings and the Rev. Mr. Priest were with me, and were entirely of the same opinion.

Beans are in common cultivation at Thelton: Mr. Havers's are dibbled in rows along ridges seven feet and a half, or ten furrows wide, on which eight rows. He ploughs the land early in autumn, and (after the frosts) in the spring only harrows, and plants immediately; two bushels of seed an acre, at the expense of 4s. and 2s. 6d. in the pound for beer. Hand-hoes two or three times, at 8s. or 10s. expense; no horse-hoeing: he gets ten or twelve coombs an acre. There cannot be better management than avoiding spring tillage, on heavy land.

Mr. Havers, and his neighbours, having kept their beans

beans clean by hoeing, plough the stubble once, harrow, and drill or dibble in the wheat seed.

Many are also cultivated at Billingford, where they are in the same system: Mr. DRAKE avoids spring tillage, ploughing before Christmas; has a set of seven small harrows, which dropping into every hollow, prepares well for the dibbles; rows at nine inches: he hand-hoes twice, keeping them quite clean. I viewed his crops and found them perfectly so, and very fine, equal in appearance to five or six quarters an acre; yet they seemed too thick, and not podded so low as they would have been with more room.

Mr. PITTS, of Thorpe Abbots, has this year (1802) drilled beans with COOK's machine, at 12, 18, and 24 inches, and he thinks the 24 promise to be the best crop. They have been horse and hand hoed, and kept quite clean. All will be eight coombs an acre at least.

Mr. KERRICH, at Harleston, lays his manure on for beans, and then drills wheat, getting great crops.

Mr. BALDWIN, near Harleston, drills many beans, and gets great crops, as well as of wheat, after them.

Through the hundred of Loddon the beans are all dibbled, one row on a flag, and all the farmers sow wheat after them, finding it the best preparation for that grain.

Mr. BURTON, of Langley, whose knowledge of husbandry is extensive and unquestioned, upon the strong land at Hempnal has always had excellent wheat, perhaps the best, on bean stubbles.

At Langley and the vicinity, average produce ten coombs. Mr. BURTON never less than that crop at a medium; this year above twelve.

At Seething and Mundham many beans, and great crops: wheat after them. Here are excellent farmers.

Admiring the fine sandy loams at Caistor, in Fleg, and combining

combining the circumstance of soil with the small variation of their crops, I proposed to Mr. EVERIT the culture of beans: he said that he had tried them, and they would not do; they run away to straw, and will not pod well.

Mr. FERRIER, of Hemsby, has known them tried; but they would not do.

Mr. SYBLE, of South Walsham, sowed five acres, and they ran away to straw, and yielded very little corn.

These articles of information were rather discouraging, but having heard that Mr. CHRISTMAS, at Billockby, had made some experiments on beans, I called on him; unfortunately he was from home, but I examined his bean crop. Passing to Ludham, the ninth mile-stone is against the field, which I note, that others may, if they please, examine the field. It is a large field dibbled one row on every flag: the soil fair Fleg land, but rather stronger than the very pale lands: the crop had been hoed, but not sufficiently, for they were rather foul: the conviction on my mind that the country will do for beans, was compleat: I guess the produce ten coombs an acre. They are rather low than high. There is a large pit of clay marle in the field.

Through the Flegs, Walsham, Blowfield, and Happing hundreds, I have been calling out for beans, and surprized at finding only one crop; but at last, at Happsborough, I found that Mr. WISEMAN had them for three years, and with good success, getting 14 or 15 coombs an acre, though in no better mode of culture than that of spraining the seed in every other furrow, and hand-hoeing them twice. He got as fine wheat after them as the best in the country. Inquiring how he came to make such an experiment, I found he had been in Kent, and seeing the effect of beans there, induced him to try them.

Mr. BIRCHAM, at Hackford, tried beans three or four times,

times, and left them off because he could not get wheat after them; but did not hoe: *could not.* I advised fresh trials in the horse-hoeing system.

Mr. STYLEMAN, at Snettisham, has had two crops of beans on a sort of marsh, stiff and strong soil. His management of the second crop, which I viewed, was to plough a wheat stubble in autumn, leave it well water-furrowed, and in the spring to dibble in the seed.

Mr. PRIEST, of Besthorpe, has drilled beans with COOK's machine at 18 inches; five coombs of seed on 17 acres; the crop a fair one: the field yielded dibbled barley the year before, on a layer. The beans were once horse-hoed, once hand-hoed, and once weeded; designs to scuffle the stubble for wheat.

At Wigenhall, St. Mary's, in Marshland, they plough in the wheat stubbles in autumn, and stirring in the spring, sow every third or fourth furrow with two bushels of horse-beans an acre; hand-hoe the rows once, and plough sometimes between them. Last year (1801) they got ten coombs an acre, but the average not above five or six. They burn the straw in ovens, &c. For the following wheat they plough once, and harrow in the seed; and if the weather is good, twice: this they reckon best on account of the *white snail,* a slug which abounds on bean stubbles: it eats the young plant of wheat the moment the seed shoots, and sometimes destroys the crop; steeping in arsenic no prevention—they eat the seed itself also, by some accounts, but this is doubtful. Their best wheats follow beans: the fallowed crops never average five coombs, but those on bean stubbles do. Such is Mr. DENNIS's account.

Mr. COE, of Islington, has generally one-sixth of his arable in beans; in autumn he ploughs the wheat stubble, and again at Candlemas, and harrows in two bushels and a half

half of seed per acre; some he dibbles, and hand-hoes in either case twice, at 5s. and 4s. an acre, and reckons a fair crop at six coombs; he takes wheat after them on one earth, and generally good; the best, if it escapes the slug, which in a wet season attacks the wheat in autumn: it is about an inch long, the size of a tobacco-pipe, and of a bluish white colour.

In going round from Islington by Tilney and Terrington to Lynn, I saw many crops of broad-cast beans full of weeds.

Mr. THORP, on Governor BENTINCK's estate, sows beans in furrows; keeps them clean, and gets fine wheat after.

1776. At Walpole, drilled in every third or fourth furrow, and kept clean by both horse and hand-hoeing; four quarters sometimes gained. Mr. CANHAM, of Southry, near Downham, had 60 acres thus cultivated in 1769, which yielded five quarters and an half round: in 1770 I viewed the wheat after them, and found it as clear and as fine as any after fallows. He has often had five quarters an acre after beans.

This husbandry continues: in 1802 I found the same; and I am sorry to say without improvement.

Produce at Walpole six coombs.

Many are dibbled in rows, and kept clean and well managed; Mr. PHILIP GRIFFIN cultivates them in this manner.

Mr. SWAYNE, of Walpole, gets the best wheat after beans, if it escapes the slug: if the land is lightish, it is common to put them in in every third or fourth furrow, for ploughing between; but on strong clay it is difficult to get the plough in. Some dibble in, but others object to it, on account of their not being sufficiently buried. It is more common to sow from two to three bushels broad-cast, and plough them in; but they hand-hoe all twice. The crop in general six to seven coombs; last year ten.

In the old lands adjoining the Smeeth, the best farmers sow them in every third furrow, but slovens at random; the rows ploughed and cleaned yield by far the best crop, even to ten or twelve coombs per acre in good years: broad-cast six or seven.

Some beans in the vicinity of Downham broad-cast, and a few farmers hoe: some furrowed and ploughed between, produce seven or eight coombs.

Mr. PORTER, of Watlington, sows them in every third or fourth furrow; if the latter, he ploughs between the rows; if the former, he hand-hoes. Has dibbled some, and they answered well; he left a wheat stubble till the spring, then set two rows on every other furrow, and hand-hoed: got ten coombs an acre; and eight or nine coombs of wheat after them: after beans, always as good as any other wheat.

In discourse during this journey with Mr. MINGAY, of Thetford, on the subject of beans, he mentioned that on his farm at Ashfield, in Suffolk, he has a field of seven acres and a half, which has been thus cropped:

1797 Beans.
1798 Wheat.
1799 Beans.
1800 Wheat.
1801 Beans (Mr. MINGAY's; and he sold 110l. worth from the seven acres and a half.)
1802 Wheat; estimated by the bailiff at from ten to twelve coombs an acre.

Soil, strong loam on clay marle.

Observation.—The not extending the culture of beans is a great deficiency in the husbandry of Norfolk. In the

rich

rich land of the eastern district there can be no question; but they ought to be much increased in that of various loams, wherein turnips are often found on land improper for that crop. But this plant would form a valuable acquisition where none are to be found in the good sands of the North-western district, because beans are, in general, found only on clays, or strong loams; the notion is general, that such only are well adapted to the culture; but on all the better soils of the last named district they would thrive to great profit, and prove a valuable variation in their course of crops. I have seen very great crops in Suffolk on rich sand, and without doubt they would do equally well in the neighbouring county.

SECT. XII.—BUCK-WHEAT.

Mr. Francis, of Martham, has sown buck-wheat after turnips, and got eight or nine coombs an acre, and wheat after the buck: the reason of this uncommon course was, because part of the field was coleseed, and it brought the whole into wheat; the crop nine or ten coombs an acre.

Mr. Cubit, of Honing, finds that nothing cleans land so well for wheat as sowing buck; he gets seven or eight coombs an acre; but oftener ploughs it under, putting a bush under the beam to sweep it down for the plough—a poor succedaneum for the skim coulter. He finds it as good as a mucking; but this only on strong land. He sows it on a second year's layer, as soon as barley sowing is over, from the 15th to the 20th of May; about the 1st of August ploughs it under, harrows well in September, then throws on the seed, and ridges for wheat: always good crops.

The

The same practice takes place about North Walsham, and it is reckoned by some, on the heavier soils, as good as half a mucking; but Mr. MARGATESON remarks, that if done on light sand it makes it too loose and puffy.

Mr. PETRE, of Westwick, sows some on ollond, for a crop, and then sowing wheat, is sure of a good produce.

Mr. DYBLE, of Scotter, has a good opinion of ploughing in green crops: he once in a first year's layer ploughed in the second crop of clover, spreading no muck there, the rest of the field mucked, and the wheat was as good as the rest.

It was much cultivated when I was at Aylesham, 32 years ago; they ploughed three or four times for it; sowed the beginning of July five pecks an acre: the average crop four quarters an acre: sometimes they got six. Esteem it as good as oats for horses. Wheat always succeeded it, and rarely failed of producing good crops: sometimes they ploughed it in on cold springy land, using a small bush faggot before the ploughs, to lay it in the right direction for turning in: it answered for two crops better than dung.

Mr. REPTON, sows buck on ollonds, and ploughs it in for wheat: it answers well.

Mr. REEVE, of Wighton, has ploughed buck in for manure, but thinks his land hardly strong enough for this husbandry; but on any piece subject to wild oats, by sowing buck after barley, on four earths, and ploughing it in at the beginning of August, in full flower, he has freed land from that weed most completely.

Mr. STYLEMAN, at Snettisham, has this year ploughed in buck for manure; it was sown the middle of May, and ploughed in in full blossom the middle of August; stirred in September, and is ready for the wheat-seed earth.

He

He has practised the husbandry before, and received material benefit from it.

Mr. THORP, on Governor BENTINCK's estate, in Marshland, last year ploughed in buck as a manure for wheat, and it answered greatly, improving the crop as far as it extended to an inch: and this year he repeated it, fortunately; for in such a drought as succeeded he had not been able to have ploughed the land at all.

At Felthorpe, buck is considered as superior to all other crops with which to sow grass-seeds—it does not rob them—it shelters better than any other from the sun—it is late sown, and consequently offers an entire spring for a second or third destruction of weeds: these are valuable circumstances, and merit the attention of those who wish to vary the methods of laying down land to grass.

SECT. XIII.—TARES.

THE culture of this plant has increased very considerably in Norfolk of late years. Within my memory they are multiplied at least tenfold.

Mr. OVERMAN begins sowing winter tares about Michaelmas, once more before Christmas, and sometimes twice and thrice more, with spring tares for succession. After mowing he does not plough the land; but runs sheep over it till wheat sowing.

But the cultivator who has made by far the greatest exertions in this husbandry that I ever met with, is Mr. PURDIS, of Eggmore, who has 300 acres every year, seeding no more than is necessary to supply himself: they are fed by his sheep; used in soiling his numerous horses; and immense quantities made into hay. His crops of 1802 I viewed

viewed with much pleasure, and found them very great indeed. In such an extent they are necessarily sown at different seasons for succession, both winter and spring. Some he has dibbled; but intends drilling many in future.

Mr. M. HILL only winter tares for soiling his horses.

Sowing tares for summer feeding sheep, Mr. HENRY BLYTHE remarks, is an absolutely new improvement in the husbandry of West Norfolk, and he thinks it a very great and important one; he does not know who first introduced it: he is largely in the practice himself, sowing both winter and spring sorts, and in succession.

Mr. DURSGATE is in this husbandry.

I found Mr. HART, at Billingford, feeding off his spring tares, the winter ones, first growth, being done: the sheep were in pens moving regularly forward; brought in at night, and in the day on the layers: a piece of 20 by 30 yards given every day to 400 breeding ewes. They draw back on the cleared land, not resting on the unfinished tares. They were sown on a two years ollond, and as fast as cleared, the land was *tempered* for wheat. Winter tares fed off in May in the same field, are now, in August, a fine crop again for the sheep, when they finish the spring ones.

SECT. XIV.—CABBAGES.

THE Rev. J. FORBY, of Fincham, was a most successful cultivator of this plant. Two acres produced 28 tons per acre, carried off the land, a strong wet loam on clay; two adjoining acres of turnips were fed off with sheep: the whole sown, after three earths, with oats, to the eye perfectly equal, and the whole produce 90 coombs, or 12 quarters per acre. Seeds took well, and he cut nine tons of

CABBAGES.

of hay. No manure for either cabbage or turnip. Cabbages never exhausted his land, which always worked better for barley or oats than his turnip land. No cattle could do better than his cows when on cabbages, and the cream and butter free from any disagreeable taste. The seed was always sown as early in the spring as possible, on land well sheltered, dunged, and dug. The moment he perceived the fly on the young plants, he sowed the beds with wood-ashes, which instantly destroyed the fly, and so far from hurting the plants, that it was astonishing to see how they were invigorated by it. They were planted out the third or fourth week in May. Mr. FORBY always mucked the land intended for the crop soon after Michaelmas, which he found far preferable to doing it just before planting. In a very severe frost which destroyed all the turnips, Mr. FORBY's cabbages escaped, and were of immense use.

Subsequent to this communication he further informed me, that he had never seen any piece of land at Fincham planted part with cabbages and part with turnips, where the former did not exceed the latter four-fold at least. Doubts having been expressed whether they did not impoverish the land, he formed an experiment upon a course of crops to ascertain that point, by a rotation which would prove it, if the fact were so. The land middling, let at 12s. an acre. He spread 14 loads per acre of dung on a wheat stubble, and ploughed it in soon after Michaelmas. Three ploughings more were given, and cabbages planted in June: then oats: then wheat. No person in the parish had cleaner oats or wheat, nor any such large crops; 15 coombs an acre of oats, and six coombs one bushel of wheat; and two years in five, 19 coombs an acre of oats, and nine coombs one bushel of wheat. No sheep or cattle fed

fed on the land at any time, nor other dressing given, except the third year, as above, for the cabbages.

This gentleman tried the red garden cabbage, and found them very hardy, and come to 14lb. but they demand more time for growing than green sorts; of which those streaked with red veins are best, and most durable. He hung those up for seed for two months after Christmas. Did not approve of setting the stalk only, as the side branches were apt to break off; each good plant yielded 1lb. of seed: he dried it on hurdles raised on stakes: and if the ground was fine under them and dunged, it became a seed-bed.

Mr. COKE has cultivated the large cattle cabbage 13 or 14 years, and got very fine crops, which yielded more food than turnips, on good sands: he has had up to 20 acres per annum. He sowed the seed in February, and transplanted as soon as the plants were large enough, continuing till the beginning of July. Manured as for turnips, 10 loads per acre. The expense the same as that of turnips.

Mr. REEVE, of Wighton, has every year a few acres, to use in frosty weather, finding a load of great use for his cows in the morning: he sows the seed in February, and plants in June, the rows three feet by two and an half: he is clear that an acre produces double the food of an acre of contiguous turnips; and does not observe any defect in the barley that follows.

Mr. RISHTON has cultivated cabbages, and has known as good barley after them, though carted off, as after turnips fed late on the land. His method of culture was to sow the seed the first week in February, once to remove the plants, and set them out by the 4th of June: hand-hoed them, earthing up: then ran the double earth-board plough between the rows, and the hand-hoes after to draw the earth still higher up. He planted by a marking rake,

with

with three teeth, at three feet asunder, drawn along and across the lands, and set the plants at the intersections.

Mr. FOWEL cultivated Scotch kale in 1801. The seed was sown in a bed in April, and transplanted the middle of June, in rows at three feet, and the same distance from plant to plant, on three acres. They were kept clean by hand-hoeing, and fed off by sheep. They yielded abundance of food; but he thought the sheep did not take kindly to them; the use, however, was very great, as his turnips rotted. This year he intended a crop, but the seed was bad and failed. The barley was as good as after turnips.

In 1784, I remarked in the *Annals of Agriculture*, vol. ii. p. 365, the great superiority, on my own farm, of borecole, &c. drilled and left without transplanting, to that which was transplanted, and thence recommended the practice; which was, *I believe*, the first public hint given of it.

COLE-SEED.

Mr. EVERIT, of Caistor, in Fleg, always hand-hoes his cole-seed once, with seven inch hoes, giving the same price (6s. 6d. an acre) as for the two turnip hoeings. The benefit is very great, not only in cleaning the land, but also in the growth of the plants: they stand the winter much better than unhoed crops, spreading on the ground, instead of running up, and being exposed and cut by frosts.

Cole-seed, in the fens of Marshland, on paring and burning, runs to thick stalk, and that quite brittle, and is excellent for stock; but at Walpole it is inferior.

SECT. XV.—CARROTS.

IT is a remarkable circumstance, that carrots should have been an article of common cultivation, in a district of Suffolk, for more than 200 years, and yet that so valuable a crop should not, in all that time, have travelled into this neighbouring county, where there are such great quantities of land so perfectly adapted to the husbandry.

Forty years ago, Mr. FELLOWES, at Shottesham, cultivated carrots with much success: he got 600 bushels per acre, which were used for horses and cows: the former never did better, and the butter from the latter, of superior quality.

They were pretty much cultivated more than thirty years ago, between Norwich and Yarmouth—nearer to the former; the farmers trench ploughed for them, and sowed in February; hoed in two months after; thrice in all, at a guinea an acre, and barley succeeded.

Mr. WOODBINE cultivated carrots at Packsfield, near Rainham, a farm of Marquis TOWNSHEND's, successfully; giving his labourers 5s. 3d. per last of 84 bushels, for digging, topping, heaping, and loading: he tried them in fatting 30 hogs; and weighing one fairly chosen, and again after twenty 28 days, found he had gained three stone 1lb. paying 2s. a week. They were sold to drovers for the London market: he then put up 40 more, giving some pease with the carrots, which made better flesh.

Mr. FORBY, of Fincham, for some years kept carrots without suffering from the severest frosts, by forming a platform of earth, six inches above the level and two feet and a half wide; on this a sprinkling of dry straw, and then a row of carrots, with their tops all on, and turned
outwards,

outwards, the tails lapping over one another; so that the width covered with carrots was about two feet; the small ones topped and laid in the middle: on every two or three rows a little dry straw, and thus to the height of four feet, the tops well covered with dry straw; another row parallel, with room for a person to walk between: these alleys at last filled with straw, and the outside guarded with bundles of straw, staked down, or set fast with hurdles, to prevent the wind blowing the straw away.

But the person who made the greatest exertions in this husbandry, was ROBERT BILLING, a farmer at Weasenham, who had them on a scale of 20, 30, and 40 acres per annum, during four or five years; gained some premiums from the Society for the Encouragement of Arts, and published a pamphlet on the subject.

SECT. XVI.—MUSTARD.

MUCH cultivated from March to Wisbeach, and about the latter place. A good crop will yield five or six coombs per acre, and it sells at from 8s. to 21s. a bushel. It is, after being in full blossom, subject to a fly, which damages it greatly.

In the newly-inclosed lands of Marshland Smeeth, mustard is the chief crop. They ploughed the old grass of that rich common once, and after one or two harrowings, sowed a quarter of a peck of seed per acre, from Candlemas to the end of March; hand-hoed the plants once or twice, as wanted, thinning and setting them out at nearly equal distances. The crop is reaped the beginning of September, and tied in sheaves, leaving it three or four days on the stubble: it is stacked in the field, and these stacks are called *pies*. If it gets rain in the field, it turns grey, and
lose

loses half its value. The Smeeth is now full of these stacks, and the season has proved highly favourable. It is threshed in the autumn, being left for a sweat, which improves the colour. A good crop, such as they have got this year, amounts to six or seven coombs an acre, and the present price at Lynn is 20s. a bushel. From this account it is not surprizing that the Smeeth of 1500 acres, letts at 3l. The price, however, is sometimes so low as 7s. 6d. to 10s. a bushel. They intend, according to the common practice near Wisbeach, &c. to sow four crops in succession; the second is usually as good as the first: and after four years mustard, a crop of wheat, then fallow.

In old cultivated lands, four or five coombs a good crop.

SECT. XVII.—HEMP AND FLAX.

Mr. ALGUR, of Buckenham, had a three-acred piece under this course for many years:
1. Hemp, dunged for,
2. Wheat;

and the wheat was always very good; more apt to be too great than too small a crop. Most of the cottagers in that vicinity have a patch of hemp, which is a great relief to them: it is very profitable, and finds the families much employment.

As an instance of the inefficacy of the bounty, Mr. ALGUR, though long a hemp-grower, never applied for it.

This culture, in the vicinity of Diss, has greatly declined; there is scarcely one-tenth grown of what there was some years past: this is chiefly attributed to the high price of wheat. Forty stone from the break, an average crop; bunching and heckling now 20d. a stone, and a good

good hand will do one stone and a half in a day. The spinners earn from 4d. to 8d. a day.

Mr. Rainbeard, Governor of the House of Industry of Forehoe, cultivates hemp every year, to supply a part of what is manufactured in the house; and in the watering of it, has a contrivance which I do not remember seeing elsewhere, by means of which it is deposited in the pit, without any necessity of a single person being wet, a pretence in common for making the farmer's tap bleed pretty freely. The pond is an old marle pit, with a regular slope from one side (where the hemp is prepared) to the depth of eight feet on the other side: on the slope above the water, the hemp is built into a square stack, upon a frame of timber, of such a height as will float and bear a man without wetting his feet: this is slid down upon the frame into the water, and when floating, drawn away, a person on the opposite bank drawing the floating stack to the spot where it is to be sunk, and on which it is built to the requisite weight.

He finds it does soonest at bottom, and would not object to sixteen feet depth of water.

By means of this very useful contrivance, he can put in a waggon-load in an hour.

The sheaves are taken out in the common manner, sheaf by sheaf: here wants a further improvement, easy to be effected.

In the parish of Wyndham, a farmer had, about seven years ago, some land (four acres and a half), dunged and ploughed, and designed for fallow for wheat; Mr. Rainbeard seeing the field, advised him to sow hemp before taking wheat, which he did, on one acre and a half; he bought the hemp at 8l. 8s. an acre, as it stood, and the farmer sowed the whole with wheat; and the crop was so much cleaner and better after the hemp than after the fallow,

328 HEMP AND FLAX.

low, that the farmer told Mr. RAINBEARD that he lost 20l. by not following his advice upon the whole piece.

Account seven miles round Diss, Lopham, &c.

EXPENSES:

	£.	s.	d.
Rent	2	0	0
Tithe	0	5	0
Rates	0	5	0
Ten pecks of seed, at 1s. 6d.	0	15	0
Sowing	0	0	6
Three earths	0	12	0
Five harrowings	0	2	6
Ten loads of farm-yard dung	2	0	0
	£.6	0	0

On an average, sells as it stands (hemp selling from the break at 6s. 6d.) for £.8 8 0
Expenses — — 6 0 0

Profit — — £.2 8 0

Suppose the crop 40 stone, dew-retted:
Pulling and spreading, turning and tying — 1 5 0
Breaking, 1s. 3d. a stone — — 2 10 0

£.9 15 0

Forty stone, at 6s. 6d. — £.12 10 0
Expenses — — 9 15 0

Profit — — £.2 15 0

Price of hemp, from the break, in
              s. d.
1789 Hemp bought at — 5 0 per stone.
1790 Ditto   —   5 6

1791

		s.	d.				
1791	Hemp bought at -	5	0	per stone.	per acre.		
1792	Ditto - - - -				£.4	15	0
1793	Ditto - - -				6	0	0
1794	Ditto - -	7	6		6	6	0
1795	Ditto - -	7	0		7	7	0
1796	Ditto - -	6	6		7	7	0
1797	Ditto - -	7	0		7	7	0
1798	Ditto - -	6	0		6	6	0
1799	Ditto - -	6	0		7	0	0
1800	Ditto - -	8	0		9	9	0

The difference of price between dew and water-retted used to be reckoned 2s. a stone; but is now full 3s. on the average.

Forty stones of hemp from the break will make 40 stones of tow, it being weighed from the break at 14¼ lb. to the stone, and that of tow is only 10 lb. At 6s. 6d. a stone from the break, it is worth 8s. the stone from the heckle. The bunching is valued at 8d. a stone, and the heckling 10d. In the House of Industry they have a very simple machine for bunching: men, by turning a winch, move semi-circular cogg'd wheels of iron, which lift, every moment, one of four perpendicular beaters, and let them fall on the hemp coiled under them to receive the stroke.

Estimation of the quantity of hemp raised in the hemp district of Norfolk.

		Acres.			Acres.
Old Buckenhan	-	20	Roydon	-	5
Carleton	-	8	Bressingham	-	10
Banham	-	20	Fersfield	-	10
Winfarthing	-	12	Diss	-	10
Kenninghall	-	6	Shelfhanger	-	5
The Lophams	-	30	Burston	-	5
Garboldsham	-	8	Gissing	-	5
			Aslacton		

	Acres.			Acres.
Aslacton	5	Starston	-	1
Forncet	3	Schole	-	3
Waketon	2	Brockdish	-	7
Tibenham	2	Needham	-	8
Titshall	1	Fretton	-	7
Shimpling	1			
Drickleborough	3			202
Harleston	5			

It may amount to about 200 acres.

Mr. GEO. EATON, linen-weaver, informed me, that about Diss there is not so much hemp grown as before the price of wheat was so high, by a third or fourth. It is affected also by the high price of turnip seed; for the cottagers, &c. sow turnips on their hemp grounds, and if seed be high, they let them stand for a crop, instead of sowing hemp every year in the common manner. The necessity of manuring for every crop of hemp, impedes much the increase of it under any circumstances.

ACCOUNT OF AN ACRE.

	£	s.	d.
Rent, tithe, and rates	3	0	0
Manuring 20 loads	5	0	0
Five earths	1	5	0
Harrowing	0	2	6
Seed, 10 pecks, at 2s.	1	0	0
Pulling	0	15	0
Dew-retting spreading, 2s. 6d. turning, 4s. 6d. getting, 5s.	0	12	0
Breaking 40 stone, at 14lb. 1s. 6d.	3	0	0
	£.14	14	6

Value

HEMP AND FLAX.

	£	s.	d.	£	s.	d.
Value then, 8s. 6d. a stone	17	0	0			
Expenses — —	14	14	6			
Profit — —				£.2	5	6
Bunching and heckling, 1s. 6d. on 40 stone				3	0	0
				£.17	14	6
Value then 40 stone, at 10s. 6d.	21	0	0			
Expenses — —	17	14	6			
Profit — —				£.3	5	6
Spinning 10 clews, at 8d. per stone, 6s. 8d.				13	6	8
Half bleaching, 40 stone, at 2s. 1d. (chiefly labour) — — —				4	6	8
Winding on bobbins, 40 stone —				1	0	0
Weaving, 40 stone — —				8	0	0
Bleaching the cloth (nearly all labour),				3	6	8
Total culture and manafacture of every acre				£.47	14	6

A good spinner earns 8d. a day.

A middling one, 6d. a day.

A bad one, or child, 4d. a day.

The weavers, 7s. to 14s. a week.

Women ditto, 10s. or 12s. One girl, 12s. but an extraordinary instance.

The price of dew-retted hemp has risen from 6s. 6d. to 8s. 6d. in two years. The war increases the demand for their linens, while the culture of the raw material falls off: this obliges them to mix Russian flax with it, the warp of one sort, the woof of the other: but nothing equals their own hemp for duration and strength. The bounty
of

of 3d. a stone did not much good; but the discontinuance of it did mischief. Mr. EATON thinks, that if a bounty of 6d. were given, it would have a good effect when wheat is cheap.

Mr. RICHARDSON, of Fritton.—Most hemp in Lophams, Garboldsham, Kenninghall, Banham, Buckenham, Diss, Bressingham, Palgrave.

The soil a rich mellow loam, on a clay bottom; worth 25s. an acre; friable, mixed, and easy working.—Manure, 20 good loads, a compost; three or four earths given; the best time, the beginning of May; 13 pecks of seed. No selection. They pull in 12 or 13 weeks from sowing; here they water-ret, in other places, dew-ret, the former best for fineness, but equal in strength: in the water, from three to five or six days; dew-retting, from three to six weeks. Dew-retted is seldom finer than 2s. a yard; water-retted, to 5s. Showers, whether for water or dew-retted, are necessary. As a crop, hemp is profitable, but yields no manure; Mr. RICHARDSON has 100 acres of land that would do for it. On a layer it is good, but another crop first, and then without manure; an average crop, 40 stone: now at 12s. a stone water-retted; dew-retted, 9s. The common price of seven years, 6s. dew, and water-retted 8s. This price will not increase the culture, because wheat is so high; nor is profitable, for want of manure: if he did not want it for his fabric, he would not sow any.

	£.	s.	d.
Rent, tithe, and rates	2	0	0
Seed, 13 pecks, at 2s.	1	6	0
Manure	5	0	0
Three earths	0	15	0
Harrow and sow	0	5	0

Pulling,

Pulling, retting, spreading, lifting, turning, binding, and housing - -	3	0 0
Breaking 40 stone, at 1s. 6d. -	3	0 0
At 7s. 6d. it only pays - -	£.15	6 0

Most is grown by cottagers in bits, roods, and half acres of land; few have so much as an acre: Russian hemp is fit only for coarse goods, being deficient in fineness, strength, and colour. A relation, at Lowestoff, made nets of it at the same price as Russian good hemp; but it would not last in salt water so long as the Russian; but Russian linen is far inferior in strength to our own home-made. They could always sell more home hemp sheeting than is possible to make, and those who cannot get it, and buy Russian, find a vast difference. But coarse sheeting is made of the refuse of all; fine sheeting, 3s. 6d. a yard, five-quarters wide.

If encouragement was given to the growth, it would be a very good thing for the poor; for it is of great benefit to them; and would answer better to them than potatoes: a rood pays their rent, and keeps them in employment. The 4d. a stone bounty was not worth going for; to encourage it by 6d. a stone, instead of 3d. might do a little good. Mr. RICHARDSON never claimed the 4d. though he grew for three years. Not half the hemp raised now, that he remembers, though the price is doubled. Where they used to have hemp, they now have wheat. And little farmers had more than large ones at present.

A small quantity is cultivated around Downham, by farmers who sell it to cottagers, at 5l. to 10l. an acre, as it grows: they sow wheat after it, and always have good crops: they muck for the hemp.

The following is an answer I sent to some Queries transmitted to me on the subject of the increasing the growth of hemp.

The only account to which I can at present have recourse, to discover what is the import of hemp, is a detail of imports of merchandize from all the world, from the 5th of January, 1792, to the 5th of January, 1793, in which rough hemp is set down at 614,362 cwt., value 522,207l. Whether this includes the import for the Royal Navy, I know not, but in the following calculations, must presume that it does include the whole. It makes 30,718 tons, at 17l. a ton, or 2s. 1½d. a stone of 14lb. from which price I conclude that this hemp is in its very earliest stage of dressing. Hemp sold in England, in 1796, at 61l. 10s. a ton.—*Annals*, vol. xxvii. p. 79.

The hemp lands of the counties of Lincoln, Cambridge, Suffolk, and Dorset, yield nearly the same crops; from 35 to 50 stone per acre; 40 may be considered as an average crop; in which ratio it takes exactly four acres to produce a ton, and 120,000 acres for the quantity imported of 30,000 tons.

I cannot conceive that it is impracticable to add 120,000 acres to our hemp grounds, even in England alone; certainly not in England and Ireland together. But it would be impracticable to add a tenth part of that space, without giving a most decided and effective encouragement.

The former bounties per stone, had no effect whatever— they did not occasion a single acre extraordinary to be sown; and I personally know, that many persons who cultivated hemp, did not even think them worth applying for; the forms were so difficult and tedious.

I have not the act to recur to, to know exactly their amount, but it was either 3d. or 4d. a stone; at the latter, 40 stone gave only 13s. 4d. an acre bounty; totally insufficient as an encouragement for an article which we know is not extended under a bounty of from eight to ten times that

that encouragement in the price of the commodity. The rise cannot be estimated at less, on an average, than 2s. a stone from the breaker. Yet this great advance does not extend the cultivation. In all these articles there is an apprehension of the price falling, if generally gone into. Two shillings a stone is a bounty of four pounds per acre—but being liable to a fall in price, it has little or no effect.

There are other circumstances which unite to prove, that a bounty, to be effective, must be very large. It would not be easy *very greatly* to extend the culture without ploughing up grass-lands for the purpose; nothing prepares for hemp so well as the land lying under grass; but landlords do not admit grass to be ploughed by tenants; and as the latter would reap all the profit, of course they would give no indulgence where they would reap no advantage. This reduces the object to the grass-lands in the hands of the proprietors, unless the parties were enabled to divide the bounty.

The land which will produce profitable crops of hemp, is applicable to various other beneficial purposes; and to induce men who possess such land to vary from their common objects, would demand a very powerful impulse.

There seem to be but two ways to give this impulse; one by a general bounty per acre (per stone would be more liable to fraud) on all hemp that produced 35 stones and upwards per acre: the other, to dispatch intelligent persons through all the hemp districts, and others where the soil is rich enough, to form contracts with all persons willing and able to raise the commodity largely; engaging to take all they produced, not less than   tons, at a given price, for five years.

To engage for such a period would be absolutely necessary, for undoubtedly a man would not plough up good grass for any inducement of a shorter duration.

I do not conceive that a less bounty than 5l. per acre would

would have the desired effect; and the only limitation that would be admissible, would be to limit it to a certain sum, for instance, 600,000l.; and if the claimants exceeded it, then to be divided among them proportionably.

If the other method of contracting were adopted, the price must, in negotiation, be raised, till it satisfied them of the profits of the undertaking: 10s. or 10s. 6d. a stone might prove sufficient: and for five years also.

I am inclined to think that by either of these measures the requisite quantity of land might be brought into this cultivation; and it would have another beneficial effect, of adding to the culture of wheat—hemp prepares well for that grain, which is commonly sown after it.

The parts of the kingdom most fitted for the culture of hemp are, the rich lands of Lincolnshire; the skirts of the fens of Cambridgeshire, great part of the fens themselves; all the upland-lands enclosed in the fens; the coast lands of Essex and Sussex; the coast, and various other tracts in Dorset, Somerset, and Devon; and much of the convertible soil laid to grass in all the central counties. In Ireland, large districts in Limerick, Tipperary, Roscommon, Meath, &c.

Ireland producing much flax, is not a proof that it would do for hemp; as much flax is sown in the north of that kingdom on soils not adapted to hemp.

The private advantages which would result in Ireland from applying public money to the encouragement of the growth of hemp, would depend on its being a measure not of the moment. To induce them to break up their grass-land for this object, and then by withdrawing the bounty leave them to common husbandry, would be injurious, supposing them unwary enough so to be led astray.

In regard to pointing out individuals able to contract for the delivery of hemp, to the amount of 50 or 100,000 acres;

acres; I could not name them. The proper way would be to take a journey into the right districts, and apply to great numbers, with powers to make such a contract; a person well skilled in husbandry, to explain the benefits, and answer objections and doubts; could by posting, traverse a great extent of country in a month, if he knew the right roads to take, and a good scattering of individuals to apply to immediately. Such an experiment would enable Government to ascertain pretty nearly what may be expected. I am of opinion that many thousand acres might thus be secured for sowing next spring.

As to giving any assistance by advancing capital, I do not conceive it to be in the least degree necessary. The culture is not expensive; and those men who possess or occupy the proper land, are generally in sufficient circumstances for the undertaking. They want only a sure price for a proper number of years. Men who are in want of so small a capital are not likely to effect the business. But attention should be paid to procuring large quantities of seed, which individuals, in case of a great extension of the culture, might not be able to procure. From ten pecks to three bushels are sown per acre.

The plans above suggested would take effect upon a very large scale next year.

As there are many allusions to the old bounty in the queries, it is necessary to observe, that they require so many forms, and such delay in payment, that no extension would make them answer. And if a bounty should be given per acre, though certain forms would be necessary, yet these should be as simple as possible.

One objection to a bounty is, that it would equally encourage and reward, at the expense of 5l. per acre, all the hemp which would be cultivated though no such measure was adopted, as well as new undertakings, alone in contemplation. Contracting avoids this: it also admits a progressive experiment in which to feel your way; but a bounty plunges at once to the full extent of the measure. —A. Y.

FLAX

Is much cultivated about Wisbeach, and in the cultivated parts to Downham, at Outwell, &c. Four acres one rood was this year sold as it stood, for the purchaser to be at all following expenses, for 16l. 5s. 6d. per acre; and as high as 17l. has been given.

SECT. XVIII.—SAINFOIN.

ONE of the most valuable plants that were ever introduced into the agriculture of this kingdom, and it may be asserted, without danger of contradiction, that it has been too little cultivated in Norfolk, especially in periods when the price of corn has been low.

In 1784 Mr. COKE had 400 acres; an extent to which he carried this excellent grass, in order to be able to feed off all his turnips by the end of March, avoiding thereby the common evil of permitting them to run up to blossom, for supporting the flock in April, by which the succeeding barley is always materially damaged: he found the object well answered. The hay was given in well contrived racks on wheels, so that both the hay and the sheep's heads while feeding, are under cover from rain. A space sown with winter tares received the sheep in May, if the layers were not ready for them.

The same excellent cultivator then made an experiment on this grass: ten acres were sown with an addition of white clover;

clover; ten with trefoil; ten with red clover; and ten with sainfoin only. The result was, that the white clover division turned out the best; the sainfoin alone the next best; then the trefoil, and the red clover much the worst.

Some years after that period he tried ray-grass, and approved of it much better than any other plant for this purpose, observing, that it kept down the blubber-grass, so apt to come with sainfoin: but from the time of the drill husbandry being adopted he has left off all additions: the regularity of the crop being so great that all spaces are filled, and the cleanness of the land being trusted to against the blubber.

Mr. ALLEN, at Stanhow, had very fine sainfoin on a rich loamy sand, on a clay marle bottom, worth 16s. an acre. The common crop was two tons per acre.

In 1792, Mr. BEVAN, at Riddlesworth, sowed two bushels an acre of sainfoin, and six pound of clover and trefoil, to give a crop the first year.

1802. It got full of rubbish, did not answer, and was ploughed up after four or five years. The method he intends to pursue is, to turnip his land for two years, both fed off with sheep, and to lay down with buck-wheat in June; sowing four bushels of sainfoin per acre: he did this some years ago, and it is now the best sainfoin on his farm. He has now 28 acres under the second crop of turnips, to be thus laid down.

Mr. BEVAN has sown sainfoin with rye to good effect. Mr. WARD, of West Harling, has done the same, and got fine crops.

Mr. M. HILL sowed sainfoin on a deep gravel, and it gave, in the third year, one load and a half per acre; from the repeated failure of seeds, he has been induced to lay down this spring 48 acres. The barley drilled first at six inches,

inches, and then sainfoin at the same distance: three bushels an acre each: looks very promising.

The Earl of ALBEMARLE made an experiment which, though not in the bounds of Norfolk, is in sight of it, and therefore I shall mention it here: taking into his own hands an immense farm of 4000 acres, with 3000 sheep, and wishing to provide all sorts of food as early as possible, he ventured to sow a field, in extremely bad order, with sainfoin alone in June: the foulness of the land such, that his Lordship's hope was not sanguine. This was in 1801, and the crop this year, 1802, was among the very finest he had ever seen, at least two tons and a half per acre. The result is remarkable, and will certainly bring to the reader's mind the husbandry of sowing sainfoin among couch in Gloucestershire, mentioned by Mr. MARSHALL.

Mr. OVERMAN, of Burnham, has broken up sainfoin layers, and suffered such losses by the red-worm, that he was fearful of sowing corn upon the last field he ploughed. He therefore ploughed it before winter, and summer-fallowed it for turnips without manure. After these he sowed barley, which crop I viewed, and found very fine; not having suffered the least attack. He remarked, that the ploughing before winter probably contributed to the destruction of the worm, as well as the summer tillage.

Mr. OVERMAN broke up another field by thorough fallowing for turnips; the crop very good: he then sowed pease; the crop middling: turnips again, bad: then barley, which was much eaten by the red-worm. This followed by winter tares, now (1802) on the ground, and a bad crop.

Mr. COKE, of Holkham, pared and burnt a sainfoin lay, intending turnips, but changing his mind sowed wheat, and the red-worm ate half of it. This shews that this
operation

operation is not in all cases effective against that destructive animal.

At Holkham there is much gravel, in some places to the depth of above twelve feet, the surface a thin covering of sand; no chalk or marle beneath, at any known depth. I was solicitous to ascertain whether this soil would do for sainfoin: Mr. COKE assured me that it succeeded very well on it; but it is to be remembered that the whole country has been marled.

Mr. OVERMAN and Mr. COKE drill sainfoin before drilling barley, in the contrary direction to that in which the barley is to be drilled, the rows at six inches and three quarters. Seed three bushels an acre, and perfectly regular; much more so than the generality of broad-cast crops.

Mr. PITT's father and grand father both tried sainfoin upon very deep and dry gravels; there are some pits 15 or 20 feet deep; but it did not answer.

It is much approved about Binham: Mr. ENGLAND, in breaking it up, would summer-fallow for wheat, as after laying so long the land is generally foul.

Mr. H. BLYTHE, of Burnham, has a very fine plant three years old, from drilling, three bushels of seed per acre, at six inches three-quarter rows; it is common to sow from four to five. This field before produced little or nothing, and was over-run by rabbits and game. He gave a clean summer-fallow for drilled oats, and drilled the sainfoin across the oats.

Mr. RISHTON, at Thornham, broke up sainfoin 11 years old by a bastard summer-fallow for oats; the wyerworm ate up the crop: he then laid on 10 loads an acre of dung, and sowed turnips, which came to nothing (eaten also I suppose by the worm); then barley, and got 15 coombs an acre; then pease, 10 coombs an acre; and this year (1802) a great crop of wheat.

Mr.

Mr. BLOOMFIELD, at Harpley, broke up a sainfoin layer in 1802, by one ploughing, and got a fine tilth for wheat by much scarifying.

Sir M. FOLKES has given much attention to sainfoin, but not with success: he has prepared for it by two crops of turnips in succession, both fed on the land by sheep, yet twitch and blubber-grass over-ran the crop in a few years, and some in that case has been spoiled in three or four: he has been careful not to feed it with sheep; and he has harrowed it the third year severely, cross and cross, and from corner to corner, yet without destroying the blubber. By reason of this circumstance the culture does not flourish at all. When broken up it is fallowed for wheat, and the crop generally good.

SECT. XIX.—LUCERNE,

WAS cultivated very successfully, thirty years ago, by Sir JOHN TURNER, at Warham, on a turnip sandy loam, of the rent of 7s. 6d. an acre: he sowed it broad-cast after turnips, and without corn. Every spring he harrowed it severely, till it carried the appearance of a fallow; manuring, at the same time, with six loads an acre, of rotten dung. It was cut regularly every five weeks, and has been often known to grow from 22 to 26 inches in 28 days. It maintained five horses per acre, from the middle of April to Michaelmas: this, at only 2s. per horse per week, is 13l. an acre—at 4s., 26l.

Mr. ALLEN, at Stanhow, had drilled lucerne at one foot, which succeeded well on a good loamy sand, on a clay marle bottom.

Mr.

Mr. FOWELL, *of Snetterton, to* ARTHUR YOUNG, *Esq.*

I beg you will communicate to the Board of Agriculture, the following experiment, on feeding horses with green lucerne.

In the summer of 1797, I fallowed a piece of land of seven acres, being part of a farm belonging to the Right Honourable the Earl of ALBEMARLE, situated in the parish of Snetterton, in the county of Norfolk, and in my occupation. The soil a sandy loam, upon a clay marle bottom, worth to rent, about 15s. an acre, exclusive of tythe.

The above land, in June of the before-mentioned year, was sown with turnips, which were well hoed; the following spring it was properly prepared, and the last week in April, was sown with six pecks an acre, of barley, and also with 20lb. of lucerne and 4lb. of clover an acre, broad-cast, which were harrowed in with the corn. In the summer of 1799, the grass produced by the lucerne and clover-seed, was mown and made into hay, and the after-grass fed with cattle. The following summer (1800) the clover disappeared, and left an abundant crop of lucerne, a part of which was given green to horses, and the remainder mown twice, and made into hay ; but as no register was made, I cannot state any particulars.

In the summer of 1801, I determined to keep an accurate account of a certain quantity, the result of which was as follows:

On the 11th day of May I began mowing 4 acres 1 rood and 24 perches of the above lucerne, which I had purposely divided off, and applied it by feeding ten cart-horses, in a walled-in yard. There was neither hay nor corn

corn given to the horses, except the first two weeks they were taken into the yard, when two pecks of oats, with some chaff, were allowed each horse per week, to prevent any ill consequences from the too sudden change from corn and hay to green food, and they were fed entirely from the aforesaid 4 acres 1 rood and 24 perches of lucerne, till the 21st of September following, making exactly 19 weeks from the period of their first going into the yard. I cannot state with accuracy the quantity of work done during the above 19 weeks by the ten horses, but, as near as I can estimate, eight out of the ten went to plough or other work, nine hours every day (Sundays excepted), and were in excellent condition during this experiment. The yard in which the horses were kept, and which they never quitted except when at work, was littered with refuse straw from other yards, green weeds from borders or waste land, or any other refuse litter that could be conveniently procured. The dung was turned over after the horses were taken from the yard, and after remaining about a month in heaps, produced 62 loads, at 36 bushels to the load. The grass had a slight top-dressing of peat ashes the first winter, and has received no manuring since, except a small part, which did no material good. The first mowing commenced the 11th May, when it was fifteen inches high; second mowing 6th July; third mowing 18th August.

I was induced to adopt the above method of summer-feeding horses, from the inconvenience I had previously experienced, from having but a small quantity of pasture land, and the consequent difficulty of preventing them from breaking the fences, and getting into and damaging the growing crops of corn.

Calculation of the value of the 4 acres 1 rood and 24 perches of lucerne, above-mentioned:

Keeping

Keeping 10 horses 19 weeks, at 6s. per horse
per week - - - - £.57 0 0
Sixty-two loads of compost, at 3s. per load - 9 6 0
                                              66 6 0
Deduct 2½ coombs of oats, at 12s. £.1 10 0
Chaff   -   -   -   0 6 0
Refuse straw and litter - - 2 0 0
                                     3 16 0
          Total   -   -   62 10 0
             Or per acre - £.13 17 4

With the greatest respect, I remain,
SIR,
Your most obedient humble servant,
RICHARD FOWELL.

Snetterton, near Harling,
Norfolk, 1st Dec. 1801.

I viewed the crop described in the preceding account, and found it very fine, regular and clean: Mr. FOWELL had twelve horses soiling on it, and some pigs in the yard, which had nothing else, and which were in very good order. He has seven acres; has sown ten more with this year's barley, which I examined, and found it had taken perfectly well, and promised to be a fine crop: he intends eight acres more next year, meaning to mow it for hay. The ten acres are the half of a field, the other half sainfoin, for comparison. He thinks the lucerne will beat the sainfoin.

Mr. BEVAN sowed at Riddlesworth, in 1793, thirteen acres broad-cast, with barley; seed ten pounds an acre, and also six pounds an acre of red clover, on good sand, worth 12s. an acre. In 1794 he mowed half for soiling and

and half for hay; the latter two tons per acre; the lucerne was predominant, rising four or five inches above the clover.

Mr. BEVAN's is now nine years old, and is still very profitable, and had it not been attacked by the parasitical plant which infests it, would have been now in full perfection. He has sown twelve acres more, which is now in the second year, and promises to be very productive. He approves highly of the culture.

Mr. BEVAN, in order to get rid of the parasitical plant which is so apt to destroy his lucerne, half ploughed it in the spring of 1802, and harrowed in spring tares, which gave him a very good crop: it did not damage the lucerne, and checked the weeds. A very good thought.

Sir MORDAUNT MARTIN has cultivated lucerne 36 years; has tried it, transplanted, drilled, and broad-cast, but, from much experience, finds the last by far the best way. Sows 16 lb. an acre; it has lasted 17 years, and when ploughed up, it was like ploughing horse-radish. Three roods have given a good noon-meal to six horses, for seven years together, the racks being filled. Has had eight acres of it, but scalding in the summer, ploughed up six of them. He thinks a fair growth of it exceeds a crop of tares: has only gravel for it: it will not do on wet land.

The Rev. Mr. CROW, of Burnham, has cultivated it for many years; and found that nothing could recompense the expense of cleaning the rows of the drilled, and if the alleys are very clean, then the lucerne is dirtied, so that broad-cast beats it greatly. It has lasted seven years; he has had four cuttings.

Mr. PRIEST, of Besthorpe, copied Mr. FOWELL's example, and sowed three acres near his stables, broad-cast—he has got a very fine plant: this the first year.

SECT.

SECT. XX.—MANGEL WURZEL.

Sir Mordaunt Martin has cultivated it from its first introduction; and generally with great success. Has usually four or five acres. He has drilled it for the purpose of better cleaning; drills it with Cooke's wheat cups, so reduced by putty, as to sow half a peck an acre. But intends it, in future, broad-cast, as the seeds are apt to be buried. Sows about the middle of May, to the 20th, being apt to run to seed if earlier. He is persuaded that an acre will carry more cow stock than an acre of turnips. It will keep hogs perfectly well, but better for large than young ones, as apt to scour: if he wanted them for hogs only, he would still have them, as very beneficial for that stock. He usually sows it after barley. They keep in the same way as potatoes: *his* method is to open a furrow and fill it, and then turn two furrows over the roots, observing that the ridge thus formed, is left sharp to shoot water off. He has a very high opinion of them, and found that they will fatten a beast faster than any root he has tried: on this gravelly loam, they yield a much better crop than either carrots or parsnips.

1802. He has this year a flourishing crop. I was unfortunate in his absence when I called at Burnham.

SECT. XXI.—POTATOES.

The quantity of potatoes in the south of Norfolk, is very inconsiderable; and in Fleg, I saw but a thin scattering of very small pieces.

Mr.

Mr. EVERIT has found that they exhaust the soil more than any thing.

Mr. CUBIT, at Catfield, had an acre two years ago, which produced 48l. at 6s. a sack: the rest of the field was turnips; then wheat after the potatoes, and barley after the turnips, seeds with both; and now, where the potatoes grew, is a bed of rubbish, the rest clean.

Mr. REPTON has raised potatoes for the consumption of his farm; but not when he has any prospect of buying them, which he has done at 1s. 6d. to 2s. 6d. a sack. He steams them for young cattle, &c. having a very complete apparatus for the purpose; boils five coppers, which steam 50 or 60 bushels a day, and answers well; also turnips, and pours their liquor on to cut chaff, giving the whole mixed together; it answers extremely well; the cattle licking it up with great avidity, and doing perfectly well on this food.

This root is much in use amongst the poor at Thornham, &c. their gardens are full of them.

A good many near and around Downham, and wheat after them; but Mr. SAFFORY sows barley, for he has found they make the land too light for wheat.

SECT. XXII.—OF THE DRILL HUSBANDRY.

THE introduction of this culture has given a new face to the fields of West Norfolk; and a new sphere of inquiry to the agricultural reporter. Some notes respecting this subject, have been introduced in the detail of particular crops; but more general observations have been reserved for this section.

Mr. DROZIER, of Rudham—To raise a great crop of wheat, how would you put it in—would you drill it?—No. I would dibble. Mr. DROZIER, last harvest, had dibbled and drilled; the dibbled beat. In 1800 he had none drilled; but in 1802, I heard that he drilled again.

Mr. HARING, of Ash Wicken, put in wheat, dibbled one row on a flag, and it beat the drill.

Mr. HOLLAND, of Bircham, in 1799, lost half his crop of barley by the drill; and in 1800, his drilled wheat was much too thin: he then determined to drill no more. His stubble did not give any signs of bad management. He remarked, that he has seen no drilled stubbles that do not shew gaps much more numerous than ought ever to be seen.

Mr. WHITING, of Fring, has been long in the practice of drilling, which he prefers much to broad-cast sowing: drills all his corn. Has a high opinion of dibbling, and for produce, knows not which to prefer; but drilling admits the hoe, which is a material point.

Mr. BRADFIELD, of Heacham, never drills. As to saving a ploughing by drilling barley, he will not admit it to be a saving; it is an earth that pays well. One-horse ploughs for putting in barley on one earth, the best instrument that has been invented. He also observes, that in a long course of experience, he had never seen a thin crop of wheat, but if any mildew happened, it was sure to be struck: drilled wheat is the thinnest of all, and most subject to mildew. In 1800, all the drilled crops he viewed were much mildewed. His neighbour, Mr. NORTON, had in 1799, a very thin drilled crop; and in 1800 he drilled none.

In 1792, I found the chief part of Mr. OVERMAN's farm drilled, and in very beautiful order and cleanness; his

stubbles

stubbles carried unequivocal marks of good husbandry. He was then in the third year of his drilling; the experiments of the first encouraged him. In the second, his drilled pease yielded twelve coombs and a half an acre; the dibbled ten, and the broad-cast eight. In 1792, the third year, all his pease, and the greatest part of his wheat, were drilled, and superior to what he had broad-cast in 1792. He has attempted more than once to drill barley and oats, but was then convinced it would not do, and did not intend to try it any more; but he changed his mind afterwards· he drills six pecks an acre of wheat; sows two bushels; he horse-hoes once, doing an acre per hour, if the furrows are long; and hand-hoes twice, each time at 20d. per acre. I viewed his machine at work, drilling wheat; a suspended marker forms the line by which the horse walks: the boy who guides, does not lead, but rides upon him, and goes quite straight by seeing the mark always between the horse's ears. The man that holds, directs the plough by the wheel mark; and as he holds for that purpose only one handle, the pressure is counteracted by a leaden weight hung on the other handle. The work completely straight. He considers the saving of seed as something. In 1791, this saving, after paying all the expenses of hoeing, gave a balance of 28l. 10s. yet the prices of corn were wheat, 24l. a last; barley, 11l. 10s.; beans and white pease, 15l. Women hand-hoe the wheat at 20d. an acre; 3s. 4d. for twice; and it is then earthed up by horse-hoeing. The rows nine inches asunder. By drilling barley he saves an earth; and thinks that the drill supersedes the use of one-horse ploughs. He never hoes it, but harrows in seeds.

 Mr. OVERMAN is a gentleman of such clear and intelligent

gent abilities, that great deference ought to be paid to his opinion.

Putting some questions to him on broad-cast husbandry, his reply was: " Pray, don't ask me questions about the broad-cast husbandry; for I wish to answer you from practice, and I have been so convinced of the superiority of the drill, that I have had no broad-casting for some years, and never shall again."

Sir MORDAUNT MARTIN, Bart. at Burnham, has practised the drill husbandry ten years, for barley, oats, and vetches, for soiling, at six inches; but no hoeing: has had fourteen coombs an acre, of barley and oats, and has had a last an acre of Tartarian oats, broad-cast. Has never observed that the drilled crops ripened more unequally than others.

Sowing on one earth in putting in barley, he considers as belonging to the drill. Sir MORDAUNT is very well satisfied with the result of his experience; he sows the seeds broad-cast, and harrows. Barley tillage is, first to scale in the tath, harrow, and then plough deep. Then harrow once for drilling, and once after; then sow the seeds broad-cast, and harrow twice, and with the harrows drawn backwards, an iron being fixed for the purpose, by which means the seed is not drawn out of the ground.

In the period of Mr. COKE's broad-cast husbandry, 1784, I found that he had two years before carried eleven coombs and one peck of clean barley to market, over 162 acres; and that year he had 300 acres, estimated by all who viewed it, part at twelve coombs, and the whole at ten.

In 1792 I was at Holkham, and found Mr. COKE then drilling on a considerable scale. In 1791 he drilled 76 acres of barley, which produced 34 lasts 13 coombs three bushels, or nine coombs two bushels one peck per acre:

93 acres sown broad-cast produced 44 lasts 18 coombs two bushels, or ten coombs two pecks per acre.

As it was a common practice in the district of Holkham to break up a second year's layer (and sometimes of three years) at Midsummer, to give a bastard-fallow for wheat; a husbandry still common in Wiltshire, where they begin with *raftering*, it should seem that one great benefit of the drill is the saving this tillage on all light soils. But it ought certainly to be remembered, that this saving belongs to the dibbling husbandry as well as to drilling.

It was with much pleasure I viewed Mr. COKE's farm at Holkham in 1800: every sort of corn was all drilled, and in a masterly manner. The wheat, however, not great that year, being apparently too thin; and I pointed out to Mr. WRIGHT many gaps in the rows, of nine inches, a foot, and even two feet in length. If a general thinness is a fault, such gaps add much to it. I made the same remark in other person's crops. Mr. COKE's distance of rows nine inches; always hand-hoes twice at 1s. 8d.; 2s. and sometimes much more per acre.

Some of the crops were immense, particularly barley; and all the barley I saw was extremely good; one acre certainly produced, as Mr. WRIGHT, the bailiff, assured me, 19 coombs one bushel.

Mr. COKE had that year some drilled turnips, but his broad-cast ones far exceeded them: these were capital. In drilling corn, the distance for barley six inches; never hoed, but drilled after harrowing on the stirring earth; then the seeds sown broad-cast, and harrowed again across. Thus the only advantages attending the drill in this crop, is depositing the seed at a more regular depth than the harrow will do, and saving an earth ploughed by one horse. And one man and one horse putting in an acre a day of barley,

barley, will reduce the saving of the earth to something not very considerable on these soils.

The seeds of two years are once ploughed very carefully for drilling wheat; and this operation is trusted only to the best ploughman on the farm; that the furrows may be so evenly and neatly lapped together, as to enable the drill to be safely used longitudinally, which is Mr. COKE's and Mr. OVERMAN's practice both for wheat and barley. The wheat is hand-hoed twice, in March and April, at the expense of 4s. an acre; the barley not at all. The wheat is at nine inches, the barley at six and three quarters.

Several gentlemen present at the sheep-shearing at Holkham were of opinion, that in the district of that seat, extending one way to Hunstanton, another to Swafham, and East to Holt, 15 acres in 20 of all the corn sown, were this year put in by the drill.

In 1792 Mr. BEVAN had made some experiments with COOK's drill, with DUCKET's, and with the drill-roller, and at that time much preferred the last.

1802. From 1792 to this time he has had no drilling; but this year began again.

Mr. BEVAN, in 1800, had six acres of wheat drilled at nine inches, in the middle of 50 acres; 44 put in with the Norfolk drill-roller. Those six acres were mildewed, and the 44 quite bright. The drilled appeared also very inferior in every respect. Mr. DAY, the bailiff's expression was, *the broad-cast will beat out and out.*

In discourse with Mr. BEVAN, in 1802, after viewing the very fine and clean crops at Holkham, he said that he had for some years been in doubt upon the question; he had tried it several times, but the result was not favourable: but this year having been desired by Mr. COKE to give it another trial, he had done it, and the barley drilled

drilled was now certainly superior to the broad-cast. His neighbour FOWELL, of Gastrop, drills every thing.

Mr. FOWELL, of Snetterton, has drilled all his farm for twelve years, and has not the smallest doubt of the superiority of the husbandry to the broad-cast; not only in yielding superior products, but also in being cheaper; he lays much stress on saving tillage, putting in his barley and pease on stale furrows scarified: this is, however, clearly applicable to the broad-cast. He will not by any means admit that the scuffle, or the one-horse plough, will bury seed-barley at so equal a depth as the drill. In Snetterton there are five farmers: four have drills; the fifth is a small one, for whom Mr. FOWELL drills. They have come about to this husbandry within two or three years, from seeing Mr. FOWELL's crops. In Harfham there is some drilling. Mr. GOOCH, at Quedenham, drills all. In Illington there are two farmers, and both are drillers. In Little Hockham one, Mr. KITTON, and he drills all, and scarifies a stale furrow for barley. At Wilby three farmers; and two, Mr. PALMER and Mr. BOWLES, drill very well. At Gastrop, Mr. FOWELL drills all.

The Rev. Dr. HINTON, at Northwold, has made some interesting comparisons: on a mixed loam he drilled, at twelve inches, with COOK's drill, one bushel, four quarts, and half a pint per acre; broad-cast one bushel and 29 quarts. Hand-hoed the drills thrice: weeded the broad-cast. The former produced 44 bushels two quarts; the latter, 38 bushels 18 quarts: weight of the drilled 62 lb. 2 oz. per bushel; of the broad-cast, 61 lb.

He drilled barley at 12 inches, and hoed in clover seed 25 days after: the crop better than broad-cast, and the clover took well; which, however, is not common, for it often fails thus put in.

His

His drilled turnips were at least equal to those sown broad-cast.

Dr. HINTON does not drill spring corn, as it is not a practice, except out of course, on account of the seeds, which will not do in drilled crops: has been tried by others, and the crops failed; but for winter corn, better than broad-cast on sandy loams; but it will not do on strong wheat soils as a general practice.

Mr. SALTER, at Winborough, whose soil is a wet loam on a clay bottom, which requires draining, does not drill, as he thinks his land too stiff and difficult for it; but dibbles largely.

In discourse with Mr. SALTER, on the drill husbandry, and inquiring how far it could be made applicable to his difficult land (note, however, that he has several fields, the surface of which is an unquestioned sand, upon a strong under-stratum, yet called *strong land* fields), he gave it decidedly as his opinion, that it would not do: yet he understands drilling, and has practised it at Snarehill. He appealed to his vast crops of dibbled wheat, and immense ones of barley, this year, 1802, promising nine or ten coombs an acre of the former, and fifteen, sixteen, and even more of the latter, and demanded whether drilling could, or did any where beat them? A respectable party of Norfolk farmers were present, and two drillers among them, but they were silent, and all equally struck with the uncommon crops we were then examining.

The Rev. Mr. MUNNINGS, at Gorget, is upon a most unkind sharp flinty gravel; red gravels are usually good soils, but his are blackish, from a mixture of black sand; nearly the worst of all soils: on this land he drills turnips at eighteen inches; also oats at nine inches; pease at nine and twelve; and once he tried wheat.

Mr.

Mr. Munnings drills his poor gravels with his barrow-drill. I viewed his turnips at eighteen inches: a very regular and well executed crop.

Mr. Hart, of Billingford, in 1802, drilled turnips at six inches, cutting up in part of the field every other row, and in part cutting away two rows and leaving one: part also broad-cast, for comparison; done with Cook's machine, and part horse-hoed with his tools; part with Mr. Munnings' expanding hoe-plough. Last year he drilled all his barley, and it answered better than the broad-cast; and this year all his corn is drilled, except the *ollonds*; these drill rolled, a practice pretty general here: but the small farmers dibble.

It is gradually coming in around Dereham, and dibbling rather going out, for want of dependence on the droppers.

I viewed Mr. Collison's farm at East Bilney, and found several crops drilled, which made a very fine appearance: one field of wheat by the barn promises to produce ten coombs an acre; some barley also very fine, and the crops in general clean: all at nine inches, and the wheat horse-hoed thrice; barley twice if no seeds, but with them not at all, as the seeds are covered by the harrow which follows the drill: Mr. Collison, however, has horse-hoed in some seeds, and did not fail. He drills seven to ten pecks of seed-barley. Mr. Collison prefers the drill very much: he forms his lands to be worked at a *bout* of the drill, the horse walking only in the furrows, and the same in horse-hoeing; a point he justly esteems essential for all heavy or ticklish land. I put the question home to an intelligent labourer who has worked 40 years on the farm, and he assured me that if he had a farm of his own, he would drill all the wheat, and horse-hoe it likewise, for when land is bound in the spring, to break the surface makes the plants grow well; but

but as to drilling barley he spoke very doubtfully, but said it was less liable to be *laid:* and in a dry time it comes best drilled.

Drilling on light lands which are laid flat, is no difficult operation, but on strong, tenacious, *clung* soils, it is no such very easy matter, without much attention. The Rev. Dixon Hoste, however, at Goodwick, has effected it with a commendable regularity: one method he practises, is that of drilling athwart the ridges; to prepare which he sometimes trench-ploughs.

Drilling is very little practised in the neighbourhood of Norwich. Mr. Crowe had a machine of Mr. Cook's, which he gave to Mr. Sillis, of Hertford Bridges, but he never used it. Mr. Crowe is so satisfied of the use of dibbling, that he desires nothing better.

Many drills in the vicinity of Thelton, a heavy land district, and much used of late.

Mr. Kerrich, of Harleston, drills or sets all his corn. Many drills in Earsham hundred; and several farmers drill for their neighbours at 2s. 6d. an acre. There is, however, as much dibbling as ever; the drilling having chiefly been substituted for broad-cast sowing.

Mr. Burton, of Langley, approves much of drilling, and has seen Cook's machine beat dibbling: he, however, does not drill himself, observing that he is absent too often.

Mr. Drake, of Billingford, near Scole, drills both wheat and barley, he has three fields of drilled barley this year, with part broad-cast; and the latter beats in all three: his drilled wheat good, and he prefers the drill for that crop much more than for barley.

Mr. Thurtell, near Yarmouth, has for two years drilled almost all his barley, by that means saving one ploughing.

No drilling in Fleg that I could hear of. Very little in Blowfield and South Walsham, but it is coming in, and Mr. SYBLE is clear that it will spread.

It cannot be said that drilling is quite unknown in the hundred of Happing, but I heard of very little, and saw none: a man last year travelled with a drill, for drilling at 2s. 6d. an acre, and some farmers employed him. Some farmers use the drill roller.

Very little practised about North Walsham. Mr. MARGATESON approves much of COOK's drill. Mr. LUBBOCK, of Lammas, and Mr REPTON, of Oxnead, are the only drillers, they practice it much.

Mr. PALGRAVE, of Coltishal, has drilled wheat, barley, and oats, for 17 years; has tried four and a half and six inch rows, but finds nine the best for all. The merit of the husbandry he thinks consists in the equal depth at which the seed is deposited, and were this equally effected, does not conceive there would be any difference in the crops, whether put in by dibbling, drilling or broadcast.

Mr. REEVES, of Heveringland, has drilled most of his corn for three years, and is clearly convinced of the merits of the practice, and determined to continue it.

Mr. ENGLAND, of Binham, has drilled all sorts of grain for five years past, and is perfectly satisfied with the practice; nor has he the least doubt of the superiority of it to the broad-cast method; there is little dibbling in his neighbourhood, drilling having superseded it.

Mr. REEVE, of Wighton, has been a driller four years, and for all sorts of corn: he has no doubt, and makes not the least question of its superiority to the broadcast husbandry. I viewed his stubbles with singular pleasure, and a more beautiful spectacle of cleanness I never beheld

beheld—absolutely and positively clean; not a weed to be seen in those of wheat, commonly so foul.

Mr. M. Hill began drilling about five years since, and ever since practised it as his general system; and is clear in its superiority, for every thing except barley, at six inches; in that he has some doubts. He has tried it for turnips, but none at present.

Mr. Henry Blythe, of Burnham Westgate, has drilled every sort of corn for twelve years, beginning in 1790 with wheat, and has continued it ever since. He is well convinced of the superiority of this husbandry.

Mr. Dursgate, of Summerfield, who, if not the greatest farmer in Norfolk, is nearly so, drills every crop except turnips, and of course has no doubt of the superiority of the method to the broad-cast husbandry. On his strong land farm at Palgrave, however, the drill is not so universally used as at Summerfield, Sedgford, and Docking.

Mr. Rishton, at Thornham, drilled all sorts of corn, and has no doubt of the method much exceeding the broad-cast husbandry.

Mr. Styleman, of Snettisham, was one of the first drillers in Norfolk: he began the use of Cook's machine sixteen or seventeen years ago, and has continued it regularly to the present time, keeping three machines in use. He is perfectly convinced of the superiority of drilling to the broad-cast husbandry.

Many farmers around Houghton drill much corn, and approve of it exceedingly.

Captain Beacher, at Hillingdon, drills all: the practice increases much, and promises to be universal.

Mr. Beck, of Castle Riseing, drills largely, and has done so for seven years: the husbandry meets with his entire

tire approbation. I viewed some of his stubbles, and found them very clean: on his sandiest soils, he, however, does not drill, because he cannot so put the seed in deep enough.

Drilling is a little practised about Downham, and is preferred by several farmers.

Mr. PORTER, of Watlington, drills both barley and wheat, and this year all, except on strong land, and finds the crops a great deal better than broad-cast. He hoes all; thus covering the seeds amongst his barley; but they do not take quite so well as in the common way. He hoes, at the expense of 2s. 6d. to 4s. an acre, once, and is clear that the benefit to the crop is very great: has had 13 to 14 coombs an acre, of barley.

Mr. MARTIN, of Tettenhall, this year drilled 190 acres of barley and 20 of wheat; is convinced of the superiority to the broad-cast.

Mr. ROGERSON, of Narborough, was a very great driller, but not at present.

About Wymondham, drilling coming in, and has been so more or less for several years. Mr. WELLS, of Hethel, letts his drill by the acre. Mr. CHURCH, of Flawdon Hall, drills all.

Mr. PRIEST, of Besthorpe, drilled largely for eight or nine years, at Shropham, on sand, and continues the practice on strong land: he is much inclined to think that it is a superior method to the best broad-cast. He has drilled all sorts of grain.

Mr. TWIST, of Bretenham, drills all the corn he can, except rye; and has no doubt of its answering.

Observations.—From these notes it appears, that, notwithstanding some failures, and probably many prejudices, the drill culture has very completely established itself in West Norfolk, and is spreading into the other districts of that extensive county. The success appears, on the whole, to be very flattering.

But there is one singular circumstance which should, so far as Norfolk only is concerned, check the unlimited panegyrics sometimes too generally heard in conversation, and that is, there being, at least to my knowledge, but one farm (Mr. HOSTE's) on strong or clay land, where this practice is thoroughly introduced. Suffolk affords multitudes; but Norfolk is at present our business; and here the farmers on strong land, have hitherto rejected it. This is remarkable, as I have heard some very able drillers give it as their opinion, that this husbandry has greater merit on strong than on light land.

DIBBLING.

Mr. BURTON, of Langley, remarked, that good as this practice was in some respects for the poor, there are inconveniences flowing from it. Girls, old enough for service, are kept at home by it. Gleaning is their employment in harvest, which gives them idle habits in the fields, then dibbling follows; and the girls lying about under hedges with the men, produces the natural consequences on their manners; bastardy flourishes, and maid-servants are uncommonly scarce.

Dibbling flourishes very greatly in Fleg, both wheat and pease, and oats.

Mr. HORNARD dibbles part of his crops, and sows part, and he is not yet convinced that the dibbled exceeds the sown. About Ludham, and through a great part of Happing, it is not so general as in Fleg. About North Walsham it is by no means general, except for pease.

Mr.

Mr. PETRE, of Westwick, puts in about 100 acres of wheat, of which he sows and drills 80, and dibbles 20.

Mr. JOHNSON, of Thurning, makes the same observation on the ill effects of dibbling as Mr. BURTON. The great girls do not drop so well as children, nor is the work so well done as formerly: they now drop between the fore-finger and thumb, which is much inferior to doing it between the fore and middle finger.

Dibbling is common around Wighton, for wheat and pease; and Mr. REEVE thinks it a great improvement upon the broad-cast husbandry, but that drilling is a step further.

There is at Snettisham much dibbling, pease and wheat on flag; and Mr. STYLEMAN thinks it never will be abandoned, as there are seasons that do not suit drilling.

In Marshland, the practice obtains every where for wheat on clover, and some on clean stubbles; 10s. 6d. an acre. It increases.

Practised about Downham, and with good success.— Mr. SAFFORY dibbles all he can, and thinks it a great improvement.

SECT. XXII.—ON THE NORFOLK ARABLE SYSTEM.

For the last four or five and thirty years that I have examined West Norfolk with the eye of a farmer, the change in the tillage system has not been great. At that period the course was, 1. Turnips; 2. Barley; 3. Grasses for two, or, in a few cases, three years; 4. White-corn; on the better soils wheat; on others, rye, &c. The only change that has occurred has been in the grasses: the variation,

ARABLE SYSTEM. 363

riation, which I believe first took place from forty to fifty years ago, was shortening the duration, from three years to two: in both cases giving what may be called a bastard fallow the last year, by a half-ploughing, soon after Midsummer. Above thirty years ago, I contended, both in print and in conversation, against it, but was held cheap for entertaining any doubts of the propriety of the practice. I have lived, however, to see this change also in a great measure take place amongst the best farmers, who now give only one ploughing for the winter corn, whether wheat or tares; or in the spring for pease. That it is an improvement, cannot be questioned. The argument for it, founded on the invention of the drill-roller, and on the introduction of the drill-plough, is good, but not singular, as the practice of dibbling is likewise far more adapted to a whole than to a broken furrow: and for broad-cast common sowing, if we are able to cover the seed by harrowing on stiff soils, once ploughed, assuredly the same practice might be better followed on sand. The other reason for the former system, spear-grass getting a-head in a layer, is quite inadmissible: for I must agree entirely with Mr. OVERMAN, that no weeds, the seeds of which are not carried by the wind, will be *found* in a layer, if they were not *left* there.

The variations which have taken place in the crop put in upon layers, are neither great, nor are they peculiar to Norfolk: the principal one is taking pease on the flag, and then the wheat, &c. an admirable system, which has long been practised by good farmers in Suffolk, and I believe, earlier still in Kent. Mr. PURDIS's substitution of tares, holds on the same principle. Considering the very great value of white-pea straw, well got as sheep food (no where better understood than in Kent) there is

no

no husbandry better adapted to a sheep-farm, than this of pease or tares preceding the wheat crop.

A great and a very important change has, however, taken place in the application of crops to sheep instead of bullocks and cows. Formerly the farmers *consumed* much of their straw by cattle: now the best tread it all into dung.

Sheep are the main grazing stock, and no more cattle kept than for *treading*, not eating straw, while feeding on oil-cake, &c. This is an important change, which has had considerable effect, and has depended not a little on the introduction of South Down sheep.

The grand object in the whole system, is the singular steadiness with which the farmers of West Norfolk have adhered to the well-grounded antipathy to taking two crops of white corn in succession: this is talked of elsewhere, but no where so steadily adhered to as in this district. It is this maxim which has preserved the effect of their marle, on thin-skinned wheat lands, in such a manner that the district continues highly productive, under an almost regularly increasing rent, for more than 60 years, or three leases, each of 21; and by means of which great tracts have been marled a second, and even a third time, with much advantage.

This system has been that to which the title of *Norfolk* husbandry has been long, and is now peculiarly appropriated; and by no means the management of the very rich district of East Norfolk, where the soil is naturally among the finest in the kingdom, and consequently where the merit of the farmer must be of an inferior stamp: barley there very generally follows wheat; an incorrect husbandry, deserving no praise. The celebrity of the county in general was not heard of, till the vast improvements of heaths, wastes, sheep-walks, and warrens, by enclosure,

enclosure, and marling took place from the exertions of Mr. ALLEN, of Lyng House, Lord TOWNSHEND, and Mr. MORLEY, which were in the first thirty years of the preceding century. They were happily imitated by many others; an excellent system of management introduced, and such improvements wrought, that estates which were heretofore too insignificant to be known, became objects of public attention in the capital. The fame of Norfolk gradually expanded, and the husbandry of the county celebrated, before East Norfolk was heard of beyond the conversation of Norwich and Yarmouth.

Without a continuance of cautious management and persevering exertions, West Norfolk would again become the residence of poverty and rabbits. Let the meadows be improved; irrigation practised wherever it is applicable; the remaining wastes cultivated, and this district will become a garden.

Mr. MARSHALL, who considers the practice of East Norfolk as alone deserving the title of *Norfolk* husbandry, mentions 40 or 50 bullocks, on turnips, as a matter of exultation. In 1768, I registered the fact, that Mr. MALLET, of Dunton, had 280 bullocks fatting on turnips, on a farm almost wholly arable.

This ingenious writer appears, in various passages of his work, to consider East Norfolk as deserving the most attention. In this, I am sorry that I must differ from him greatly; and I think, that had he *resided* on the other side of the county, he would have adopted a different opinion himself.

" In West Norfolk, no general plan of management has yet (1780) taken place." " Viewing the state of husbandry in West Norfolk collectively, it is much beneath that of East Norfolk." In West Norfolk, the most steady and regular plan of management had then, and for

many

many years taken place, that was to be found in the kingdom; and at that time, probably, for the soil, the best. It is to be regretted, that so very able a writer did not examine it with more attention.

There are some circumstances in the husbandry of Fleg, &c. which deserve attention, before the merit of it can be duly appreciated. 1st, The soil is certainly amongst the finest in the kingdom. 2d, They are within reach of marle, by water carriage, to every part of the district, comprehending, besides the Flegs, the hundreds of Happing, Tunsted, Blowfield, and South Walsham: and they have, in addition, great plenty of a fine clay marle for variation, under many parts of the whole. 3d, They have an enormous quantity of marsh and fen, and low rough waste, and rushy grounds, which yields a very considerable bulk of coarse hay and rushes, with which to make yard-dung, as well as to support great herds of cattle. Examine the map, and it will appear that near half the district is marsh, fen, and water. These are circumstances so favourable to the arable part of the country, that I am more surprized their products are so small, than that they are so great. In discourse with Mr. HORNARD, of Ludham, on two years layers, he said, that he was forced to have recourse to them, for since the sea, eight or ten years ago, broke through the Marram banks, and flowed over the top of the marsh banks, destroying the rushes and coarse fodder, it has made a difference to him of 60 to 70 loads per annum of rushes, that were four feet high: now he has none, and therefore must leave his layers two years for want of manure. What advantages has West Norfolk to be contrasted to such a circumstance?

View the two districts in another light: I conceive that no farmers in England would make more of West Norfolk, than those of the district actually make at present.

present. But I have little doubt that East Kent and Isle of Thanet men, would make 20 per cent. at least, more of East Norfolk than is at present made by the occupiers; they would not be long before they shewed what beans would do in such a soil.

IMPROVEMENT.

In discourse with Mr. BIRCHAM, at Hackford, on the public benefit of tillage; and the effect of the landlords restraining their tenants from breaking up grass-land; he asserted it as a fact, of which he had not the least doubt, that tillage, well managed, would support as much live-stock on the seeds, turnips and straw, as the same land would do all under grass; consequently, the corn is all gain to the public. *I am certain it would:* he spoke of moderate pasture, that keeps two beasts of 40 stone per acre in summer.

Great improvements have taken place in Earsham hundred, in twenty years. The number of horses much lessened, by not ploghing so often for barley; scarifying instead of it, and even putting that crop in on one earth.—Mr. PAUL, of Sharston, has even trusted to scarifying only, and thus got the best barley on his farm. Besides this practice, general management is better; and far more weeding done than formerly.

Mr. THURTELL is clear, that in five and twenty years past, the general average produce is, at least, one-fourth more, probably one-third.

Mr. FERRIER, of Hemsby, is sure, that in husbandry there is great improvement in 30 years. The seeds were then left three years; and they did not raise half the corn that is produced now. Summer-fallows were common then; now, no such thing known, unless by chance, when no manure for turnips.

Mr. SYBLE, of South Walsham, is clear that husbandry

dry is much improved of late years: summer-fallowing, heretofore so common, is quite done away, yet the land cleaner. Small trifling enclosures are thrown together, the hedges and pollards grubbed, and the sun and air admitted. Dibbling has spread very greatly.

Mr. PALGRAVE, of Coltishal, has no doubt of the improvement that has taken place in 17 years; every thing is better done, and business carried on with more animation: drilling is spreading, and dibbling increased.

Mr. PARMENTER, of Aylesham, has not the least doubt of husbandry being greatly improved: more land in cultivation, and a greater expense in manuring, and every other article bestowed.

Mr. STYLEMAN has improved his Ringstead farm from 5s. to 15s. an acre. A farm in Snettisham, which he has let, from 11s. in 1783, to 17s. in 1798 ; and has laid out, improved, built, and let seven farms, at a very considerable improvement; and this he considers as the most profitable object of a gentleman's husbandry. I viewed several of his new farms, and found the buildings on a rational scale; so small, yet convenient, that the expense was no formidable objection, even for small farms. He has also accommodated several tradesmen in the village, with closes of land for their horse or cow, for which they are glad to give a very high rent, even to 3l. an acre: this should universally be done ; and to the poor also, though at a lower rent.

This gentleman has no doubt of the husbandry of the vicinity being much improved in 20 years, exclusive of new inclosures: the crops were then disfigured by weeds, but now, every man is ashamed to have such seen on his farm: drilling and dibbling have done much.

Mr. SAFFORY, of Downham, has no doubt of husbandry being much improved in 20 years; they plough better,

better, manure more, and have carried all rough banks and hills on to their fields.

Upon the heavy lands of Goodwick and the vicinity, husbandry is very little improved in the last fifteen years.

Mr. HILL thinks that the husbandry around Waterden has wonderfully improved in the last fifteen years. He attributes it chiefly to drilling, and the various conversations which have taken place upon that topic. Another essential point, is the increase of sheep; cows much lessened, and consequently fewer turnips being drawn for them: if the two greatest blood-suckers of an arable farm are to be named (I use Mr. HILL's terms), they are a dairy of cows, and the sale of lambs from a merely breeding flock.

Mr. FOWELL, of Snetterton, a very intelligent and observing cultivator, is clear that the husbandry of all that vicinity is much improved in the last twenty years; for besides the introduction of drilling, the rotation is improved. At that time, if their seeds laid two years, they took two crops after them; a practice of some few now; but the better farmers, never more than one. Less tillage is now given, yet the crops greater; and they are much improved in better exertions; in hand-weeding, &c.

Mr. ROBINSON, of Watton, has no doubt of husbandry, in general, being very much improved in 20 years; in almost every circumstance.

CHAP. VIII.

GRASS.

NO person can have been in Norfolk without quickly perceiving, that in this branch of rural economy the county has very little to boast. No where are meadows and pastures worse managed: in all parts of the county we see them over-run with all sorts of spontaneous rubbish, bushes, briars, rushes: the water stagnant: ant-hills numerous: in a word, left in a state of nature, by men who willingly make all sorts of exertions to render their arable land clean, rich and productive. To make many notes would be useless, for through nine-tenths of the county, they would consist of disgusting repetitions—the same objects continually recurring, to be condemned in the same terms.

Improvement.—It is, however, with great pleasure that I have it in my power to mention under this head, one of the most original discoveries (for such I esteem it, in common with many excellent cultivators) that I have any where met with in the improvement of grass-land. Mr. SALTER, of Winborough, near Dereham, upon his large farm o above 800 acres, found 3 or 400 acres of old meadows entirely poisoned by springs, which, from every sort of impediment that neglect could cause, had formed bogs and moory bottoms, famous for rotting sheep and miring cows; with blackthorns and other rubbish spread over large tracts. His first operations were, to grub and clear the land, and open all ditches to the depth of four or five feet, and to cut open drains in almost every direction for laying them dry; burning the earth, and spreading the ashes on the ground: so far, all was no more than common good husbandry; but he applied a thought entirely his own: as he found that the flinty gravel, marle, and other earths, but especially the gravel, was very beneficial to the herbage, he thought of sowing winter tares and white clover upon the places wherever any earth was spread, or any other operation had laid bare the surface, harrowing in those seeds. I had the pleasure of seeing several of these crops growing: the success has been uncommonly great; for the land thus sown not only has given large and very profitable crops of hay, but has also received a rapid improvement in the herbage; the cover and shade of the tares, so beneficial to all land, mellowed the surface, and seemed to draw up as well as protect such of the old plants as received improvement from the manure, and exhibited a much superior fleece of grass to any spots where this singular management had not taken place. So that nothing can be clearer, on viewing this large tract of meadow, than the superiority of the improvement resulting from the growth of the tares: the effect of the manure is much accelerated and rendered greater. The idea is certainly applicable to many of the grass-lands of the kingdom, especially such as are improving by the addition of chalk, marle, clay, loam, sand or gravel: 40 loads an acre of any of these bodies will much improve coarse or wet, or moory grass-lands; and then to add tares secures an immediate profit, and makes the manure work much sooner and more powerfully. He sows some so late as the middle of May. An idea here strikes me, which I shall venture to add; that if I was to scarify any mossy, hidebound or poor pastures, &c. it should be with a drill scarifier, drilling in winter tares by every tooth of the scarifier, and I have no doubt but the tares would take well, and effect a considerable improvement, even without manuring.

nuring. Mr. SALTER has practised the tare husbandry on meadows for 10 years, but his first beginning was 17 years ago, at Ellingham: the cockchafer-grubs had destroyed a part of a meadow; he harrowed in tares and seeds, and the success was great.

Tare-seed running short, he this year sowed pease and oats mixed on some spots, and they do well; and this husbandry he pursues, whether he intends mowing or pasturing.

I will venture to add, that if making known this single discovery had been the whole result of examining the county, the Board would not have failed in the object of ordering the district to be reported.

Mr. BEVAN's arable sand, at Riddlesworth, joining to his low boggy meadows, gave him the power of carting sand down hill at an easy expense; and thus he improved some parts of those meadows to great effect: from 100 to 150 loads an acre were spread at the expense of 4l. or 5l.

	£.	s.	d.
A team of five horses, 30 loads a day, and wear and tear	0	12	6
Driver	0	1	6
Filling, at 2d.	0	5	0
	0	19	0

1802. It has answered very greatly: these meads were then not capable of irrigation, but one meadow has since been watered, and the water has taken much greater effect on account of the sanding, than if that operation had not been performed. The sand has all been laid on the most boggy meadows.

There is a great range of meadow and marsh all the way from Norwich to Yarmouth. Those at Thorpe are very good, and let at 50s. an acre; but at Whitlingham they

they are neglected, and therefore only 20s. At Surlingham are boggy marshes, and to Buckenham ferry, Rockland and Cautley, they are in a bad state, wanting much improvement, by draining and embanking, and clearing from rubbish; these are not more that 8s. an acre. To Reedham better, and let at 20s. At Langley there are 600 acres newly enclosed, let at 12s. but will soon, from the improvements introducing by Mr. BURTON, steward to Sir THOMAS BEAUCHAMP PROCTOR, be much advanced. A circumstance not uncommon was found here: a great range of marsh, but with no safe means of communication with it. Mr. BURTON made a solid road banked and gravelled, above half a mile long, so traced as to communicate with all Sir THOMAS BEAUCHAMP's allotment: a great and most useful work, equally well imagined and executed. Five miles of banking against the river, &c. are also made: part drained by mills, and part by sluices. From Langley to Yarmouth the marshes are good; all 20s. and upwards, and some to 30s.

At Dawling and Gestwick Colonel BULLER has from 7 to 800 acres of pasture, which he esteems worth 30s. an acre, but let at about 20s. Some is let by Mr. COUSSMAKER, a relation of the writer, at 30s. to 36s.

At Tasborough, between Wymondham and Stratton, some very fine well-drained meadows, with a fertile aspect.

Mr. JOHNSON, of Thurning, classes with the very few in Norfolk, that have made any considerable exertions in the improvement of permanent grass; he has converted 200 acres, which were rough, moory, and of small value, into a tract which does credit to his husbandry. He cut off the springs which poisoned them from the adjoining hills, by outside fences, acting as drains, which worked a vast improvement: moved great quantities of earth in levelling

levelling inequalities, to the depth of five feet, to free the land from water. Under-drained to a very great amount. The soil moory, and under it a gravel; part of it bog. Some parts were so bad, that no stock ever went upon it; his predecessor let 15 acres for 5l. a year, clear of all town charges, and at that rent they were flung up, and let to another person at the same rent. At present these worst parts graze bullocks well, and are very valuable lands. These works, which were only preparatory to various other articles of good management, cost him, over a large part of the tract, 10l. an acre. It is no easy matter to dwell sufficiently on exertions of this sort, in a county where they are so rare: the efforts of years, and the expense of thousands to the improver, are dispatched in a few lines—but let those who best know the deficiency of the county, estimate such works as they deserve, and hold in due honour and veneration the men who have thus laudably stepped beyond the common practice.

In these most meritorious works of this active farmer he has but one defect: a perennial brook runs through his meadows, which he thinks erroneously would not be beneficial in irrigation: this will be his next undertaking: he will not continue to let it run waste.

Mr. PALGRAVE, at Coltishal, improved some meadows very capitally, by spreading sea-sand from Yarmouth, which killed all sedge and rushes, and brought up a fine sheet of white clover. The soil boggy. In 1780 they were valued at 7s. 6d. an acre, and lately by the same person at 3l.

Mr. REPTON, at Oxnead, drained a large tract of meadows, and did all they would admit in good grass husbandry, but a water-mill belonging to his landlord is a nuisance to all such endeavours, and keeps the water so high that all exertions are vain: when will landlords have

just

just ideas on this subject of mills? There is scarcely one to be found, but does mischief to an estate to ten times the value of the rent these wretched erections let at. Mr. REPTON's father, on the same farm, improved a bog 25 years ago, as well as the mill would permit, laying a great dressing of gravel on a boggy part; but from being kept by the mill a saturated spunge, the gravel is now got down two feet deep, and overgrown with the spontaneous rubbish of the bog. If you will have mills, you must give up all ideas of true meadow improvement.

Mr. REEVE, of Wighton, may perhaps be considered as the prince of grass-land improvers in Norfolk: he has very few rivals that have come to my knowledge: one great improvement consists of 45 acres, effected without irrigation; the other of 50 acres, by means of many exertions, finishing in irrigation; the latter will be mentioned under another head. The former tract is situated between a line of chalk hill, on one side full of springs, and a mill river pent up on the other, so as to poison and turn to bog all the land below it: the springs from the hills thus meeting the soakage of the river, it may easily be supposed what the effect must be: the land was a quaking dangerous bog. His cure was effective, and such as could not fail; he turned an arch, traced by a level, 160 yards long, for catching the soakage of the mill-pond, and continued it an open drain to the length of three quarters of a mile parallel, and near the river, gaining such a fall as now to keep the water in the drain four feet below the surface of the land, in places where before it was never more than six inches. To cut off the springs from the chalk-hill he run a deep ditch at the foot of the hill, varying in depth according to the level of the line, but effective in cutting off the springs.

In the parts where it must be of the greatest depth, as
that

that of eight feet, he turns an arch of brick-work, as the expense of sloping down the sides would exceed it; and that the bricks of the wall against the hill may not impede the entry of the water, he has made them each with two open grooves for admitting it.

He has yet much to do to finish the improvement, as he intends carrying great quantities of earth from the drains to spread on the boggy parts, to finish the consolidation, and improve the soil. From the harvest of 1801 to that of 1802, he has had constantly at work from 20 to 30 men, on this tract and that irrigated.

One of the richest tracts of grass in Norfolk, is the district of marsh lying to the south of Lynn and east of the Ouze. Sir MARTIN FOLKES has there 700 acres, which let at 42s. an acre short measure, besides 2s. tithe. The tenants are restricted from mowing two years together; a bad covenant; for mowing and feeding should be on distinct lands; and for every load of hay they sell, must bring three loads of muck. Some tracts in the hands of butchers are never mown, which has made them very superior to the rest. In general these marshes, like all others in the county, are hired by the upland sand farmers, and not stocked regularly, but merely as convenience occasions, to ease their farms. They will carry, when so fed, a large beast to two acres, and a few sheep besides. They never have too much water, and can let in fresh water at pleasure.

ROUEN.

In 1792, I found that Mr. BEVAN had not read the *Annals of Agriculture* in vain; he had a fine field of 28 acres of excellent *rouen* saved for the ewes and lambs in the spring.

1802. His present system of allotting ewes to different tups, in separate fields, prevents his being able to avail himself

himself of this article of food, as every enclosed pasture is, from the 15th of September to the 15th of November, forced to be fed each with a lot for this purpose. But his opinion of the great value and use of rouen, is as complete as ever: the object of improving his flock, alone occasions the change.

Mr. MASON, of Necton, near Swafham, keeps grass from the end of July, and does not turn into it at all, till early in the spring of the following year, when he puts in his fatting bullocks and sheep, which have had hay in the winter. The old grass nurses up a great bite of young growth, and both together carry on the bullocks well: and it is excellent for sheep; nothing at that season equals it.

Mr. OVERMAN, of Burnham, in 1799, kept 13 acres of grass, from Midsummer, an exposed piece, open to the sea and N. E. wind: turned into it 10 score and 16 ewes and their lambs the 27th of March, and it kept them well a month. They would have been half starved without it; but were well supported, to the surprize of many who saw them feeding. The piece was equally *tathed* in every part.

Laying down.—Mr. COKE, at Holkham, has laid down various pieces with good success, and he is decidedly of opinion, that the best method is that of a fallow, till about the middle of August, and then sowing the seeds alone; keeping off all stock in the autumn, and sheep feeding for two or three years.

Mr. DENNIS, of Wigenhall, St. Mary, in Marshland, lays down with barley, or oats after fallowed wheat, sowing 10lb. of white clover, 10lb. of trefoil, and a sack of hay-seeds. Manures the stubble of the corn amongst which they were sown, in autumn. Sheep feeds the new lay the first year; and much better to observe the same for two or or three years; by which means it comes to a good pasture soon; but if mown, it is a long time before it gets a good covering.

Break up.—Mr. REEVE, of Wighton, broke up an old pasture in 1800, for oats, which he dibbled in on the flag; he then scaled the oat stubble, and on a second earth, drilled in another crop of oats: the wyer worm did some mischief to both. He then laid on 60 loads an acre of marle, and 10 of muck, and sowed turnips, which are now one of the finest crops I have ever seen.

At Snettisham, 600 acres of common being enclosed by act of parliament, much of it was broken up, in 1801, for oats, by mere ploughing: they did not succeed at all. In 1802, they ploughed and took oats again, and the crop still worse.

Mr. CRISP broke up a four-acred field of old grass behind his house at Dereham; the four last crops in which, paid him 100l. a year; one of which was coleseed; one wheat; another year hemp, and after it turnip seed; the former producing 149l. 12s.; and the turnip seed, 18 coombs, at 36s. a bushel, 129l. 12s. It is now (1802) in wheat, and the crop very fine: he has railed off a walk around this field, and has laid it down to grass: such a piece of land affords entertainment, and might yield instruction in experiments;—but who has not a grass field to walk into, which affords nothing of the sort?

Mr. SALTER, of Winborough, upon his finely cultivated farm of about 800 acres, of which he had permission to break up a large portion of old and miserably bad grass, poisoned with springs, and over-run with bushes, and all sorts of aquatic rubbish, first surrounded every field with ditches five feet broad and four deep; then hollow-drained every acre completely; and broke up for a crop of dibbled oats; took a second crop of oats, and on the stubble spread

spread 100 loads of marle (called here, as every where in Norfolk, clay, and in much of it there is a large portion of clay), and then took turnips. His success was various; many oats he lost by the wyer worm; and his first crop of turnips was, in some fields, very indifferent. In one large field the two crops of oats failed entirely, and the turnips the same. He has been there seven years: this year (1802) his turnips were the finest I had seen; seventeen hoers in one field; and all his corn an uncommon spectacle of greatly luxuriant crops.

He is of opinion that all the pastures and meadows of the farm ought to be broken up. I think he will make them excellent

Mr. HEATH, of Hingham, in 1796, broke up seven acres of old pasture; the soil, a turnip loam on a marle bottom, sowing oats, the crop 21 coombs per acre.

1797 Wheat, much damaged by the wyer worm.
1798 Oats, 23 coombs.
1799 Oats, 21 coombs.
1800 Barley, 16 coombs.
1801 Clayed near 100 loads an acre for turnips.
1802 Barley, and the crop very great indeed.

Mr. ROBINSON, at Carbrook, in 1795, broke up an old pasture, dibbling in oats, and got 17 coombs an acre.

1796 Oats again, 16 coombs.
1797 Clayed for turnips.
1798 Oats, 15 coombs.
1799 Clover.
1800 Wheat, six coombs.
1801 Oats, 16 coombs.
1802. Turnips.

Sir THOMAS BEEVOR broke up an ordinary pasture, and dibbled in pease; the crop five quarters an acre.—Then he sowed buck-wheat; followed by wheat, which

produced

produced six quarters an acre; succeeded by turnips, and barley, with grasses.

The instances of the great profit of breaking up old grass, on whatever soil, and under whatever circumstances (but on dry land being every where superior to wet), should be combined with the arrangement of a farm in the course of shifts. Upon the poor sands of the southern half of West Norfolk, ray-grass and trefoil are their only dependence: their grand-mothers trusted to these plants, and the farmers of the present day do the same; after a year and a half they produce little; after two years, scarcely any thing. Thus the land does not get rest enough: when broken up, it has not been sufficiently impregnated with the dung and urine of sheep, and the crops consequently are poor; but if the farmers were persuaded that their future crops would be exactly proportioned to the stock kept, from the case of a layer of five years to one of 50, they would set themselves to find more durable plants, and bring their waste arable under shifts that should secure very different products.

The failures in this husbandry of breaking up, have all been for want of paring and burning.

CHAP. IX.

WOODS AND PLANTATIONS.

THE modern spirit of planting took place as early in Norfolk as in any other county of the kingdom; and in some cases, upon a very considerable scale. The exertions of Sir ROBERT WALPOLE, at Houghton, and of Lord TOWNSHEND, at Rainham, were followed by many other persons spread over the whole county. I have, on various occasions, noted several cases in planting in this county; but rather than transcribe here, I wish to refer the reader to Mr. KENT's Report of Norfolk, who has treated this subject in a satisfactory manner.

There are some large woods at Billingford and Thorp Abbots, where hurdles and hoops are the principal object. Hurdles, 12s. the dozen. Admiral WILSON has 60 acres at the former place, let at 10s. an acre; and Marquis CORNWALLIS 144 acres at the latter, at a higher rent. The Billingford wood abounds much with hornbeam, which is made into hurdles, but is inferior in working to hazel.

At Catfield, in Happing, there was a wood of 150 acres, belonging to the Earl of ABERGAVENNY, that was grubbed ten years ago, and it has produced very fine crops ever since; it was not, while a wood, let at above 50l. a a year; the produce faggots.

Mr. ALLEN, at Stanhow, took potatoes for the first crop, 150 bushels an acre. Then carrots very fine; then Poland oats, a last (84 bushels) an acre. Fourth, winter tares. Fifth, turnips. The success on the whole very great.

A List

A List of Trees planted at Holkham, from the year 1781 to 1801.

Acres planted—718 acres, 2 roods, 11 perches.

Oak	336,700
Ash	420,200
Sycamore	179,600
Beech	166,280
English Elm	178,260
Cherry	92,800
Spanish Chesnut	45,430
Horse ditto	16,360
Birch	49,490
Lombardy Poplar	51,720
Canada ditto	34,950
Abele	20,150
Aspin	2000
Acacia	6600
Hornbeam	4700
Plane	4110
Witch Elm	3700
Lime	900
Larch	23,820
Scotch Fir	172,850
Spruce ditto	57,850
Silver ditto	4950
Weymouth Pine	950
Pineaster	900
Evergreen Oak	17,900
Holly	10,950
Mountain Ash	1860
Maple	2100
Crab	600

Service

Service	260
Alder	2900
Willows	27,300
Weeping ditto	12,500
Sweet ditto	2990
White Thorn	65,600
Hazel	68,280
Portugal Laurel	1250
Laurel	300
Juniper	1900
Laurustinus	300
Privet	15,300
Laburnum	700
Elder	6300
Golden Osier	1300
Sea Buckthorn	2700
Virginia Dogwood	2700
Guelder Rose	300
Sweet Brier	1200
Spindle Tree	320
	2,123,090

Mr. BEVAN, at Riddlesworth, has planted 966,000 which have, at present, a very flourishing appearance.

Marquis TOWNSHEND has been long in the practice of feeding cattle, sheep and deer, with the trimmings and thinnings of plantations: half an acre thickly planted thus fed 35 bullocks, 20 cows, 7 young cattle, 200 sheep, 300 deer two weeks, with the assistance of three tons of hay, at 4l. The browse saved seven tons, or 28l. Sheep are fond of the bark of Scotch fir and ash, as well as the trimmings of those trees, and it is a cure for the *scour*. The stock prefer ash, Scotch fir, and oak; but seem to dislike beech, birch, spruce, fir, and larch.

Mr.

Mr. CROWE, of Lakenham, is a great advocate for trees in grass-land; he has a pasture with many large oak, ash, and other trees, with spreading tops, and he is fully persuaded that he has to the full as much, and as sweet grass under them, without waste, as if no trees were in the field. This is novel, and an idea I have not met with before; it deserves much observation: Mr. CROWE's rule is to admit air and light, for if the trees are close, or the branches hanging to the ground, in such case the grass becomes coarse and sour, and is refused by the sheep; but all is fairly and closely eaten under Mr. CROWE's trees.

LARCH.

Colonel BULLER shewed me a circumstance relative to this tree, which merits noting: old sows, if allowed to get at them, will bark them for the sake of rubbing themselves in the turpentine: he had some killed, in this manner, before he knew by what cause; when informed, he ordered the sows to be watched, and had it confirmed by his own view.

OSIERS.

Planted in small spots, and along some of his hedges, supplied Mr. FORBY with hurdle-stuff enough to make many dozens every year, as well as a profusion of baskets. The common osier cut at three years; the yellow bark at four.

CHAP. X.

WASTES.

NOTHING can cause more surprize in the minds of many strangers on their first visiting Norfolk, than to find, on entering the county by Brandon or Thetford, a long stage of 18 miles to Swafham, through a tract which deserves to be called a desert: a region of warren or sheep-walk, scattered with a scanty cultivation, yet highly improveable. This is a capital disgrace to the county, and has been the result of an absurd prejudice in favour of these old heaths for sheep. They have been let for 1s. 1s. 6d. and 2s. an acre for many years; have been valued at 2s. 6d. of late years, the best at 3s. and 3s. 6d. and while left at such rents they are not likely to be improved. Something, however, has been done; better ideas are slowly creeping in, and some men have begun, though *good* clay be not found their farms. Chicory would treble the rent of these lands.

Mr. BEVAN, after trying several methods of bringing old heath-lands into cultivation, gave the preference to the following: sow oats and seeds on one earth after the drill roller; after harvest feed hard with sheep for two years, in order to rot the old turf; then prepare it for cole-seed, by repeated ploughings and harrowings; feed off the cole with sheep, and prepare for rye, with which sow seeds again, and let it remain as a layer till it can be clayed or marled with 60 loads per acre, when it may be brought into the regular shifts of the farm. This process will give the old flag time to rot, and will not exhaust the soil, so frequently done with new lands.

1802. The crops on the fields thus managed have been very good, and are now great. He continues of the same opinion; paring and burning he has not tried, but has advised a friend to compare it with this method. Mr. BEVAN now generally takes two successive crops of cole, both fed off with sheep (the latter greatly superior to the first), and sows seeds with the rye.

I crossed 400 acres of thick fern, called Eccles Common; half in that parish and half in Snetterton: Lord ALBEMARLE has much property in both; and being a good farmer, it is to be hoped so fine a tract of land will not long remain in such a horrid state, exhibiting in its spontaneous produce, its great capabilities of yielding corn and turnips most amply.

The commons are immense at Attleborough; Turnmoor, Westear, Broad Moor, Fen and Row, Lyng, Bacon's Thorpe, Decoy, Bunrough; these are all above 100 acres, and some above 200, with many smaller; I was assured that they amount to between 2 and 3000 acres.

Mr. FARROW, of Shipdam, purchasing 200 acres of Sayham common, under the act of enclosure, pursued, in breaking up, the practice common in Norfolk. He took two crops of oats, and then clayed for turnips: some on this common, and on that of Ovington, have begun with pease, and got very great crops; then oats, and then clay for turnips: for two years past the pease have answered better than the oats; the crops very large: the second crop of oats the best, and have produced in many instances 2d coombs per acre. I viewed various fields, both at Sayham and Ovington, in 1802, which promise that produce at least. I saw wheat also, which must be 12 or 13 coombs: barley exceedingly great. In a word, all the products immense.

immense. The barley follows the clayed turnips: Mr. FARROW had 17 coombs an acre of barley big: with that crop clover: some with ray and trefoil.

Among the improvers of Norfolk, Mr. OVERMAN, of Burnham, ranks very high. When he first took the farm, the land surveyor employed by the landlord pronounced that it was a dangerous error to think of wheat, as the soil was adapted only to rye and light oats. The great success he has had in raising very fine crops of wheat, proves that his husbandry has been conducted on sound principles.

Improvements on heaths and sheep-walks, which bring them from the state of desert wastes to be productive of corn and grass, are certainly in the very first class; but there are others which, though less striking, manifest abundance of exertion, vigour and perseverance. Mr. OVERMAN took a farm of Mr. COKE, at Michaelmas, 1800, which was, as I could easily judge from a part, the improvement of which was not finished, in a very bad state: and this spirited farmer not being of a temper to dream over any thing, determined to bring the whole into order as soon as possible: very little remained to do when I saw it in June 1802, and that little would be finished by wheat sowing. Besides paying the outgoing tenant 5l. 10s. per acre for desisting from sowing the lays, and summer-fallowing lands which many would have sown, largely manuring, and other more common exertions, he is going to enclose the whole farm at his own expense, throwing down the old ragged fences, and arranging the fields anew according to his intended *shifts:* but planting quick he postpones, till all spear-grass and weeds are quite destroyed: all corn sown is drilled, and as clean as a garden. He brings rape-cake in his own vessel; has a numerous flock of

South

South Down sheep; and, in a word, will carry the productiveness of the land to its ultimate degree of perfection.

There are 600 acres of good land in Sparham-heath, that calls loudly for enclosure.

Sporle common, near Swafham, has much of excellent land.

South Creke commons, 1000 acres; four great farmers, and four sheep-walks; passed it by West Basham enclosures, where it is covered with thick fern, yet this is the worst land of the four. Mr. SMITH's common belongs to Mr. COKE, and is very fine land, worth 30s. an acre, rent, tithe, and rates.

Mr. CROWE broke up a warren at Ash Wicken, of 300 acres, and re-laid it for a sheep-walk, with the greatest success.

Mr. REEVES, of Heveringland, broke up 165 acres, an old sheep-walk, on a poor sandy soil: he began with turnips, claying and mucking for them; the crop very good, and fed on the land by sheep: then he took oats, which were likewise very good: after the oats, wheat, and a fair crop: now turnips. The improvement great and profitable, and with this excellent farmer's management will be durable.

One thousand six hundred acres of wastes at Holt; they have talked for five years of enclosing, but nothing yet done.

Commons and unimproved marshes abound much in Fleg; but many are enclosing and draining: 400 acres of common, and as much open field at Hemsby, and an enclosure just agreed upon.

In the parochial notes entered under the Chapter of Enclosures, are numerous other cases, as well of wastes as of their improvement.

CHAP.

CHAP. XI.

IMPROVEMENTS.

TO examine the county of Norfolk with a single eye to this object, and explain in full detail the causes, progress, and consequences of the improvements which have taken place, would demand at least a year's travelling, and would require a large volume to contain the notes necessary for such an undertaking. The subject is of such importance as to demand, in every work that concerns the agriculture of this county, a particular attention. The methods more especially to be treated are:

1. Draining,
2. Irrigation,
3. Manuring,
4. Paring and burning,
5. Embanking.

SECT. I.—DRAINING.

MR. FREEMAN, of Swanton Morley, possessing a tract of meadows on the river, at Billingfold, poisoned by the water being pent up by the mill at Elsing, and no fall to be gained on his side for draining it, laid a truck under the river, and, by permission, cutting a drain on the other side, gained a fall, and by it drained 120 acres, to his great profit, and also to lowering the soakage of his neighbours' meadows. The improvement doubled at least the value of the land; the truck delivers a good stream now (in August).

There

There are numerous facts which shew that water is, in certain cases, so confined within the earth, that if the reservoirs of it are pierced into, it has a force sufficient to rise to certain heights. At Fincham, a man complaining that his well was often dry, Mr. FORBY advised him to bore at the bottom of it. The well was 28 feet deep; and on boring, the borer suddenly dropped down to the head, and being drawn up, the water gushed after it, and has ever since ran over the top of the well.

The exertions which Mr. SALTER, of Winborough, has made in draining his great farm of above 800 acres, have much merit. The first year of his coming he made a straight cut for the brook which runs through it, 342 rods in length of seven yards; 1116 rods of open drains in the meadows; 2937 rods of ditching, five feet broad and four deep; and 4871 rods of hollow-draining: these works he continued, and in 1801 did above 4000 rods of hollow-draining. In another farm he has at Carbrook, of 400 acres, he did in 1791-2, 798 rods of ditching, and 788 of draining; in 1792-3, 371 of ditching, and 1562 of draining; in 1793-4, 571 of ditching, and 897 of draining; in 1794-5, 201 of ditching, and 687 of draining; in all, 1941 of ditches, and 3931 of drains. His drains are in general 30 to 36 inches deep, some to four feet. He uses any sort of wood, chiefly the bushes that were a nuisance to his fields, but of late has been forced to buy great quantities.

In this note of the exertions of the master, it will be fair to minute those of one of his labourers, who did 1300 rods of hollow-drains between Martinmas and harvest.

I was rather surprized to find that this improvement was necessary on Waterden farm, in a very dry country: but Mr. HILL, finding that the springs were injurious to much of his land, made very laudable exertions in freeing his

his farm from them. He digs hollow-drains, from two feet and a half to four feet deep, filling them very carefully with stones, hand-picked from the heaps by women, to prevent any earth going into the drains and impeding the current of the water. The effects of the improvement are great.

In going from Waterden to Rainham, passed by Sculthorpe Mill, and there enter a region that must make a farmer's heart ache. Of the nuisances that a country can be plagued with, certainly water-mills class very high in the black catalogue: for the sake of this beggarly mill, which apparently cannot be worth more than from 20l. to 30l. a year, here is a noble tract, from a furlong to a mile wide, of what, ought to be rich meadow, poisoned with water, and producing rushes, flags, sedge, and all sorts of aquatic rubbish. Who would not suppose the two sides of the river belonging to little proprietors, as beggarly as the mill, who could meet over their tankard to wrangle, but never agree? No such matter. Marquis TOWNSHEND on one side, and Mr. COKE on the other. It would not be amiss to couple the two stewards of the estates up to the chin in one of these overflowing dykes, till they settled the matter, *for the benefit of the public.*

The Rev. DIXON HOSTE, who has done much hollow draining, twists three sticks or poles together, which he lays in at the bottom of the drain, and then fills, six inches deep, with stones; as he has found, that when the drains are filled with stone only, they do not run quick enough. The smaller drains he digs 24 inches deep, and the leading ones 30. The price 4¼d. a rod for the one, and 5d. for the other.

Mr. HAVERS, at Thelton, drains attentively; the distance from eight to ten yards; the depth, in general, 30 inches, but of leading drains, 32; fills with bushes and

straw;

straw; the expense 5s. a score. He has tried the draining plough, at the depth of 14 inches; but the bailiff (Mr. H. not at home himself) thought that it did not answer. Mr. SMITH, at the inn at Scole, has done much. On every farm the improvement is very great.

There is some wet land in Attleborough; and to and around Hingham, draining is well established and much done; but, strange to say, none on pastures, be they wet as they may.

About Watton, much done, and the effect such, that one crop has paid the expense.

The father of the present Mr. KERRICH, of Harleston, began hollow-draining at Redenhall 27 years ago; his drains were filled with bushes, and they *work* now. He also drained much pasture land, which was the better for three or four years, but worse afterwards, by being too dry; moles and rats have now stopped many of the drains, and done good by so doing. I had this fact from the present Mr. KERRICH. Much has been done by Mr. PAUL, of Starston, who recommends greatly the culture of sallows in hedges, as that wood lasts longer in drains than any other, and is as good for the purpose as the hazel.

In Loddon hundred, Mr. CRICKMORE, of Seething, began hollow-draining above 20 years ago: he is an excellent farmer, and has been much imitated in this great improvement.

Mr. BURTON, of Langley, has made great exertions in this husbandry at Hempnal, laying out 300l. in one year.

Mr. JOHNSON, at Thurning, has made a great improvement in draining meadows, described in the following extract from a letter he favoured me with:

" The track of land I took of the late Mr. ELWIN, is about

about 26 score acres, consisting of five farms, jumbled together, for more than 80 years; three of them have not had a resident occupier during memory. The lands were chiefly small pieces and large borders; the meadows a long strip, of about nine score acres. It is not in my power to represent the bad state the whole of the lands were in, nor can I give an exact account at what prices the work was done. The fences are chiefly drains for the land, which always should be the case, if possible. I began the work on the chief of the meadows, the year before I took the farm: the account I annex was all done the first year of the lease, and I have since expended much on all the lands. I have always had more labourers than I wanted for harvest: I began the work in all directions, which made it look in a confused state. My neighbours said at the time, it would never be made a job of; but before the year was expired, Mr. DUGMORE, who knew the state the land was in, looked it over, and paid me the compliment of saying, it was the greatest and best work he ever saw in our county. In draining land, the main object is to form the main cut in the bottom; the fences for the next drains; then begin on the great springs, and see what effect they have: but draining small springs seldom has any effect on the large ones.

S. JOHNSON."

Thurning, Jan. 31, 1803.

DRAINING.

Work done on the late Mr. ELWIN's estate in Thurning, from Michaelmas 1796 to Michaelmas 1797, by SAM. JOHNSON.

Rods.		s.	d.		£.	s.	d.
14	at	3	0	per rod	2	2	0
99	at	2	6	- -	12	7	6
401	at	2	4	- -	46	15	8
225	at	2	3	- -	25	6	3
598	at	2	0	- -	59	16	0
326	at	1	10	- -	29	17	8
593½	at	1	9	- -	51	18	7
138	at	1	8	- -	11	10	0
348½	at	1	6	- -	21	2	9
28	at	1	4	- -	1	7	4
185	at	1	3	- -	11	11	3
53	at	1	2	- -	2	18	10
589	at	1	1	- -	21	18	1
500	at	1	0	- -	25	0	0
152	at	0	11	- -	6	19	4
15	at	0	10	- -	0	12	6
66	at	0	9½	-	2	12	3
150	at	0	9	- -	5	12	6
1864	at	0	8	- -	62	2	8
262	at	0	6	- -	6	11	0
36	at	0	5	- -	0	15	0
119	at	0	4	- -	1	19	8
187	at	0	3	- -	2	6	9
920	at	0	2	- -	17	13	4
7868					£.420	16	11

		£.	s.	d.
7868 rods	— — —	420	16	11
Stubbing, clearing, levelling by the piece —		178	8	8
Mould filled on three-wheel tumbrels by the day, on a valuation of 1d. per load, 36,000 tumbrel loads: two tumbrel loads make one cart-load		262	10	0
New barn, stables, bullock-houses, &c. —		500	0	0
Twenty extra horses, keeping; 1200 loads town muck; rape cake; labour for extra jobs, &c.		1200	0	0
		£.2961	15	7

Mr. PRIEST, of Besthorpe, digs his drains 30 inches deep and seven yards asunder, price 6s. a score.

SECT. II.—IRRIGATION.

THIS improvement is of very late standing in Norfolk: the experiments made are few, but they are interesting enough to promise a speedy extension.

In 1792 I found Mr. BEVAN had made some progress in watering his meadows. In 1794 he had completed some. He purchased them at the rent of 4s. an acre only, and his tenant had now offered him 40s. an acre rent, for all done. That year he fed them with ewes and lambs till the 15th of May; the rushes were then swept over, and the produce of hay two tons per acre, though watered only for 48 hours the middle of June, and cut the 15th of July. The expense of making them was 5l. 10s. per acre; the produce leaving 36s. per annum interest for that sum, or about 26 per cent. profit.

1802.

1802. After some years experience of these meadows, Mr. BEVAN found, contrary to his expectation, that the rushes would not give way to the water; and Mr. BROOKS, from Gloucestershire, viewing them, and having the favour of a visit from the Rev. Mr. WRIGHT, he employed Mr. BROOKS to new form the works, by altering the direction of the beds, and reducing them from ten and twelve yards to seven, and the immediate improvement made was very great: and he has also made several new meadows; the first done were 46 acres, and 14 more are now adding.

Mr. P. GALWAY, at Toffts, has watered 20 acres: having read Mr. WRIGHT's treatise, he made application to that gentleman, who procured for him a man (Mr. BROOKS) well skilled in the Gloucestershire method. I viewed the meadow which was fed. I have no doubt of the improvement being exceedingly great, though the expense, by contract, was only 4l. 4s. per acre; but I have great doubts of the method followed. I think the beds or panes too flat, and that, consequently, the water has not a motion sufficiently nimble; and this, I conceive, results from the error of taking the water from a level of too small an elevation. Mr. LUCAS has done a meadow of eight acres, still lower, at Lyndford, upon which the improvement, owing to the same cause, may perhaps disappoint him.

Having crossed the Bridgham river, going from Riddlesworth to Thetford, and observing that it brought down (in July) a most copious stream, I made it a point to examine it higher up; and under the conduct of Mr. FOWELL, of Snetterton, traced it from East Harling church to Bretenham, being throughout that line chiefly bounded by the property of Sir JOHN SEBRIGHT; generally on the left side, West Harling; but interrupted by the commons and some

some smaller properties in Bridgham, on the right side. How far this circumstance may operate in preventing a system of irrigation, depends on the rights of various persons, but this precluded, here is a very fine field for a capital improvement by watering; for some distance the stream is sluggish, and therefore may not give so good an opportunity as lower down, upon the lands between the Hall and Bretenham; but a large tract is evidently below the level, and consequently capable of a very important improvement. The rental of the West Harling estate, on old tenures, is 2387l. Roundham is a very fine farm of 1600 acres, including 500 of ling heath: it is 500l. a year at present; and contains much fine sandy loam.

I crossed a fine stream at Chapel mill, in Gressenhall and Hoe, which runs to Wendling; another in Elmham and Beteleg; much water in August, and a falling valley with it, capable of much irrigation; but not a thought of it. At Billingford, it is a fine river: passed above three miles over Mr. BLOOMFIELD's farm, and again examined the river and meadows to Elsing mill: they are in a sad state for many miles, caused by mills keeping up the water as high as the adjacent lands, and in some places higher; ruining the lands, which it would convert to gardens, were the waters applied to irrigation instead of grinding. Messrs. BLOOMFIELDs made a weir of timber and stones to discharge the water at a certain height, sufficient for the mill, which I examined, and could not but commend; yet this work was opposed by the miller, and was near causing a law-suit. It is the same story over half the kingdom; and were the extent of the mischief known, would prove how necessary it is to apply other powers for this purpose, steam, wind, &c. rather than suffer a trifling rent of a mill,

mill, to prevent fifty times the amount being gained by the improvement of meadows.

Mr. BLOOMFIELD, on the recommendation of his landlord, Mr. COKE, has irrigation in contemplation; he has a small stream at command, and has begun by a straight cut, which is thus to be applied.

Note, in 1792, the stream at Cley offers so fine an opportunity for watering, that I stopped my horse, repeatedly, to view it with regret. Poor sand-hills might be converted to rich meadow.

The river at Bodney, and the poorer arable lands, which come down below the levels it affords for watering, struck me as offering an uncommon field for irrigation; which calls aloud for the exertion of that spirit which has just begun to awaken in Norfolk. The two streams between Swafham and Buckenham, join at Bodney.

Mr. REEVE, of Wighton, has made an exertion in irrigation that has uncommon merit. Having a long, rough, and very coarse meadow, of above 20 acres, through which a small stream runs, which is nearly dry in any time of drought, but has sufficient water in any other season, he cleared a straight channel for it through the centre of the meadow, and taking the necessary levels, threw it at pleasure into carrier trenches, under the two hedges that bound the meadow, and accompanied those trenches with drains: the first part he did, remains too flat; but as he advanced, he corrected that error, and raised the surface into arched lands, running the carriers along the centre, from which the water flows down the sides to the drains. This part is very well done. The experiment having been lately finished, the full effect is not yet seen; but it is evident enough to determine, that the value of the

meadow

meadow is trebled. The expense of the first part was about five pounds per acre; but by filling up holes, taking up turf, and laying it down again, &c. some of the latter part cost him, it is said, above 20l. per acre. Gentlemen who attempt this improvement, cannot be too much impressed with the idea of the necessity there is, that water, applied in irrigation, should always be in nimble motion; the effect is greater and more certain: another circumstance, oftentimes not sufficiently attended to, is that of the carriers being so levelled, that the water will flow over the edges in every part, by which means it is much more equally delivered than when let out by small cuts.

Mr. PURDIS, of Eggmore, shewed me a mead of eight acres, which he had very lately renewed; employing Mr. BROOKS, from Gloucestershire: upon examining the spot where it would be proper to fix the sluice for throwing the water of the river into the main carrier, the foundations of an old sluice were found, in a sound state; and the whole immediately renewed: on further examination, the carriers and drains in the meadow were all traced, opened afresh, and thus an irrigation formed upon very nearly the plan of old works, which had been utterly neglected for at least 80 years: upon further inquiry, it was found that this former irrigation was obscurely known to have existed, but no records gave any information of the time when it had been formed: it is extremely curious thus to trace former exertions in so excellent a husbandry, followed by so long a period of darkness and ignorance, as to suffer such immense advantages to sink into a state of neglect and ruin. Mr. BROOKS approved the former mode of irrigation. I may observe upon it, that the water through-

out

out the meadow flows through little cuts: not one carrier *overflows:* equally varying in breadth as it advances, for the equal delivery, as explained by Mr. BOSWELL; the delivering trenches are not on a very gentle, but on steep declivities: and the benefit is unequal: at the end of every little cut, there is a great bunch of grass, with spots on the sides, inferior: I pointed this out to the waterman, and he admitted the defect, observing, that those little cuts must, another year, be greatly multiplied. It ought to be apparent on the first view, that the equal distribution, by a universal overflowing, without any *cuts* for the purpose, must be superior. The company with whom I viewed this and Mr. REEVE's meadow, I found strongly impressed with the idea of the great superiority of the water first taken from the river, and reasoned as if all the benefit arose from a deposition of certain particles, which being dropped, the water became of little value: these ideas, in certain cases and to a certain extent, are just; but they seem to be carried here much too far, and may have ill effects in causing a small value to be assigned to water taken at the second and third hand.

Mr. BROOKS forms his works by the eye, and without using a spirit level: the consequence is, that the distribution by overflowing is, in that manner, impossible to be attained.

The improvement, however, as in all cases of irrigation, is very great: the meadow had been watered but three weeks, and the growth was very luxuriant; the benefit is unquestioned; and the exertion does great honour to Mr. PURDIS, who has the merit of effecting a most valuable amelioration, which so many of his sleeping predecessors utterly neglected.

April

IRRIGATION.

April 7th, 1802, he turned in his cows, and they found full feed during the rest of the month. May 1st, shut them up, and in nine weeks cut two tons per acre, and they have been fed since: these meadows were full of the broad-leaved plaintain, which has disappeared, and are now clothed with good grasses.

I have rarely seen a finer opportunity for irrigation than at Hillingdon: Sir MARTIN FOLKES and Mr. COKE have the stream as a boundary for two miles together. Captain BEACHER walked with me on the banks of it for a considerable distance: I found the declivity every where so rapid, that it cannot be doubted but that it may be carried over a large tract of poor arable slopes, to the effecting a most profitable improvement. Nothing yet done: but this subject is well started in Norfolk, and the age of dreaming passed. If these proprietors muddle themselves in the low flat lands, on the river banks, already of a good, though inferior value, instead of running levels as high as possible for floating the dry arable, they will make a shilling where they might make pounds. For carrying irrigation to the highest improvement, the levels should be taken for two or three miles before a spade is in hand: and then, if there be some lousy miller below, he yelps at the undertaking. Three or four proprietors should unite, and buy, or burn, the mills, before they think of beginning.

Uncommon opportunity for it from Sedgford to the sea, through Heacham: two mills denote a fall of ten feet in three miles, besides the rapid motion of the water every where: after a long drought I found ample water in the stream for great improvements, and the circumstance of many dry arable fields under the level, will, by-and-bye, be found of great value. Mr. STYLEMAN has engaged Mr. BROOKS to make a trial.

SECT. III.—MANURING.

This is the most important branch of the Norfolk improvements, and that which has had the happy effect of converting many warrens and sheep-walks into some of the finest corn districts in the kingdom.

1. Marle.
2. Lime.
3. Gypsum.
4. Oyster-shells.
5. Sea-ouze.
6. Sea-weed.
7. Pond-weeds.
8. Burnt-earth.
9. Stickle-backs.
10. Oil-cake.
11. Ashes.
12. Soot.
13. Malt-dust.
14. Buck-wheat.
15. Yard-dung.
16. Leaves.
17. Burning stubbles.
18. River-mud.
19. Town-manure

MARLE.

Thirty years ago, being in Norfolk, I was informed by the late Mr. CAR, of Massingham, that 25 years before that period, 70 loads an acre had been commonly spread; after which, many farmers tried 30 more, but without success: his own practice was to lay on 35 to 40 load, and in three or four years after, as much more, by which means he found that it incorporated better with the soil.

Thirty years ago the quantity spread from Warham to Holt, was 60 loads an acre, which lasted 15 or 16 years in perfection, then they laid on 25 or 30 loads more, which lasted ten or twelve years longer; repeating it still; so that previous to 1770, that country had much of it been marled thrice at least.

Mr. BEVAN marled (clayed, as it is called) his whole farm

farm, from 60 to 80 loads per acre: I found his team at work at it, and measured the carts: length, six feet; width, four feet; depth before, two feet; ditto behind, one foot; contents, 36 cubical feet; price of filling, 28s. per 120 loads, filling and spreading, and pumping out water from the pit, if necessary; six horses, two tumbrels; four men fill 30 loads a day. The expense per acre for 60 loads:

	£.	s.	d.
Filling and spreading - -	0	14	0
Two days work of six horses, allowance of oats, two coombs a week, at 10s. which, for two days - -	0	5	8
Hay - - - -	0	9	0
Decline of value, 5l. 12s. 6d. per horse per annum, at such hard work, or 4½d. a day for 300 days; and for six horses, 2s. 3d. a day - - -	0	4	6
Interest of his purchase, 30l. or 30s. a year, 1¼d. a day, and for 6, 7½d. -	0	1	3
Driver, at 1s. 6d. - · -	0	3	0
Wear and tear of carts, and interest of first cost, 25l. say 15l. per cent. 3d. a day	0	0	6
	£.1	17	11
Sundries, shoeing, harness, &c. -	0	1	3
Per acre for 60 loads - -	£.1	19	2

He scuffles the marle after spreading.

Examining a team (belonging to one of the tenants of Mr. COLHOUN) at clay cart, I found the tumbrels four feet three inches long, three feet nine inches broad, and two feet three inches deep, consequently hold, if full, 35 cubical feet: the wheels five feet six inches high, and six inches broad: eight horses (sometimes nine) were employed;

four

four men filled, and did regularly 32 loads a day, eight loads per man being the stint; the distance to the heaps 334 yards. Each load made eight heaps, and to discharge it the easier, there is a false tail-board in the cart. The tumbrel does not separate from the shafts, as in our common ones in unloading; the shafts are fixed, and rise in the air, the traice horses drawing by the ends of the shafts, and the thill horse by short traices fixed at the other end of the shafts. The price given to fill and spread, 25s. for 120 loads, or $2\frac{1}{2}$d. each. I examined the heaps with particular attention, and guessed them at four bushels each, and then, asking the men their opinion, they also guessed the same. Surely it behoves the gentlemen and farmers of this country, to reconsider this business entirely; for seven horses (and there are often eight) to draw 32 bushels of clay, seems preposterous; $4\frac{1}{2}$ bushels a horse are a load, which, when you come to divide it, seems such a system of trifling, as to be worthy only of children; but viewed in that bundle of logs called a tumbrel, with great lumbering wheels, the machine and the load seem more congruous, and to the eye there appears something for the horses to draw; but calculation tells us there is nothing but a heavy, ill-contrived, unmechanical cart, lessening the power of the horses, till one draws not more than a man would push in a wheel-barrow. All this evil, and an enormous one it is, springs from this circumstance of uniting many horses in the same draught, to form a team, which is never analyzed and well examined, but you find it a barbarity worthy only of savages. Take a light cart, such as I use and have recommended, or even the little car of an Irishman; put one of these horses in it, and see if $4\frac{1}{2}$ bushels will be the load!! See if you are content with 9; try if you cannot carry 12; try again at $13\frac{1}{2}$, or the treble of what your horse does at present!!! Thus one horse

horse does the business of three. But drivers? Let us examine: I get but little there; but what do I lose? A man, at 1s. 6d. a day, drives away 32 bushels; this is a fraction more than a halfpenny a bushel: a boy at 6d. with me drives away $13\frac{1}{2}$; this is not a halfpenny. Here then is no loss in driving, with an enormous gain in team; and the measure of employing children to execute the work of men, is a parochial and national benefit, which wants no explanation. I need not observe, that with one-horse machines, of whatever kind, it is not necessary to allow a horse in the standing cart; the horse in an Irish car, or any other, if prepared with that view, is attached in an instant, as quickly as you hook a traice-horse to a thiller. But in this point, an improvement which was introduced by Mr. COLHOUN, deserves attention.

This is, a contrivance to draw the carts of any size out of the pits by means of a capstan; he uses large three-wheeled tumbrels, and to save the extra number of horses, which are used in common to get the load out of the pit, he applies a boy and a horse to the lever of a capstan, and draws up the load with so little loss of time, that the whole operation takes but three minutes and a half, and with horses in the common way, three: if it demanded more, the objection would go no further than letting there be an extra cart in the pit, which would prevent any waiting. Mr. COLHOUN's are three-wheeled carts. By means of this machinery, the pit may be dug of any depth, without impeding the raising the load; a great advantage, not only to the men in filling, but also in the quality of the clay or marle, which is usually better at the bottom of a pit than in any other part of it. I measured the depth of one pit, which was above 20 feet.

The country about Snetterton was all marled many years ago; Mr. FOWELL's farm, fifty years past, and was

done

done by the landlord, who contracted for the work, paying by the load; and this was the cause of an evil, felt to this day; they laid on a great deal too much near the pits, and too little at a distance: the soil in the former situation now tills badly, insomuch that Mr. FOWELL wishes it had not been done at all. The marle is yellow, from the mixture of clay, but ferments strongly with acids.

All the country about Watton has been marled (clayed, as they call it), and the general way has been to do it upon the first breaking up of all old grass. They take two crops of oats in succession, and then clay for turnips, 40 to 60 loads an acre.

Mr. SALTER, at Winborough, in seven years has clayed, as it is called, 100 acres, at 100 loads an acre; a quantity which he thinks necessary on his soil, which is a wet loam, or springy sand, and also brick earth; but observed, that if so much was to be laid on the dry Norfolk sands, they would be *set fast*, and it would be many years before the clay would work.

All Mr. JOHNSON's farm at Kempston is marled: he approves of doing it at twice, rather than giving the full quantity at once.

The country about Thorpe Abbots, has, in general, been clayed; and on the gravels it answered greatly. But some being done on Mr. PITT's farm a second time, it did more harm than good.

The hundreds of Loddon and Clavering have all been clayed. I observed many pits of clay marle every where. At Langley, they now bring white marle from Thorpe, near Norwich, by water; laying on 12 loads per acre, at 4s. 6d. from the *keel*, barge, and costs 5s. on the land.

Caistor and the vicinity, has all been clayed, 40 loads per acre.

Mr. EVERIT marls from Wightlingham, at 5s. a load,

load, at the water-side, of two chaldrons; lays 16 chaldron an acre; carriage, three miles besides.

At Thelton, &c. no old pasture is ever broken up without marling, called here, as every where in Norfolk, *claying*. Much done in the vicinity, especially in all the new enclosures; and the best practice that of claying one year before breaking up. Without clay, the straw on new land runs up weakly and *faint*: this manure stiffens it, and much increases the produce.

Mr. THURTELL, near Yarmouth, is decidedly against claying on a layer: he has found, that in this way, it is four or five years before it works well. He spreads it, to chuse, in the winter before a turnip fallow, by which it is thoroughly mixed with the soil: he knows several practical farmers of the same opinion.

At Hemsby they spread from 20 to 70 and 80 loads an acre, of their own clay marle; the latter quantities, if not done in the memory of man; but for renewing 20; the effect lasting 30 or 40 years. Some white marle is brought by water from Thorpe, &c. Mr. FERRIER agrees with Mr. THURTELL, that it is best spread on a fallow; it works quicker and mixes better, than when on a layer.

Mr. BROWN, of Thrigby, having a piece of land that he had over-clayed, ploughed it a little deeper, and it then did well. That parish was marled above 30 years ago from Thorpe; he now clays 35 to 40 loads an acre, and it does well.

At Martham they marle from Wightlingham; a keel costs 5l. 5s. and does two acres well; it lasts 30 years. Mr. FRANCIS has no clay on his farm; he lays on 10 cart loads per acre of the marle, equal to 20 chaldrons: 6s. a load: spreads it to chuse on a fallow for turnips; does not approve of marling on layers, because it hurts them, otherwise it is a good method, and not apt to sink so soon.

At

At Ludham they have it by water from Thorpe, Wightlingham and Wroxham, at 4s. a chaldron; and lay from eight to ten cart loads an acre; it lasts 20 years. Mr. HORNARD spreads it on stubbles scaled: none better for it than a one year's layer that is to remain another.

He also lays it in summer on turnip fallows: first marls, then spreads the dung; scales in, and then deeper for the seed earth.

Mr. CUBIT, &c. at Catfield, brings marle from Horsted, 20 miles, by water; costs at the staith 5l. 5s. a keel of 36 chaldrons, 18 to 20 loads. He lays on seven or eight loads per acre, generally on a fallow for turnips: lasts 30 or 40 years. Mr. CUBIT has some land done 40 years ago, and does not yet want renewing.

Mr. CUBIT, of Honing, spreads eight or nine loads per aere, from Wroxham, at 5s. a load at the staith, three miles off: he would give 9s. a load in his yard; no claymarle of their own, except at Happsborough. When turnips shew the anbury, it is a sure sign that the land wants marling.

They have white marle in North Walsham, and it is much used: Mr. PAYNE has done 40 acres in one year, 12 loads an acre: it lasts 14 to 20 years. It is common to make layers of mould, the marle on to that, and then the yard muck, and turn the whole over together: they also lay lime on mould, turn it over, leaving it some time, then muck on, and turn the whole over again; all chiefly for turnips, but some for wheat. Mr MARGATESON brings it five miles, from Oxnead and Lammas; lays on ten cart loads an acre: it lasts 20 years. If the land has not been marled, or wants a renewal, the turnips have the anbury, which this manure prevents entirely.

All the country about Scotto has been marled; *full of marle*

marle is the expression; ten loads an acre on strong land, and eight on light; and lasts above 20 years.

Good yellowish marle at Coltishal; they spread 12 loads an acre, and it lasts 14 years.

Much at Oxnead, &c.; Mr. REPTON lays 20 small loads an acre: it lasts 20 years: a second marling answers well: the foulness of land shews the want of marling. He spreads on the layers, and thinks that the longer it is kept on the surface the better.

Mr. REEVES, of Heveringland, who is a very attentive and spirited farmer, lays on 28 loads an acre of clay-marle, free of *callow*, which he reckons equal to 40 to take all as it comes; but as this demands a greater depth of pit, he reckons that it costs him 9d. or 10d. a load.

Mr. BIRCHAM, at Hackford, from 40 to 60 loads; and the whole country there has been so marled; the white best by far, 10 loads as good as 40 of other sorts. It prevents the buddle *(crysanthemum segetum)* which, and sorrel, are the signs that land wants marling; when done, these plants disappear. Marle does a second time, but he has found that the best way of applying it then, is by composts with dung; on layers for wheat. Marle, Mr. BIRCHAM has found bad for turnips, except when it is wanted to get a farm done as quickly as possible.

Mr. JOHNSON, at Thurning, clays, 40 loads an acre, which lasts 20 years: it is best on a layer, but hurts the grasses to the amount of half the produce: he reckons that the longer it is kept above ground the better. Much marle about Holt.

All the land about Binham is marled that wants it. Mr. ENGLAND lays on from 40 to 50 large loads an acre: it lasts 20 years; but this depends on soil. He thinks it best spread on ollonds, between the first and second year: but

this

this must be governed by convenience. After a proper time a second marling is as good as the first.

Mr. HILL's father marled 350 acres of the Waterden farm, at 80 loads per acre; he has done the rest of the farm himself, 35 per acre, and never exceeding 40, thinking it much better to do it at twice than at once. The colour is yellow and white mixed; considers the yellow as best. Spreads it to choose on a one year's layer; and in winter rather than in summer; leaves it a year, and ploughs for the first time very shallow. The first turnips are not the better for it, but the barley great, especially when it comes to the bushel. His rule, in after-manuring is, to muck the first time, in preference to folding. His father did one field, at the rate of 122 loads per acre, which was so over-dosed, that the land has not recovered it yet: the soil light. It has given but one good crop, which was wheat, nine coombs three bushels (old measure) per acre: every other crop has failed more or less. If Mr. COKE had not granted a second 21 years' lease of this farm, the benefit, after much loss, would all have gone to others.

On Mr. REEVE's farm, at Wighton, I saw an extraordinary fine white marle, not as in common, in globules, but more resembling the equal consistence and texture of white butter.

In all the light lands of Norfolk, *clay*, as it is called, but which ought to be called clay marle, from the quantity of calcareous earth it contains, is preferred much to more chalky marls; and of all others, the hard, chalky, and stony marls are reckoned the worst: when these only (called also *cork)* are found under tracts of waste or poor land, they are not deemed improveable to profit. Mr. OVERMAN has made the experiment of such, and has found

found the benefit so great, that, directly contrary to the common opinion, he prefers them. I saw such used in 1792, on a very large scale on his farm. He spreads from 40 to 50 loads per acre. Such manuring prevents the *anbury* in turnips.

Mr. H. BLYTHE, of Burnham, has no doubt of white marle, on the sands of his farm, being better than clay; it works better and sooner, adding the expression, *it will buy a horse, before clay will buy the saddle.*

Mr. DURSGATE has found white marle more profitable than clay; working much sooner: nor has he any objection to that hard chalk called *cork*. Of white marle he lays on 50 to 60 loads an acre.

Mr. WRIGHT, of Stanhow, a very attentive and excellent farmer, has found, that the longer marle or clay is kept on the surface the better. He spreads it on a lay of one or two years old (the latter best), and leaves it a year; then ricebaulks the land, leaving it so for the summer; and it gives as much food as if it had not been stirred, and thus the marle works without being buried. He marls largely, yet is clear that it is an enemy to grass and turnips.

Cork has been used successfully at Ringstead.

Marle is found under all the country at Snettisham, generally white: the farmers lay on from 60 to 100 loads an acre.

Mr. GODDISON, at Houghton, laid, in six months, 3200 loads on 44 acres of very poor black-sand heath, and broke it up; the success great, and shall do as much more this year. He pays 7d. a load, to a man who finds team and every thing.

Mr. BECK, at Castle Riseing, has clayed all his farm, 60 to 80 loads an acre, and covered the whole with Lynn muck.

muck. He thinks, however, that it is better to lay on less at first, and renew it by composts. He always lays it on ollonds, and leaves it above for a year.

Mr. SAFFORY, of Downham, is for 40 loads an acre, and then repetitions, which is a far better system than much at first; and he would always spread it on a one year's layer, to be left another year: it then does not sink so soon. Not much done near Downham.

Mr. PORTER, of Watlington, lays 100 to 140 loads per acre of clay (marle) on to his gravels; and never found it too much; the benefit of some, done thirty years ago, is to be seen now: spreads on a layer for the sake of frosts taking it: one acre done before winter, as good as two in summer.

Mr. MARTIN, of Tottenhill, lays 100 load of clay per acre on black sand and gravel, at 25s. per 120 in winter, and 30s. in summer.

About Wymondham, 60 to 100 loads of clay; 80 common; some chalky; some blue; and some yellow.

At Besthorpe, white, blue, and brown marle; all ferment in acids. Mr. PRIEST, 64 loads an acre, at 30s. per 120 cubic yards; now, 35s. to 36s.

LIME.

Mr. BIRCHAM, at Hackford, has used lime at the same time with yard-muck, very successfully for turnips; three chaldrons, at 10s. or 11s. a chaldron, spread out of the waggon on land, on which twelve loads of muck are spread: and when a piece has had part of it muck only, and the rest muck and lime, the effect is seen to an inch.

Mr. SAFFORY, of Downham, has tried lime, sixty bushels

bushels per acre, on his fen farm; but discontinued it, as it did not answer.

"In East Norfolk, lime is successfully used, even after marle. It is of the greatest efficacy on hot burning soils, and is perhaps the most effectual cure of *scalds*: hence considered as a cold manure."—*Marshall.*

GYPSUM.

Mr. ALLEN tried this manure, very carefully, at Stanhow, on clean clover.

March 31. No. 1 and 4. No manure; produce average of the two, 38 lb. 6 oz.

2. Four quarts sifted coal-ashes kept dry, 50 lb.

3. Gypsum, one quart, $54\frac{1}{2}$ lb.

The ashes, therefore, gave an increase of 11 lb. 10 oz. and the gypsum of 16 lb. 2 oz.

OYSTER-SHELLS.

In East Winch and West Bilney, and scattered for ten miles to Wallington, there is a remarkable bed of oyster-shells in sea-mud: the farmers use them at the rate of 10 loads an acre for turnips, which are a very good dressing; they are of particular efficacy on land worn out by corn. Mr. FORSTER several years ago laid 20 loads an acre on some worn-out land, and they had an amazing effect in producing grass, when laid down in seeds, giving a deep luxuriant hue like good dung: the benefit very great at the present time. They are found within two feet of the surface, and as deep as they have dug, water having stopped them at 16 or 18 feet deep. They are used again and again on the same land, and with the same effect. At East Winch, Mr. CROWE has acres together of this most

most valuable manure. They fall to powder on being stirred.

SEA-OUZE.

Mr. PALGRAVE, at Coltishal, uses much sea-mud, scraped up by the *bear* from the bottom of Yarmouth Haven: he lays on 40 loads per acre, and has thus manured 70 acres; the improvement very great. I found, on trial, that it is a calcareous mud: on scalds, or burning places of sand or gravel, it forms a cold bottom, and is an effectual cure.

Fifty loads per acre, of sea-ouze, have been used on the upland sandy loams of Warham, with very great success: superior crops the consequence.

SEA-WEED.

What other name to assign to a very singular manure on the coast at Thornham, I know not. In the great and accurate map of the county, published by Mr. FADEN, there is a mark on the shore for what is called *crabs, scalps and oak-roots*. Mr. RISHTON had the goodness to take me to view this spectacle, which is an extraordinary one: it is evidently the ruins of a forest of large trees, the stubs and roots remaining, but so rotten, that with a spade I dug into the centre of many, and might have done of all, with as much ease as into a mass of butter. Where the stumps are not found, on digging I turned up a black mass of vegetable fibres, apparently consisting of decayed branches, leaves, rushes, flags, &c.; to what depth this vegetable stratum extends is not known, but at some creeks on the very edge of the sea, at low water, there is a very fine soapy sea ouze, at two or three feet depth. The extent of

SEA-WEED. 415

of this once sylvan region, which every common tide now covers, can scarcely be less, in one place only, than from 5 to 600 acres. There is not an appearance of any tree lying at present from the stump, as if blown down or left after falling, but rather that of a forest cut down in haste, the stems cleared and hurried away, leaving the branches to rot: but this is mere conjecture. It is remarkable that there is not, as I am informed, any mention of this ruined forest in the old historians of the county; nor does tradition offer the least conjecture or report on the subject. Trees, roots, and stumps, are very common in bogs, wherever found; but here is not the trace of any thing like a bog, the earth is solid, and all a fine ouze or sea-clay.

Mr. RISHTON viewed these relicks with the eye of a farmer; for experiment, he sent his carts down for some, and spread 10 loads per acre of it, for turnips: it answered perfectly, and on comparison, equalled his yard-dung: and also rape-cake. In another experiment, he manured two acres for wheat, with a compost, consisting of nine loads of this weed, and three chaldrons of lime, mixed; one acre with yard-muck; one acre with tallow-chandlers' graves, 16 bushels, and the rest of the piece with rape-cake. The graves were, in effect, far beyond all the rest; between which the difference was not very perceptible. The expense only 1s. per load: but if a barge was floated to the spot, and anchored when the tide was in, for loading at low water, it might be procured at a much cheaper rate. This gentleman is going to quit his farm, but not before he has opened a real mine to such farmers as shall have the sagacity to dig in it: it appears astonishing that none of them should long ago have made the same experiment, and consequently have profited by so beneficial a vicinity.

POND-WEED.

Several persons in Norfolk are in the regular habit of clearing their rivers and ponds just before turnip-sowing: they cart them immediately on to the land, and plough in as muck, and load for load they are equal to farm-yard dung. Mr. COKE thus manures from 20 to 30 acres annually from the lake at Holkham.

Mr. CROWE, of Lakenham, manures four acres annually for turnips with the weeds of a river that runs by his farm; the plants are chiefly the *Phelandrium aquaticum & Sium nodiflorum* (water-hemlock and water-parsnip). He lays 20 loads of 30 bushels per acre, and ploughs in directly: are as good on sand and mixed loam as the best dung, but not equal on stiff soils.

Mr. BLOOMFIELD, of Billingfold, has been in the habit of manuring his turnip-land with weeds fresh from the river, and ploughed in quickly; they have answered as well as yard-muck.

BURNT EARTH.

Mr. SALTER, of Winborough, whose fine farm offers many proofs of excellent management in every part, burns all the turf and rubbish which comes out of the numerous open drains he has made throughout his moory meadows, as well as the first spit of many hedge rows and borders: this he spreads on the grass, with earth from hills and rows. In his turnips, observing one part of a field finer than the rest, I found it had been manured with cottagers' ashes, who burn the parings of grass-land: these ashes never fail of giving great crops.

At Summer-green, in Dickleburgh, I observed a large heap

heap of burnt earth, the ant-hills, and inequalities in the common pared off.

STICKLEBACKS.

These little fish, which are caught in immense quantities in the Lynn rivers about once in seven years, have been bought as high as 8d. a bushel. The favourite way of using them now, is by mixing with mould and carrying on for turnips. Great quantities have been carried to Marham, Shouldham, and Beachamwell. Mr. FULLER there, is reported to have laid out 400l. for them in one year: they always answer exceedingly.

Mr. ROGERSON, of Narborough, has gone largely into this husbandry, laying out 300l. in one year, at from 6d. to 8d. a bushel, besides carriage from Lynn: he formed them into composts with mould, mixed well by turning over, and carried on for turnips: the success very great.

OIL-CAKE.

From 40 to 50 years ago this was a very common manure in West Norfolk: 35 years ago I registered the husbandry of manuring there with oil-cake; then chiefly spread for wheat. Mr. CARR, of Massingham, tried it largely: he laid out 140l. for one crop, in which he received very little benefit from the manuring: on another occasion his expenditure was for lintseed-cake, to fatten beasts, and the dung thus gained answered much better than buying rape cake. About the same period this manure was much used at Snettisham, at the expense of 3l. 10s. to 4l. 10s. a ton, which quantity did for three acres; the benefit there reckoned exceedingly great, but lasting only one crop.

At that period, from Holkham to Holt they spread one ton three-quarters to three acres; bringing it from Holland and Ireland, but they found the Dutch cakes best, from not being pressed so much. It lasted strongly only for one crop, wheat; but of some use to the following turnips.

In 1784 I found Mr. COKE in the regular practice of using this manure, at 5l. per ton. He found it more forcing to a crop of wheat than either dung or fold; but the turnips after the wheat not so good as after dung spread for that crop.

At present (1803) it is 8l. 10s. per ton, and he drills in with the turnip-seed a ton to six acres; and though when used in lumps, it may be better to sow it six weeks before the seed, yet, in his estimation, this is not the case when reduced thus to a powder.

In the district of Holkham rape-cake is very generally, perhaps I might say universally, used. They now spread a ton on three or four acres, usually sowing it (about eleven or twelve bushels to the acre) for wheat or turnips, or for both. Mr. OVERMAN has compared English, Dutch, and Irish cake; the latter he thinks the worst, and suspects from its breaking of a black grain, as well as from its want of that agreeable scent yielded by other cakes; that it has undergone in the manufacture some operation by fire. This manure is very effective.

The expense rising so high, induced Mr. COKE, some time ago, to recommend to Mr. COOK, the patentee of the drill-plough, to add to that machine an apparatus for sowing rape-cake-dust with turnip-seed; but his various engagements preventing the necessary attention, Mr. COKE

Coke has procured one from Mr. Burrel, of Thetford, which worked while I was at Holkham, to the great satisfaction of every one who saw it, delivering a constant stream of powder so regularly, that no doubts were entertained of the great success of the invention. It contains alternate divisions, with large and small cups for the delivery of both cake and seed into the same drills. In this way a ton does six acres, instead of three or four in the common method. In a letter I afterwards received from this very able cultivator, he informed me that the experiment answered to his satisfaction.

Mr. Hill, of Waterden, has much doubt of the benefit of this manure, and thinks that it is often used (the great expense of it considered) to loss. For the last three years it has decreased in goodness, by reason of the increased power of the mills, exertions caused as he thinks by the great demand. It should not be used in less quantity than two tons to five acres; and always for turnips in preference to wheat.

Mr. Johnson, of Thurning, finds that from one-third to half a ton per acre for turnips, will, in a wet season, beat every thing. It ought to be sown the first week in May, if possible; all ideas of putting it in at the same time with the seed, he thinks erroneous; he has seen it quite lost thus: in dry seasons it has no effect. Of this manure Mr. Johnson uses much. He thus expresses himself in a letter he favoured me with:

" Rape-cake is an excellent manure for turnips, and does not subject them to mildew; they will grow longer than from any other manure; the turnip that grows most after Michaelmas is always of the best quality. Malt combs quick to bring the turnip to the hoe; rape-cake slow; where both are used they should be sown together,

but

but not at the same time; the rape-cake should be broken to the size of walnuts, the less dust the better, and should be sown in April or May, as near the second ploughing as conveniently can be done, to have a shower on it; the malt-combs should be sowed on the last earth, and harrowed in with the seed."

Mr. ENGLAND, of Binham, uses much rape-cake, and this year his turnips, thus manured, are his best. The cake-dust should be scaled in, early in May.

Mr. REEVE, of Wighton, uses large quantities of rape-cake for his turnips, which in a wet season is an excellent manure. Mucked turnips come quicker at first than caked ones, but the latter exceed them afterwards: it is best applied three weeks or a month before sowing the seed, scaled in by the last earth but one: the deeper seed-earth then deposits the manure in the centre of the furrow.

Mr. HENRY BLYTHE, of Burnham, has this year turnips for which he spread rape-cake, at the expense of 3l. per acre, and the crop is not worth 20s.; but it answers in a wet season.

Mr. SYBLE, of South Walsham, feeds many bullocks with oil-cake, and finds that one load of the dung is worth two of any other: this he thinks by far the best, and even the cheapest way of getting a farm into condition, and laughs at the idea of buying rape-cake for manure, when compared with this superior practice. It is expensive to men who put lean beasts to cake, but if they are what is called fat before cake be given, it answers well.

Mr. BIRCHAM, at Hackford, has found that 10 load of dung from cattle fed on oil-cake, will do as well as 16 loads from turnip-fed beasts.

Mr. STYLEMAN, finishing his turnips before the grasses were ready for the sheep, gave oil-cake in troughs, with cut

cut hay, to the amount of just 6d. a week for cake, on a pea stubble; a pond in the field: it continued four or five weeks, and turnips succeeded, which were the best on his farm, and a very great crop.

ASHES.

Mr. SALTER, of Winborough, buys all the ashes he can get, of the poor people who burn *flag* parings, and the strength of them is very great on his heavy land; but on dry sand he remarks that they do little good.

Mr. STYLEMAN, of Snettisham, has manured sainfoin with coal-ashes; 40 bushels per acre, and with great success.

SOOT.

Mr. BIRCHAM, of Hackford, lays out from 70l. to 100l. a year in soot, sowing 20 bushels per acre on his wheat, in March or April.

MALT-DUST.

Mr. KERRICH, of Harleston, manures for turnips with 50 sacks of malt-dust per acre, at 1s. 6d. and gets finer crops than with yard-dung.

Mr. M. HILL uses this manure; 10 sacks per acre, at 3s. 6d. a sack.

BUCK-WHEAT.

Mr. BEVAN has been in the practice some years of ploughing in buck-wheat when in full blossom, as a manuring for cole: the success to his satisfaction.

Mr. SALTER has ploughed it in as manure, fixing a bush

bush to the beam of the plough, to brush it down for that purpose.

Upon a harsh stiff piece of marshland, at Warham, Sir J. TURNER sowed buck, and ploughed it in for wheat; it answered well as a manure, besides saving much tillage. Sir THOMAS BEEVOR tried the same husbandry on strong clay, and got five-quarters of wheat per acre.

YARD-DUNG.

Through every part of West Norfolk, from Brandon and Thetford, to Snettisham and Holkham, the farm-yard dung, after foddering is over, is turned up in heaps in the yard, or carted on to heaps in the fields, where it is turned over for mixing: Mr. DENTON, of Brandon, has made an observation on this point, which has a tendency to a change of system. It seems from the general practice, that the gentlemen and farmers, for all are in the same husbandry, do not conceive that the sun and wind have any power of extracting those volatile particles which contribute to the food of plants; and the common way of leaving the heaps when carted on to the land for some time before spreading, and again, when spread, before ploughing in, shew that this is the case. Here, however, I must make an exception of Mr. COKE, whose teams I saw so proportioned, that the dung was turned in as fast as carried out, and very completely buried; but it was *short*.

The observation alluded to is this—Mr. DENTON shewing me his beautifully improved warren farm at Feltwell, he remarked, how much better one half of a layer of seeds was than the other; occasioned by one part being manured with long dung, and the other with old turned-over short dung. The soil, quite a sand: I observed the different appearance

appearance clearly. The best was that which was covered with the long manure.

Thirty years ago they reckoned, near Holkham, that the wheat stubble ploughed in, was as good as a light coat of dung.

Mr. BRADFIELD, of Kerattishall, Suffolk, tenant to Mr. BEVAN, carries the yard-dung long from the yard, without any turning or mixing, and spreads it about six weeks before sowing; ploughs it in fleeter than for the sowing earth, but not so fleet as what is called scaling; ploughs two or three times after the manuring.

Mr. BLOOMFIELD's bailiff, in the absence of his master, gave his opinion in favour of rotten dung, rather than long and unrotted: however, this year, he says, they used the latter, and the turnips on bad land, justify an opinion different from his own.

The Rev. Mr. PRIEST, at Scarning, in 1801, on a field that had been unkind for turnips, drew it into baulks, and laid long fresh dung into the furrows, then split the baulks, covering the manure, and drilled the turnips on the tops of the ridges; the crop proved the largest and best in the neighbourhood. His man, not satisfied with this method, and thinking that he could get a better crop, Mr. PRIEST permitted him to try: he laid the land on broadlands, and sprained the seed into every other furrow, but in such a manner, that it came a broad-cast crop and good for little; *because his dung was not buried.*

Mr. PRIEST, complaining to Mr. BIRCHAM, of Reepham, that he had some land on which it was difficult to get turnips, had this answer—*Put on your dung in autumn, and your difficulties will vanish.*

Mr. SALTER, of Winborough, pointing at some dunghills, observed, that he had now got a year's muck beforehand; *over-year* muck, he thinks, far preferable to long fresh,

fresh, as the latter breeds insects, &c. and sometimes is hurtful. We soon after entered a very fine field of turnips, the crop beautiful; and long dung lying about some of the best parts of the field: his theory was here evidently condemned by the appearance of the plants.

Mr. HAVERS, at Thelton, in common with his neighbours, keeps his yard-dung in hills, and composts, called *over-year muck:* that is, kept over the year to have it old, and for use in succession, so managed.

Mr. DRAKE, of Billingford, in the same vicinity, does not approve of over-year muck for heavy land; but on light land, subject to burn short dung best: and he has observed, that when land has been over clayed, long dung helps it much.

Mr. PITTS, of Thorpe Abbots, carts clay marle on to heaps in the summer, to which he carries his yard-dung, turns over thrice, and spreads it for turnips, or wheat, or on young seeds, and he finds that on his burning gravels, it answers better than dung alone; though a second claying on the same land will do more harm than good.

The farmers about South Walsham, &c. mix dung and marle together. This Mr. SYBLE thinks a bad practise, as the marle will not *work* in the land, after it has *worked* in the dung.

Mr. BURTON, of Langley, does not approve of over-year muck: the best method, he thinks, is to spread earth over the farm-yard before foddering begins, to let it be late before it is turned up, to turn over the hill once, and in a month after to cart it for turnips. He never mucks for wheat: but very good to do it for winter tares, in order for having turnips immediately after; in this way he always gets good turnips.

Mr. THURTELL does not approve of over-year muck; he

he carts it on to heaps as soon as his turnips are sown, to rest without turning till wheat-sowing; as all his turnips are manured from Yarmouth. He has accidentally carted long *par* muck for turnips, to finish a field, and they were certainly a worse crop: for wheat it may be different.— Mr. THURTELL thinks the winter is the worst time of all for carrying out muck, whether from yards or composts; it should be either in summer for turnips, or in autumn for wheat. He does not wish any of his straw to be eaten; all trodden into muck.

Mr. EVERIT, of Caistor, never uses over-year muck: he carts from the yard, late in the spring, forming heaps; in three weeks turns over, and in a fortnight more carts and spreads for turnips: when he has fallen short in quantity, he has taken long and fresh dung, and has had as good turnips as after the short. Upon strong land, he has known long fresh dung answer very well: the chief objection to it is the difficulty of turning it in. He was much pleased at the idea of the skim coulter.

At Hemsby, Mr. FERRIER, &c. thinks short dung, from being carted to a hill, best; but no over-year muck. Mr. FERRIER gives fifteen loads per acre to turnips, and six to wheat.

Mr. BROWN, of Thrigby, has carted long stable-muck in March, without any stirring, for turnips, and had as good crops as from hilled short muck; but in such cases gives 15 loads per acre, instead of 12. If the same quantity, he thinks the rotten would prove the best. His objection to long muck is, the idea that seeds would be carried out which would not vegetate in time for the hoe to destroy them, such as docks and needles; and these, he imagines, are destroyed by the fermentation, when hilled, and for this purpose, the muck, by all good farmers, is

thrown

thrown light on to the hills, without carting on to them.—
He never turns them over. Mr. BROWN puts no value
on the dung made by straw-fed beasts. He has also tried
dung collected from commons, and found it of little or no
service. He remarked also, that the dung which in long
snows has been deposited by sheep under hedges, has
proved of very small use. Mr. BROWN, and other good
farmers in Fleg, are attentive, in carting out muck, &c.
to make the drivers keep on the head-land till they come
to the end of the land which is manuring, so as to make
each ridge bear its exact proportion of damage, if any:
for want of this attention, if the men are left to themselves,
they make roads across from the gate, in every direction,
to the great injury of the crop.

Mr. SYBLE, of South Walsham, thinks over-year dung a
bad system: he is in the common practice of the country,
but were he to farm a strong soil, he would carry out
long dung directly to the land: and on all soils it cannot
be too new, if it be in the right state.

Mr. FRANCIS, at Martham, no over-year muck, but
in manuring for wheat, some left was carried on for turnips, and there the crop not so good, though perhaps a
fuller plant. He has tried long muck, fresh from the yard,
and it does as well as any, but not quite so quick a growth
for the first six weeks. He has no objection to the practice, but the difficulty of burying it. He likes the idea of
the skim coulter. He carts on to heaps, and if the team
goes on, always turns the heap. He lays twelve loads per
acre for turnips, and likewise eight for wheat. This large
manuring, common in these hundreds, depends much on
the quantity of marsh and fen land, abounding in all this
country, and which commonly yields a great plenty of
rough coarse fodder and rushes, for thatch and litter.

At

At Ludham, 12 loads per acre of *par* dung common. Mr. HORNARD often lays on 20. He carts his muck to heaps, on moulds, not suffering the teams to go on to the hills. Some farmers turn it over. He never keeps it over-year. He lays some on for wheat, but for turnips in preference.

Mr. CUBIT, at Catfield, and others, carry out muck in frosty weather, on to mould heaps, and also when barley sowing is over. He has often carried out long fresh dung for turnips, and the effect has been very good, especially if the land has been at all strong or wet. The objections are, the difficulty of hoeing well, and the manuring being unequal, from some yards and parts of yards being better than others; whereas, in carting to hills, Mr. CUBIT takes from oxen and stables, alternately in order that the whole may be mixed and equal, when turned over.

Mr. CUBIT, at Honing, thinks that for land on a wet bottom, long dung is good for turnips or wheat; but he carts on to hills, to have it as short as possible; for on light land he has known it fail, for turnips, when long.

Throughout the hundreds of North and South Erpingham, the same management prevails: all cart out and hill, and in general turn over.

Mr. DYBLE, of Scotter, makes platforms of earth, then a layer of marle, and turns over, then adds muck, and turns again, whether for turnips or wheat. Has on many acres carted long fresh stable muck for turnips, ploughing it in at once, and gained fine crops if the season proved wet; but not in a dry time.

Mr. REPTON, at Oxnead, thinks long dung the best for turnips; however, he seldom uses it; but when he has, the turnips have generally been the best. Carts his dung on to heaps of marle, and turns over.

Mr. JOHNSON, of Thurning, thinks that muck wastes
by

by keeping to an unprofitable degree, and that the more it is turned over, the worse: he has tried long muck, fresh from the yard, for turnips, and got as good by it as by any other; and the barley also as good. The difficulty is to get it buried: he has employed boys to tuck it in: he approves much of the idea of the skim coulter.

Extract of a Letter from Mr. JOHNSON.—" Where lands are unkind for turnips, straw may be converted into muck with profit, by feeding the pigs with pease in the yards; and the muck kind for turnips; the quality of the muck depends on what the animals is fed with; muck made from turnip-fed beasts is better for grass or wheat than for turnips: if beasts have nothing but straw and turnips, it is not so kind for turnips as muck made in straw-yards from other food."

Mr. ENGLAND, of Binham, carts his yard-muck on to heaps in the winter, and turns up the rest in the yard, to get it rotten for the turnip-seed earth, and thinks it would lose its virtues if carried on long: on strong land it may do, carried on in winter, for turnips, and has done it on such a soil with good effect. He has no doubt of the superiority of rotten muck for turnips, but is against keeping it over-year.

Mr. REEVE, of Wighton, carts his yard-muck on to heaps, and turns them twice, to destroy the seeds of weeds, and the shorter the better, provided it be in a fermenting state: eight load of short are as good as twelve long; but over-year muck bad, as fermentation in that is over. He has tried long and fresh dung, but it has not answered so good a purpose. He lays on all his muck for turnips; none for wheat.

Mr. REEVE is clear that all straw should be trodden into muck, and none eaten. He has kept a large dairy of

cows,

cows, and thinks them the worst stock that can be on a farm, as turnips are drawn for them, instead of being fed on the land by sheep; and more straw is eaten by them, instead of being trodden, than by any other stock. His expression was, " *I would not have a mouthful eaten.*"

I have observed on many farms, the dung for turnips either not well turned in, or harrowed out again, and often recommended the use of the skim coulter. I should observe, however, that on Mr. COKE's farm, his dung was very well *tucked in*; whether it would have been the same had it been long dung, is a question.

Mr. M. HILL remarks, that long muck is best, if laid on in November, for turnips the following year; especially on wet cold land: but short and rotten for summer manuring on the land that has had three earths. The difference, in this case, little in the barley after the turnips, but much in the turnips themselves.

The Rev. DIXON HOSTE, at Goodwick, prefers short and rotten dung: dunghills for rotting seen in all that country.

The quantity per acre generally applied on the sand district, north of Swafham, is 10 cart-loads; and every man tills that quantity for a day's work. The price for filling and spreading, is 3d. a load, of large three-horse carts: three men spread; one to throw out, and two to break to pieces, and shake about equally.

Mr. OVERMAN desired me to remark a superiority of a part of a field of sainfoin: it was very visible. He could attribute it to nothing but that part of the field having been dunged twelve years before: the soil a sharp gravel, commonly thought to *devour* dung quicker than most other soils.

Mr. DURSGATE carts his muck on to heaps, and then turns over: he has tried it long and fresh for turnips, but
likes

likes short much better: though 100 loads carried out, becomes but 60 on the land. Mr. DURSGATE would not have a bullock on his farm, except for *treading* straw into muck: he would have none eaten.

Mr. STYLEMAN, of Snettisham, carts out his yard-muck on to platforms of marle, turns over, and lays it on for turnips. He thinks long muck might do well for strong land.

Mr. SAFFORY, of Downham, turns over the dung in the yard, and then carts it for turnips, ploughing in directly. He has seen very long fresh dung spread and ploughed in directly for turnips, and it has answered well on strong, but not on light land. Some cart out of yards, and mix with mould.

Mr. PORTER, of Watlington, turns over dunghills, to have the muck short for turnips, not liking long dung at all; it makes the land scald.

Mr. ROGERSON, of Narborough, carts earth into his yards previously to foddering, and when it is done, turns it over for turnips.

Mr. PRIEST, of Besthorpe, forms his yard-muck into a heap, and turns over: chuses to have it short: even in this way he has boys to tuck it in.

Mr. GODDISON, Steward to the Earl of CHOLMONDELEY, at Houghton, considers rotten dung as necessary for wheat on light soils; and he prefers top-dressing wheat crops to ploughing in at seed-time: he also top-folds as much as he can do by Christmas, harrowing both across in the spring: but for turnips he has a high opinion of long-muck; he carries it out of the yard without any stirring over, and ploughs it in for that crop; little harrowing up; nor does it impede the hoe: it answers greatly, and the barley after has been much better

ter than in any other way, insomuch, that if a fair comparative experiment were made, he would bet on long dung against short.

Mr. E. Scott, at Grimstone, has carted out long muck from Christmas to Lady-day, and ploughed it in for turnips, and had none better: nor did he see any difference in the barley, but believes that it does not last quite so long as short.

" Muck from the straw which is trodden only, is, by some, thought to be better than that from the straw which is eaten by lean stock." A capital farmer much in favour of fatting pigs loose in a littered yard: " What a rare parcel of muck they make, compared with what neat beasts would have made from the same straw!"—*Marshall.*

Observations.—Many of the preceding remarks are extremely interesting. The negative of so many able and intelligent men against suffering any straw to be eaten, which is the common practice of the larger part of the kingdom, deserves much attention: and the consequent practice they are in of buying oil-cake, often to loss, that their straw may be trodden into dung by fatting beasts, is a perfection of management not often met with.

In regard to the question of long and short dung, opinions are evidently much divided; and though, in the common method of the county, short dung is preferred, here is enough said upon the merit of that which is used in a long state, to prove that the inquiry deserves more attention than it has met with. Mr. Denton's experiment is remarkable, and the observations, founded on practice, of Messrs. Bradfield, Brown, Syble, Francis, Dyble, Repton, Johnson, and Scott, are all much to the purpose. Comparative experiments, very easy to make, would ascertain this point, which is certainly of considerable importance. A prevailing idea in Norfolk is, that long dung is best for strong land, and short for light soils: the general practice is that of spreading *short* in all cases. But Mr. DENTON's soil is sand*.

LITTERING.

There is a singular practice at Yarmouth, which has been common time out of mind, of littering all stock, such as horses, cows, &c. with sea-sand. A number of Yarmouth one-horse or one-ass carts, are employed to bring sand from the shore for this purpose, and it is done the more largely, that the quantity of muck to sell to the farmers may be the greater. Mr. THURTELL manures all his turnips with this dung, and it is excellent. The sand ought to be ten days or a fortnight under the horses and cows, being gradually drawn back with hoes, and fresh supplied: many thousand loads are thus made annually; and great quantities are taken into the country by the sailing barges called *keels*. Ten large cart-loads per acre are a good dressing, as much as three horses can draw. It sells at 4s. a waggon-load in the town, and six of these loads do an acre. Mr. THURTELL brings it all winter long. He observes, however, that it is not durable; the chief force of it is exhausted in the turnips and following barley.

Mr. EVERIT, of Caistor, manured a field with *par* yard-muck for turnips, but falling short two acres, he finished that part with Yarmouth sand-muck: the turnips

* Much information on this interesting question, is to be found in many passages in the *Annals of Agriculture*.

were equally good; but the barley on those two acres inferior to the rest of the field.

Mr. FERRIER's horses, at Hemsby, all littered with sand; and the manure very good.

LEAVES.

Sir THOMAS BEEVOR, more than 30 years ago, constantly swept and raked up all the leaves in his park to use for litter, at the expense of 6d. a load: the success, in adding to the farm-yard-dung, great.

BURNING STUBBLES.

I found many oat-stubbles in the new enclosure of Marshland Smeeth burning, ready to put in wheat or cole for seed: the crops had been immense in straw, and reaped, and the land quite black with the ashes; but many partially and badly done, not half burnt. Mr. JOHN THISTLETON, of Walpole, had burnt his completely: I saw the fire spread over several in an unbroken moving wall of flame, and must be to the utter destruction of many insects, and all grubs and slugs not buried in the earth. Where stubbles are stout it must be excellent husbandry, and will remind the reader of burning straw in Lincolnshire as a manure for turnips.—*(See my Lincoln Report.)*

Mr. PORTER, of Watlington, has burnt oat-stubbles for sowing wheat, with much success: is now threshing 10 coombs an acre of wheat thus gained. Harrows fine after the burning: dibbles if strong land; drills if light.

RIVER-MUD.

Mr. PALGRAVE, at Coltishal, has used moory-mud, from the bottom of the river, mixed with lime and marle, and

and spread upon the sandy uplands, and it produced a profusion of weeds, especially *persicaria*.

Mr. BIRCHAM has used the silt out of the brook at Reepham, for a moory meadow: it killed the rushes, and covered the land with white clover.

TOWN MANURE.

Mr. REEVES, of Heveringland, for five or six years kept one or two teams almost constantly at work bringing manure from Norwich, at the distance of eight miles, laying eight loads per acre: the expense heavy, but he thought it answered while the price was 4s. or 5s. for good stuff; but the price rose, and the manure became adulterated.

Mr. BECK, of Castle Riseing, for seven years kept a team constantly at work, bringing Lynn muck.

SECT. IV.—PARING AND BURNING.

MR. DRAKE, of Billingford, broke up a rough coarse pasture; the soil poor, wet and hungry, on brick-earth, worth scarcely any thing, from the kind and state of its herbage; by paring and burning, at the expense of 2l. 12s. 6d. per acre; he then ploughed it as shallow as possible, hardly more than an inch and half deep, and dibbled in oats, covering the seed with a very light harrow, bushed: the crop, which I viewed, very great indeed; it varied in parts of the field, but the produce must be eight or nine quarters per acre. He proposes to plough the furrow back in the spring a little deeper, and dibble oats again; then to work it well for barley, laying down and claying on the layer. I remonstrated against these crops, but he urged the necessity of the flag (as he calls it, though pared) rotting,

and

and the tillage for barley, mixing the ashes well with the soil: as it is very thin skinned land, he will lay 80 loads per acre. He has hollow-drained part, and intends the rest.

On another piece of the same soil he has got turnips for the first crop. His oats, however, are worth at least 6l. or 7l. per acre more than the turnips.

Mr. WYNEARLS, near Marham, on a common being enclosed, pared and burnt 200 acres for turnips and coleseed.

SECT. V.—EMBANKING.

THE tract of land in Norfolk, between the rivers Wyne and Ouze, called Marshland, is one of the richest districts in the kingdom. It spreads also into Lincolnshire, and forms altogether by far the largest salt-marsh we have. As the sea still retires from this coast, it is easy to perceive in what manner all this country has been the gift of that overwhelming element, which in other places encroaches so severely, and is, at high tides, restrained even here with so much difficulty.

The soil of the whole is the subsidence of a muddy water, with a considerable portion of what the waves, powerful in their agitation, wash from the bottom of the adjoining gulph, which forms the embouchure of two considerable rivers. It is a mixture of sea-sand and mud, which is of so argillaceous a quality, that the surface of it which covers the sand, gives it the common acceptation of a strong clay country. Is its extraordinary fertility at all owing to the marine acid, with which every particle is impregnated? That cause has every where on the coasts of every part of these islands, as well as other countries, some effect. If the sea leaves only a running sand, the saline particles are soon washed away or exhaled; the land may be barren, though never in the degree of vulgar conception. But when the sand is mixed with, or covered by a more retentive substance, such as an argillaceous or calcareous earth, then the particles, whether saline or mucilaginous, are retained, and the surface classes amongst, or rather is at the head of all, fertile soils.

I observed that the whole country has been a present from the ocean: this is obvious from numerous appearances; but those who wish to know its history particularly, should consult DUGDALE. I may remark, that there are ranges of banks at a distance from each other, which shew the progressive advances which industry has effected, eager to seize the tracts which so dreaded an enemy relinquishes. One of these banks is called the Roman, which naturally brings to our mind the vast exertions which that people made in agriculture, wherever their victorious eagles flew. The distance of this bank from the shore, if it really is Roman, and not a misnomer, is not so great as it would have been, had the sea in all ages been as liberal as it is in this. It probably varies considerably in this respect in different periods: at present it retires very rapidly, so that though Count BENTINCK's embankment has been finished but a few years, there will be, in twenty years, a thousand acres more ready to be taken in, belonging to Mr. BENTINCK, the present possessor.

The mud deposited by the sea, is at first, and for some years, bare of all vegetation: the first plant that appears is the marsh samphire; by degrees grasses rise, which, from their appearance at the time I viewed them (October), and eaten close down by cattle, seemed to be the common ones

of

of the improved salt-marsh, but not the *diadelphia* family, which come afterwards.

Long before it is raised enough by successive deposits of mud from high tides, it lets to the farmers of the contiguous improvement for 5s. per acre; some years since at 2s. 6d. Broken as it is by holes and little creeks of water, it lets, immediately after embanking, at from 20s. per acre; a few years ago to 40s.; and 42s. at present. I observed one or two pieces within Count BENTINCK's new bank, that were left in that rate for cattle, but in general they were under the plough, and the grass-fields laid down after a course of tillage.

The business of embanking to take in a new piece of marsh, is done sometimes at the expense of the farmers, who make the bank, to have the land rent-free for 21 years. Adjoining to the Bentinck improvement, is a piece of 80 acres thus taken, but the bank very ill made, at no greater expense than 40s. a rod. Those constructed by landlords, were deficient in not having slope enough given towards the water. Count BENTINCK laid out his upon a scale never practised here before; and his son, the present possessor, has far exceeded it. The former extends about four miles, and added to his old estate, 1000 acres. The base of the bank is about 50 feet. The slope to the sea, 36 feet, forming an angle, as I guess from my eye, of 25 or 30 degrees. The crown is four feet wide, and the slope to the fields, 17 feet, in an angle, I guess, of 50 degrees; the slope to the sea, very nicely turfed. The first expense of this bank was 4l. per rod, but a very high tide coming before it was finished, not only made several breaches, but occasioned an additional height and slope to be given to several parts, to bring it to the above dimensions, all which made the gross expense about 5l. a rod. The whole cost something above 5000l. The expense of the

the buildings, and other things, amounted to as much more, for five new farms, with houses, barns, and all necessary offices, were immediately raised; this was, however, going to a greater expense than necessary, for the land would have let as well in two or three farms, as it did in five. Calculating the expense at 10,000l. and the new rental at 1000l. a year, it is just ten per cent. for the capital. The expenses certainly ran too high; for the value of the marsh, at 2s. 6d. an acre before embanking, reduces it to less than nine per cent.; after which, there is still to be deducted, the almost periodical repairs, which remarkably high tides still occasion, and which may be averaged at once in ten years. So that when we consider it not as a purchase of a new estate, but an agricultural improvement of a waste, the profit is not equal to what might be made on other species of waste lands.

This is probably owing to the husbandry of these stiff wet soils being very ill understood, and managed in a manner that is reprehensible in almost every particular.

Instead of a system of miserable tillage, with weeds the chief signs of fertility, the plough ought to be introduced only as a preparation for the most perfect grass system that can be devised. These lands, when well laid down, will fatten the largest bullocks and sheep in England, which is the right employment of them; and in which application they would be better worth 30s. than in their present state 20s. Hence it should be an improving landlord's business to farm the marsh till he got it to a very fine grass, laid down himself, for I scarcely ever saw a tenant that would do that well. Ray-grass, and the weedy rubbish of a loft, which he calls hay-seeds, with, perhaps, some common clover, are what he has recourse to; and, under such management, the wonder is, that he ever gets a pasture worth even 20s. In all improvements, where

the

the previous steps are very expensive, like embanking a marsh, draining a bog, &c. it is essential to profit, that the land be advanced to the highest perfection possible, as those preparations to culture cost no more for a great than a small rental.

Count BENTINCK had one idea in the execution of his work, which had considerable merit; he planned a navigation from a quay to each of his farms, over the whole estate, by a large ditch capable of admitting long-boats, some of which he actually built ready for the business: by this means the farmers would be able to carry their corn, or bring manure from Lynn, if they chose to do it, without the least land-carriage; but his death, which was occasioned by too assiduous an attention to building the bank, living in a tent, in a bad season, and aguish situation, without the precautions of adapting his diet to those circumstances—prevented the execution.

One circumstance of folly in his neighbours, prevented the improvement from being so considerable as the Count had planned. At the further extremity, towards the Wisbeach river, there is a common belonging to the parish of Terrington, to which the sea, by retiring, makes additions similar to those by which individuals have profited.

A continuation of his bank, in nearly a right line to the Wisbeach river, would have taken in about 500 acres of that common. Mr. BENTINCK applied to the parish for their consent to do it, which would have been the means of shortening his bank. Though several individuals would have been glad of making use of so favourable an opportunity, the body refused their consent. They were even so preposterous in their opposition, that when he afterwards offered to be at the sole expense, provided they would give him a lease of 21 years of the land recovered, they still refused it. Upon which, he was obliged to follow the irregular

regular outline of his own property. The motive of the parish for refusing their consent to a proposal so advantageous to themselves, arose from this circumstance. It is of great extent; the proprietors adjoining the common, make, at present, nearly the whole advantage of it; but when embanked and let, those at a distance would come in for their share, a jealousy of which, occasioned the failure of the scheme.*

The spirit and unlimited attention, even to the loss of his life, with which Count BENTINCK planned and executed this great work, ought to render his memory dear to every lover of agriculture. His active mind had taken a strong and most useful turn towards that art; apparent, not only in this great and successful project, but in the original invention of an admirable machine for drawing up trees by the root, which executed that difficult work with expedition and cheapness.—*Minute, in* 1784.

New Embankment.—The men were paid 4s. 6d. a floor of 400 cubical feet, but they find wheeling planks, barrows, trussels, &c. &c. When formed, the front slope is sodded, for which they are paid 4s. a floor of 400 square feet, earning from 5s. 6d. to 7s. a day. And some small expense follows for beating it firmly down. The whole expense of bank, sluice, and all, 3300l. The quantity of land taken in, 273 acres of marsh, and 18 of bank. The previous value absolutely nothing; now, Mr. MAITLAND, steward to Governor BENTINCK, was at once offered 4l. an acre for four years; or 3l. an acre for six years. The former amounts to 4368l. in four years, or the whole expense, and 1000 guineas over. Some buildings, however, in this case, to be erected: the Governor

---

* This tract has been since embanked, and allotted by act of parliament, passed in 1790.

let

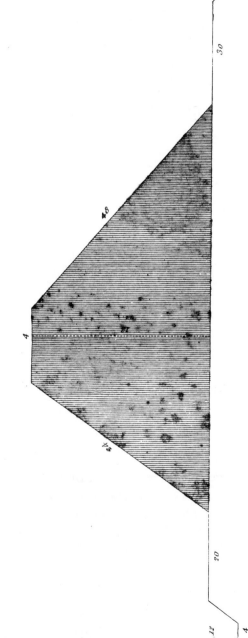

Count Bentinck's Embankment

let it to his old tenants at 40s. an acre, without any expense of building, a permanent rent, and under restrictions in cropping: confined to cole-seed and corn; of which, to take seven crops, laying down to grass with the seventh, to remain seven or fourteen years, and when broken up, to lay down an equal quantity of their old farms. This, upon the supposition of the grass failing, but if good grazing land, to remain unploughed. This system of cropping I must think much over-doing it, if rich grazing-land is the view. There is a great treasure in the land, and it should have no more tillage than necessary to prepare it for grass: if that grass does not turn out of a luxuriance and sweetness sufficient to carry a full stock of bullocks, then is the time for a tillage system; but I conceive that after seven crops of cole and corn it will not be found *fine* bullock-land. All ideas of the fertility being inexhaustible are idle and vain; you cannot make it bad land, but there is a great difference between common grazing pastures, like some in these marshes, and such as will carry an ox of 100 stone.

The Governor has the agreeable prospect of taking in 300 acres more in six or seven years. The space is fast covering itself with grass; and wants very little more than to have the cracks silted up*, which adds so considerably to the convenience and value of the land, that if this tract was taken in now, it would not let for more than 20s. an acre; but a few years hence it will be worth 40s. But to assist the operations of the tide, in thus silting up the creeks, much attention is necessary to accelerate the effect; and I found Mr. MAITLAND engaged in this operation:

* It is proper to observe, that the effect of the water on this coast, depositing its sediment, is here called *silting;* which upon the Humber and the Trent, is termed *warping;* they are the same operation of Nature. These waters are muddy to an extraordinary degree.

where the sea-water remains in a creek after the tide retires, the creek will not silt up nearly so fast as when channels are made for conveying it quickly away : by making stanks across the large creeks for giving a new direction to some of the water, and cutting channels from hole to hole for their drainage, this is effected. Mr. MAITLAND shewed me the surprizing difference in the silting up of creeks without water, compared with those where it remains. The floorings, from which the bank was raised four spits deep, are now, in only two years, nearly silted up in some places, and in others, not more than a foot deep : by these attentions, regularly put in execution, the tract preparing to be taken in, will be ready many years sooner than it otherwise would.

Governor BENTINCK has a very attentive and understanding agent in Mr. MAITLAND, as his plans sufficiently prove. These are, to build a small house close to the new bank, for a steward, at an angle, to command a view both ways; and where one or two cottagers should be always ready for executing the works necessary in assisting the silting of the tides, and any little reparations, or rather precautions, that the banks may want ; to make a road from the highway to the new bank ; this is done. To build a granary on the shore for the tenants to lodge their corn, for taking it by water to Lynn. And lastly, to plant the old banks, rendered useless, unless in the case of breaches, to supply the want of wood on this estate of fertile land.

In regard to the cultivation of this fine estate, I wish I could add, that it is worthy of the soil; but this is far from being the case. The tract taken in, in 1800, was ploughed directly, and sown that year, part with cole for seed, and part with wheat; and in the spring of 1801, part with oats. The cole-seed crops were great; and the wheat some

some of the best in the country. This year (1802) it is under cole, wheat, and oats again; the wheat and cole the best. It is let from 2l. 2s. to 2l. 12s. 6d. per acre: tithe, 4s. 6d. per acre; and rates, 2s. in the pound. The expense of the whole, 4000l.

CHAP. XII.

LIVE STOCK.

SECT. I.—CATTLE.

THE cattle predominant in Norfolk are Scotch, bought in every year from the drovers of North Britain. The quantity of these is very considerable indeed; as there is scarcely a farmer of any consideration in the county, that does not turnip-feed a lot proportioned to the size of his farm. The profit was formerly respectable, but it has been gradually lessening, by reason of the high price at which they are bought in. The breed is lessened in Scotland, and the drovers are more skilful than they used to be, in estimating what the demand will be from the state of the turnip crop, and they accordingly proportion their supply so exactly, that cheap lots are rarely procured. This circumstance has had some influence on the change that has taken place in the county respecting sheep. This stock has of late years been very greatly increased, to the diminution of bullocks, and has been effected by the superior profit derived from the South Down breed over that of Norfolks. The change has been of great importance to the farmers, insomuch, that they have by this means raised their profit, at the same time that their farms are in better heart: if they were in the habit (could it be effected on a large scale, of which I have little doubt) of treading their straw into dung by their fat sheep,

sheep, they would have still less occasion to buy bullocks.

Cattle in Norfolk of other sorts, do not offer much that is interesting: they have a breed of their own which possesses no qualities sufficient to make it an object of particular attention.

I saw upon Mr. MONEY's farm at Rainsham, a Norfolk horned cow, which is undoubtedly 35 years old; she has not had a calf for about ten years; she is old to the eye, but in good condition, and no marks of extreme age, except a stiffness in her motion, and a halting gait, as if her feet were sore.

I viewed a dairy at Mileham, Mr. CARRINGTON's, the only one left in that country of the true old Norfolk breed of cows: middle horned; some rather shorter, and tending to the Alderney horn; colour red, in some not much unlike the Devon; as loose and ill made as bad Suffolks. Mr. MARSHALL gives a much more favourable idea of those of East Norfolk.

"Small boned, short-legged, round barrelled, well loined, thin thighed, clean chapped; the head in general fine, and the horns clean, middle-sized, and bent upwards: the favourite colour, a blood red, with a white or a motsled face. The Herefordshire in miniature, except that the chine and the quarter of the Norfolk breed are more frequently deficient: no better flesh'd beasts are sent to Smithfield. This quality of flesh, and fatting freely at an early age, do away every solid objection to their size (40 stone) and form. One of the best farmers in the district apprehended that the best cross, that of the Highland Scot, would not fat so early: he is clear that a Scot does not fat kindly even at three years old, much less at two, at which age many hundred head of cattle are annually fatted in this county."—*Marshall.*

Mr.

Mr. DIXON HOSTE's are between the Suffolk and the polled Scotch; came originally from the Duke of GRAFTON's; they milk very well; 12 were in August, fatting two large calves, supplying the family with milk and cream, and giving 60lb. of butter per week.

About Attleborough, Hingham, and Watton, there are many dairies, but fewer cows than formerly: as the farmers have changed their system to grazing, there is some good fattening land all through that country.

Mr. COKE, at Holkham, has had many breeds; and he was once almost exclusively addicted to long horns. Some of FOWLER's stock are now at Holkham: he has many Devons.

Mr. PURDIS, of Eggmore, imported from Devonshire, in 1802, above 40 cows and heifers, and two bulls of the true North Devon breed, from Mr. PESTER. I viewed them with pleasure, and also 16 oxen of the same breed, which were ploughing for turnips He works four to a plough, in yokes and bows; they moved fully as fast as the horses at plough in the same field. This gentleman was before in the long-horned breed, of which he exhibited two cows at Holkham; but the part of his stock which most attracted notice was a Galloway heifer, of a most beautiful form; of a singular disposition to fatten; much admired by every one who examined her.

Mr. BIRCHAM, at Hackford, buys his bullocks in October, and puts them to turnips directly, thrown on the ollonds, but the beasts brought home at night to straw; continues thus to Lady Day; if turnips be done, he puts them to hay and oil-cake, or ground pease or barley; and he reckons that cake at 8l. or 9l. per ton, is equal to pease at 16s. per coomb: he has fed with cake when it was at 13l. per ton. His son at Reepham, on 300 acres, keeps 40 bullocks, and 100 to 150 sheep.

Mr.

Mr. Pitts, of Thorpe Abbots, is of opinion that cake at 10l. 10s. per ton, is a cheaper food than bean-meal at 14s. per coomb. He gives meal with turnips, but with cake only cut hay.

In Happing hundred, some farmers feed with cake, but the number not considerable: more do it in Fleg, but the practice by no means general.

Mr. Repton, at Oxnead, feeds with cake, and has used much linseed jelly: he has gone to 13l. 13s. per ton, for cake, but that price too high to answer, except in the manure alone.

Mr. Havers, at Thelton, has imported two bulls and several Devon cows and heifers: some are very beautiful, and must be esteemed a well-chosen selection: he has 26 in all: rears all the calves.

Mr. Francis, of Martham, thinks that his own home-breds do as well in grazing as Scots, but that it is different with such as are bought—it is uncertain. He turnips 40 bullocks a year; half in the yard, and half abroad; but the latter at home every night.

SECT. II.—SHEEP.

Norfolk and Suffolk have for ages been in possession of a breed of sheep, of which the farmers were (as they generally are, whatever the breed) extremely proud; thinking that no other sort would suit their country. Of this breed, the district of Bury, in Suffolk, possessed the best. They have often been described. I shall therefore only observe, that they are horned; bear clothing wool, the third in the kingdom for fineness: fleece, about 2lb. shape bad,

bad, loins narrow, back-bone high, chines thin, legs long, pelt good, disposition very wild and roving, not hardy, though formerly thought so: rate of stocking half a sheep per acre. Mutton 18lb. a quarter equal to any in the world in cold weather, and yields an uncommon quantity of high coloured gravy.

Mr. KENT thus characterizes them:—" When great tracts of heath-land were brought into cultivation, the Norfolk sheep gave great aid to the new improvement: hardy in their nature, and of an agile construction, so as to move over a great deal of space with little labour, folding became in high estimation. The turnip system enabled the farmer to improve his stock by better keeping, so that at this time they are become respectable and profitable in their return, and in as high estimation at Smithfield as any sheep whatever.

" I have no patience with the great farmer, in suffering himself to be lulled into so gross an error, as to prefer Lincolns and Leicesters; he will never be able to substitute any other sheep that will answer penning so well as the native sheep. It is a manifest incongruity, to cross a Norfolk with a South Down."

The following is Mr. MARSHALL's account:—" The Norfolk breed of sheep, taken all in all, appears to be singularly well adapted to the soil and system of management prevalent in this county. They may be bred, and will thrive upon heath and barren sheep-walks, where nine-tenths of the breeds in the kingdom would starve: they stand the fold perfectly well; fat freely at two years old; bear the drift remarkably well to Smithfield, or other distant markets; and the superior flavour of the Norfolk mutton is universally acknowledged; therefore the Norfolk husbandmen, in their sheep as well as in their cattle, have much to lose. The sheep appear to me, from a

knowledge

knowledge of different breeds, to be better adapted to the soil, situation, and system of management of the county at large, than any other breed at present existing in the island."

I proceed to the minutes I have taken in the county at different periods; in all such cases the opinion of strangers must fall before the experiments of the natives: *their* practice, *their* opinion, are what County Reports should contain.

BREEDS, CROSSES, AND VARIOUS CIRCUMSTANCES.

The South Down breed is getting rapidly in possession of all the country from Swafham to Holkham; but from Brandon to Swafham many Norfolks remained: I observed, however, some mixture even in that district.

In 1784, being at Holkham, I was informed and registered, that Mr. COKE " last April twelvemonth sold sixty Norfolk shearling wethers at Smithfield for 35s. each: he has killed them at two years old of 30lb. a quarter." At that time these were thought extraordinary circumstances.

1784, with one of the finest flocks in Norfolk, Mr. COKE is not so devoted to the black-face and leg as to hesitate at any experiment calculated to compare it with another breed. He purchased a number of Leicester ewes of Mr. WALKER's breed, to whom he put one of Mr. BAKEWELL's tups: he is well satisfied of the advantage attending this breed.

Mr. COKE's flock of 160 New Leicester ewes, produced in 1802, 100 lambs; his flock of 630 South Downs produced 830 lambs living in June. The same farm yields a most interesting comparison between Norfolks and South Downs: his former flock was 800 Norfolks, SELLING *all the produce:* he planted 700 acres, and now has 800 South Downs, KEEPING *all the produce.*

Mr. COKE's New Leicester hogs and theaves produced 8lb. of wool each in 1802, yet they had been hard kept on seeds fed very bare.

Of all the crosses of sheep Mr. COKE has tried, none strikes him so much as that of a New Leicester tup and a Norfolk ewe: the change is almost total, to a degree that is extraordinary indeed. I viewed the hoggits of this cross, and found them enveloped in about 7lb. of long wool; no horns; faces, some white; and the form surprizingly improved.—*(Note, some years back).*

In 1803 I found his opinion changed, from much experience; so that he prefers the cross of a South Down ram on a Norfolk ewe to that of a Leicester ram.

Mr. HOSTE has had the same cross, and they come to 32lb. a quarter, at two-shear. He put a Norfolk tup and a BAKEWELL tup at the same time to the same parcel of Norfolk ewes, and at St. Ive's fair sold the lambs fat at six or seven months old, and the BAKEWELLs brought just double the price of the Norfolks.

Mr. COKE, in April 1799, sending Norfolk, South Down, and New Leicester three-shear wethers to Smithfield, that had been fed together, the return:

	£.	s.	d.
Average per head, Norfolks	3	0	0
Leicesters	4	2	2
South Downs	3	7	2
Ditto, fleeces included, the others being in their coats	3	15	3

And in May following above 100 going, the South Downs beat the new Leicesters by 2s. a head.

Mr. MONEY HILL, at Waterden, with about 500 acres
less

less land than at present, kept 27 score breeding Norfolk ewes, and sold the produce of lambs: now he has 35 score South Down ewes, and keeps their produce, selling his wool at 5s. a tod more than the Norfolk.

In 1798, Mr. MONEY HILL sold a flock of Norfolks, reckoned a very fine one, and they brought 34l. 10s. a score, on the average. The next day he went to the South Downs, and bought 1000 ewes, at 31l. a score home: he culled 200 that were rather coarse-woolled behind, or not well made, which he sold for 400l. to a neighbour. In 1799 he lett one tup for 10l. 10s. In 1800 he lett five at 5l. 5s. and eight at 1l. 11s. 6d. In 1801 he lett twelve for 237l. and one to various persons, sending 60 ewes, at 10s. 6d. each, besides 40 of his own, being in all 50 guineas for one. In 1802 he lett ten for 254l. 3s. His mode of letting is by a table of the number and price at which they are put up at auction. In 1802 he sold 157 culled ewes for 368l.

Mr. HILL estimates the difference of stocking between Norfolks and South Downs, at one-third in favour of the latter, in number, in better condition, and of greater weight both in wool and carcass; all fairly attributable to the superiority of the breed, and free from any change of lessening cattle, &c. When his flock was of Norfolks, scarcely one in a score had a whole fleece; but now they are South Downs, scarcely one in a score is broken. His flock at Midsummer: 700 breeding ewes, 660 lambs, 45 rams. The wethers are grazed off in the spring, the last lot going before the ewes lamb; such as are short of shearlings go in their wool: culled ewes are sold in July, one, two, and three years old. Crones fed and killed in harvest. Mr. HILL thinks that South Down stock-sheep and hoggits are generally shorn three weeks too soon, when later there is more wool and better clipt: common time

time about the 20th of June; would be better the 10th of July.

In 1801 he gained the prize, a silver ladle, for the best South Down ram shewn at Swafham, given by the West Norfolk Agricultural Society.

Mr. BLYTHE, of Burnham, had, four years ago, a flock of between 5 and 600 Norfolks: he has now 1000 South Downs on the same land.

Mr. BLYTHE has double the wool from his land, stocked with South Downs, to what he clipped when under Norfolks.

March 27, 1799. Mr. OVERMAN took from turnips 24 two-year old Norfolk wethers, and 10 South Downs of the same age, having always lived together from the time they were lambed, and two hours afterwards weighed as follows:

	st.	lb.		st.	lb.	oz.
24 Norfolk, from the field,	264	7½	average	11	1	15
Do. after fasting 28 hours,	257	13		10	10	7
Difference,	-	-		0	5	8
10 S. Downs, from the field,	109	4	average	10	13	0
Do. after fasting 28 hours,	106	2		10	8	9
Difference,	-	-		0	4	7

One of each lot slaughtered:

NORFOLK.

	st.	lb.		£	s.	d.
Mutton, -	6	10	at 6d.	2	7	0
Tallow, -	1	2½	at 5d. -	0	6	10½
Head and pluck,	0	10½	-	0	0	9
Skin, -	0	9¾		0	1	0
Wool, -	0	3¾	at 17d.	0	5	4
				£3	0	11½

Blood,

		st.	lb.
Blood,	-	0	$6\frac{1}{4}$
Entrails,	-	0	11
Loss,	-	0	$0\frac{3}{4}$
Live weight,		10	$12\frac{1}{2}$

SOUTH DOWN.

		st.	lb.		£	s.	d.
Mutton,	-	6	$8\frac{1}{2}$ at 6d.		2	6	3
Tallow,	-	0	$13\frac{1}{4}$ at 5d.		0	5	$7\frac{1}{2}$
Head and pluck,		0	10	-	0	0	9
Skin,	-	0	10	-	0	1	0
Wool,	-	0	$7\frac{1}{2}$ at 18d.		0	11	3
					£.3	4	$10\frac{1}{2}$

Blood,	-	0	7
Entrails,	-	0	11
Loss,	-	0	$0\frac{1}{2}$
Live weight,		10	12

			£	s.	d.
Norfolk,	-	-	3	0	$11\frac{1}{2}$
Down superior by	-		£.0	3	11

These Norfolk sheeep losing 1lb. 1oz. more of their respective weight (taken full and empty), is a strong circumstance against them. The Downs are run much thicker on the land than the Norfolks.

Mr. OVERMAN's 600 South Down ewes have this year (1802), 645 lambs.

Mr. PURDIS, of Eggmore, in November 1801, on a farm of 1900 acres, had 70 score fatting sheep, 20 score South Down ewes and their lambs, and 15 score Leicester ewes and their lambs; in all very near 3000. About one

one and a half per acre on a corn farm! full double the ratio of black-face stocking.

South Downs, in the Holkham district, 18lb. a quarter, two-shear; and Norfolk about the same: Mr. COKE 20lb.

Fleece, South Down, three and a half on an average.

Norfolk, before, on the same land, one and three-quarters.

Mr. LONG, of Cranworth, bought in wether hoggits at 35s. clipped them twice, and sold all at 4l. without any refuse.

Seventeen years ago, being at Mr. BAKEWELL's, he mentioned to me the curious circumstance that Norfolk mutton would not keep so long as South Down, on the authority of the butcher of Eton College; I immediately desired my late brother, then Fellow of Eton, to apply to the butcher for particular information, and his own account was as follows:*

" The Norfolk mutton certainly will taint sooner than any in very hot weather; neither is there any sort (that I know) of a worse flavour at that time, though inferior to none in cool weather. Many very fine and fat Norfolks do not please on the table. The fat often runs away in roasting, if they are laid to a hot fire; and they rarely are so sweet as the South Downs. The latter are in hot weather, worth a halfpenny a pound more than the Norfolks.

" When both are *completely* fatted, it is hard to say (supposing the season cool) which, upon an average, is fattest: the flavour too, in such a season, I think is equal; and as to coarse meat, there is none in either sort. But if they are killed in cool weather, before they are very fat, the preference must be given to the Norfolks, because the

---

* I printed this paper in the sixth volume of the *Annals of Agriculture*, but as it is a very curious decision relative to Norfolk sheep, I insert it here.

meat will in that case eat better, and there is a probability of much more fat within.

"With respect to profit to the feeder, if they are fed entirely with grass, and upon good land, my opinion is decidedly in favour of South Downs; or if they eat turnips in the winter, and after that are kept two or three months upon grass in the spring, it is the same. But if they are half fat against winter, and are to be completed at turnips, I believe no sheep are more profitable than Norfolks, perhaps none so much so. But both sorts should be kept where there is both turnip and grass-land.

"JOHN VYSE, *butcher*,
Eton College."

Mr. BAKEWELL observed upon this account, that the Norfolk mutton not keeping, connects very much with the quantity, and perhaps the colour of the gravy. In all sorts of meat, that which is chosen for gravy, and which actually abounds with it most is the lean; and the freer from fat the better. That loose texture which is implied by the very circumstance of being full of gravy, is the cause of the meat tainting so soon, by the admission of air.

To this account there is one collateral circumstance to be named: the Wiltshire sheep have proved in various trials an unprofitable breed, as well as the Norfolks; but it is remarkable that for turnips, no sheep are said, by many practical and experienced husbandmen, to pay better, if so well. In Hertfordshire many who turnip-feed adhere to that breed, who admit the South Downs to be a superior sort for grass-feeding. I cannot but conceive that this whole comparative inquiry into the particular merits and demerits of the breeds of sheep, is yet in its infancy: certain important facts are gained; but when they are combined, and the causes to be assigned, we are still in great want of further observation and experiment.

Mr.

Mr. BAKEWELL, on the same occasion, gave as his opinion, that there is no comparison between the Norfolks and South Downs; that the latter are much better for any kind of food, for folding, or for any purpose, than the former, except the flavour of the mutton.

Mr. SAMUEL THORN, at Kimberly, buys wether lambs in August, at 12s. or 14s.; keeps them highly; winter-feeds on turnips; then on clover; turnips a second time; and sells, six weeks after Christmas, at 36s. each: 1791. This was mentioned as a profitable system with Norfolks; but if wool makes the profit 25s. they do not pay above 4d. a week.

Experiment by Mr. CROW.—" About Michaelmas I put ten Norfolk, ten Leicestershire, and ten South Down wether hoggit lambs to turnips, that they might learn to eat them readily, and let them remain together till the 6th of November, when the ten Leicestershire, the ten South Down, and nine of the Norfolk (one having died) were numbered, weighed, and put each sort by themselves, into three pieces of wheat stubble, of one acre each, separated by hurdles, and I provided at the same time three other pieces of the like size, and separated in the same manner, to shift them into. They were fed upon turnips, topped, and tailed, measured to them in bushel skeps, with great exactness, from that time till the 14th March, and then weighed again.

LEICESTERSHIRE.

Weight of each sheep, 6th November.				Weight of each sheep, 14th March.			
No.	Stone.	Lbs.	Ozs.	No.	Stone.	Lbs.	Ozs.
1.	4	13	4	1.	7	0	8
2.	5	4	4	2.	6	9	0
3.	5	5	0	3.	7	7	8
4.	5	7	8	4.	6	12	8
5.	5	8	0	5.	6	12	0
6.	5	9	0	6.	7	6	0
7.	5	10	4	7.	7	2	8
8.	6	0	0	8.	7	9	0
9.	6	2	8	9.	7	8	8
10.	6	4	0	10.	8	1	0

Average weight 6th of November, about 5st. 9lb. 3oz. each.

Average weight, 14th of March, about 7st. 4lb. 1 oz. each.

Average increase of weight in eighteen weeks two days, about 1st. 8lb. 14oz. each.

SOUTH DOWN.

Weight of each sheep, 6th of November.				Weight of each sheep, 14th of March.			
No.	Stone.	Lbs.	Ozs.	No.	Stone.	Lbs.	Ozs.
1.	4	13	0	1.	6	13	12
2.	5	0	0	2.	6	11	8
3.	5	1	8	3.	6	13	8
4.	5	2	12	4.	7	2	0
5.	5	4	12	5.	6	9	4
6.	5	5	12	6.	6	12	0
7.	5	7	0	7.	7	2	0
8.	5	7	0	8.	6	10	8
9.	5	7	4	9.	7	3	8
10.	5	8	8	10.	7	0	8

Average weight, 6th of November, about 5st. 4lb. 5½ oz.

Average weight, 14th of March, about 6st. 12lb. 7 oz. each.

Average increase of weight, in eighteen weeks two days, about 1st. 8lb. 1½ oz. each.

NORFOLK.

NORFOLK.

Weight of each sheep, 6th of November.				Weight of each sheep, 14th of March.			
No.	Stone.	Lbs.	Ozs.	No.	Stone.	Lbs.	Ozs.
1.	6	4	8	1.	8	2	0
2.	6	6	0	2.	7	10	8
3.	6	6	4	3.	7	4	0
4.	6	6	8	4.	7	5	8
5.	6	7	0	5.	7	11	8
6.	6	8	12	6.	7	9	0
7.	6	11	8	7.	8	2	0
8.	6	12	8	8.	8	0	0
9.	7	2	0	9.	8	9	8

Average weight, 6th of November, about 6st. 8lb. 9oz. each.

Average weight, 14th of March, about 7st. 12lb. 4oz. each.

Average increase of weight, in eighteen weeks two days, about 1st. 3lb. 11oz. each.

Quantity ate in eighteen weeks two days:

The ten Leicestershire ate 588 bushels of turnips.

The ten South Down ate 589 bushels of turnips.

The nine Norfolk ate 607 bushels of turnips.

The offal turnips were, at different times, collected, measured, and deducted from the account of the quantities given to them.

If nine Norfolk consumed 607 bushels, ten would, in the same proportion, have consumed above 674 bushels, or 85 bushels more than the South Down, and 86 bushels more than the Leicesters.

The Norfolk and South Down are about the same age; Leicestershire about six weeks younger.

I should

I should have observed, they were both times weighed, after having stood some time in pens to empty themselves.

The Leicestershire and Norfolk were bred by Mr. COKE, of Holkham; the South Down by Mr. ELLMAN, of Glynd; and as they were all picked out for the purpose, I doubt not but you will allow they are prime stock of their kinds. The Norfolk were chosen out of about 700 lambs; the Leicestershire and South Downs, out of about 100 each.

South Downs are coming in about Watton: it is no sheep country; but on the commons Mr. ROBINSON shewed me his own little flock, bought of Mr. BRADFIELD, of Knattishall, who parted with them because he thought they would not do for ling walks.

A Gentleman remarked on the Norfolk breed, that their pelts were more valuable than any other, being worth 4l. 4s. a dozen (1791) to the London butchers, owing to the singular quality of being separable into three flakes, or skins.

Mr. WRIGHT, of Stanhow, an excellent farmer, and very attentive to his flock, kept on 800 acres of very good land, four hundred breeding Norfolk ewes, 80 or 100 of which went to his marsh (exclusive of 800) in the spring, and stayed till after Michaelmas; three hundred ewes in summer; one hundred hoggits ditto, on the marsh in winter. Little more than one-half a sheep per acre. Never sold wether lambs higher than 16s.—*(Note, some years past)*.

Mr. SALTER, of Winborough, keeps only Norfolks; whatever success may attend other breeds on dry land, he is sure they would not do with him: they would not travel through the mud of his gateways.

In 1792, I found a South Down flock, of 30 score, on Mr. BEVAN's farm, and having a flock of Norfolks on an adjoining farm at Knattishall, he had an opportunity of comparing the wool exactly: 34 score of Norfolks produced 43 tod at 28 lb.; and 34 score of South Downs produced 61 tod; which 61, kept till November, became 64, but the summer very wet.

South Down	-	1708 lb.
Norfolk	-	1204 lb.
Superiority, just ¾ lb. each	-	504 lb.

In 1791, the shepherd would not let his own Norfolk ewes take the South Down ram; but in 1792, he was ready enough. He said, they would eat *harder* than the Norfolks; and would eat what the Norfolks would not: that they are more quiet and obedient than the Norfolks; so that he has done with them what he could not do with the Norfolks; fold them almost to an inch without hurdles.

A neighbouring farmer bought three rams of Mr. BEVAN, at 5l. 5s. each; but afterwards repenting, because they would *stain his flock*, Mr. BEVAN offered him 6d. a head, for all their lambs, more than he sold his Norfolks for, in the same flock, at Ipswich fair. The offer was accepted; the price proved 6s. 3d. for the ewe lambs, and 9s. for the wethers. Mr. BEVAN re-sold the ewes for 9s. and the wethers for 10s. 6d. or 2s. 1½d. a head in favour of the half-breds.

When his sheep were Norfolks, he kept 500; but in 1794, he had 960 South Downs.

BREEDS, CROSSES, &c.

Produce of 116 ewe lambs, bred by Mr. BEVAN, at Riddlesworth, 1792:

		£.	s.	d.
Wool, 12 tod 16lb.	-	26	8	0
48 Lambs, sold for	-	32	2	0
5 Ram lambs, ditto	-	8	8	0
6 Refuse ditto	-	2	10	0
7 Refuse shearlings, ditto	-	5	10	0
10 Good ditto	-	10	10	0
1 Ditto	-	1	0	0
87 Ditto	- - -	91	7	0

		£.	s.	d.
105		£.177	15	0
9 Died	Cost	57	14	0
2 Dunt				
116	Actual profit £.120		1	0

Mr. BEVAN was early in trying South Down sheep, but finding them tender at lambing, went into a new Leicester cross: these he abandoned, and got back to South Downs, but still esteems them a tender breed, and that they ought to have yards sheltered and littered for lambing in bad weather; remarking, that all the farmers he knows on the South Downs, have these yards for that purpose.

Mr. DRAKE, of Billingford, buys ewe-lambs in August, selling them that time twelvemonth, folding the whole year; a great improvement on his hot gravelly lands: wool included, he more than doubles his money.

Mr. HART, of Billingfold, all South Downs, except a few Leicesters. He and Mr. BLOOMFIELD were gone, when I called, to Mr. SCRACE's sale in Sussex.

Loddon and Clavering hundreds are no sheep countries; there are a few, and South Downs and Leicesters have been creeping in.

In Fleg one scarcely sees any sheep: I was told that Mr. CHRISTMAS, at Billocby, has some Norfolk ewes, but the number inconsiderable. Bullocks the general stock.

This

This year (1802), there were but two pens of Norfolk lambs at a fair in this neighbourhood; not many years since there were no others: at present chiefly *half-bred*.

Mr. PETRE, of Westwick, keeps 17 score in a country where flocks are not common; he has some South Downs but more Norfolks.

Mr. REPTON, at Oxnead, keeps fifty breeding ewes of the New Leicester kind, which answer greatly: Norfolks are so mischievous to fences, that he does not like to have any thing to do with them.

Mr. REEVES, of Heveringland, buys in lambs, and sells them shearlings from the fold. Tried half-bred South Downs from Mr. DURSGATE, which paid him better than Norfolks, and he likes them so much better, as to intend continuing to buy this sort in preference.

Mr. BIRCHAM, at Hackford, declares against having any favourites; he has generally bought Norfolks, and half-bred lambs; some few South Downs, but they did not answer: has had some Leicesters: any sort he can get worth his money. Little farmers who keep a few sheep, find the polled breeds very convenient from their quietness, and therefore prefer them. Norfolk lambs bred near Cromer, were bought by Mr. G. JONES at 14s. were run on stubbles in the autumn, and put to turnips at Christmas, then to layers of the first year, probably as the best food for sheep, and sold shearling wethers at Michaelmas at 55s. each; 20 to 24lb. a quarter.

Colonel BULLER, at Haydon, is convinced that Norfolks answer better than South Downs: shearlings come sometimes to 20 and 25lb. a quarter, and have had $19\frac{1}{2}$lb. of tallow; he has a breeding flock of 400: sold his wether lambs at 26s. and his ewe lambs at 24s.

Mr. JOHNSON, of Thurning, has 40 score of South Downs, which he has been rearing these six years, having bought many ewes and got good tups. He has, however,

ever, a good opinion of Norfolks, and will not be surprized to see them come into fashion again. In May 1792, he sold two shear Norfolks at Smithfield, for 3l. each. He admits their rambling disposition, which is much against them; and he is clear that he cannot keep so many on his farm as of South Downs. The South Down wool is not, on *good keep*, so good as Norfolk wool, but the fleece is heavier. Five years ago he got a lot of Yorkshires from the Wolds, white faces, polled, and the wool very coarse, but they throve wonderfully; never having had any sheep that did better, insomuch, that he was sorry when he parted with them. Norfolks, he thinks, will bear folding better than South Downs. The latter will, however, come to hand rather sooner, but not on ling: has had three shear South Downs of 28lb. a quarter. *Mr. Johnson, shall you go back to Norfolks?*—" Certainly not, for my lands lie wide." I like this discriminating attention; it is a sign of accurate observation, and the balance is enough in favour of South Downs.

Mr. ENGLAND, of Binham, got South Downs last year, and approves what he has seen of them: he thinks they may be run thicker on the land; and if as much profit is gained from five as from four, it is a better system.

Mr. REEVE, of Wighton, keeps only Leicesters: while he was in the Norfolk breed, his flock was 18 score breeding ewes: he has now 10 score Leicester ewes, and all their produce, amounting in the whole to never less than from 28 to 30 score, lambs included; but the account taken at any time of the year, the amount is 10 score more in number. He has had this breed six years, and prefers them to South Downs. Mr. REEVE's account is, however, candid, for he admits that they are apt to go barren; and the highest bred, the most so; clips 6lb. on an average, ewe and hog: his Norfolk ewes 12 to 14 to a tod. He letts from 35 to 40 tups annually, at from five to ten guineas, and a few from fifteen to twenty.

Mr. H. BLYTHE, of Burnham, has been for three years entirely in the South Downs; clips 1000; a stock greater than ever he kept of Norfolks, the comparison fairly made: his flock averaged $3\frac{1}{2}$ lb. of wool this year; his Norfolks never exceeding $2\frac{1}{2}$ lb.

Mr. DURSGATE has had South Downs six years, and is clear that, free from all change in husbandry, or other circumstance that would unfairly affect the comparison, the number kept, compared with Norfolks, has been as five to four. The carcasses as heavy as the Norfolks; more wool, and at a better price. He does not fold; but the South Downs would bear it better than the Norfolks. At Palsgrave, he folds the South Downs, because there is a sheep-walk—a Norfolk flock changing gradually to South Downs.

Mr. RISHTON, at Thornham, South Downs, and approved very greatly of the breed; from 250 acres of land, sold off a thousand pounds worth on quitting the farm.

Mr. DODMAN, at Thornham, South Downs.

Mr. STYLEMAN, at Snettisham, keeps 2000 of various breeds, South Downs, New Leicesters, and half and half; in number considerably more than when, on the same land, he kept Norfolks: his farm may, and probably does, produce more sheep-food than it did at that time; but he is perfectly clear in the great superiority of the number, this circumstance deducted, and that the profit is considerably greater. Clear also in the superior hardiness and kindliness of feeding of the new breeds. Of all cross breeds, he thinks the first cross of the Leicester tup on the Norfolk ewe the best, and that wool now (1802) sells at 46s. a tod; fleeces 4 lb.

Mr. GODDISON folds Lord CHOLMONDELEY's flock

of

of Norfolk and South Down, all the year, except while in turnips; and Mr. BECK, at Massingham, who has 35 score South Downs, folds them as regularly as any Norfolks.

At Hillingdon, all either Norfolks or half-breds, a Leicester tup on a Norfolk ewe. Captain BEACHER thinks there are no sheep in the island which the Leicester will not improve. He has grazed many Wiltshires, and thinks them the best of all for cole-grazing in the fens.

Mr. BECK, of Castle Riseing, has had South Downs 13 years, beginning with some from Mr. TYRRELS, of Lamport, and has imported three or four times since. He has now 800, and is quite convinced of their superiority to Norfolks: when he was in that breed he had not half the number; but after abating fully for improved husbandry, and every other circumstance, he is clear that there is a superiority of four to three. His fences are and must be bad, and in such a farm quietness is a vast object: his farm 486 acres. He gained the first prize for ewes both the last and this year at Swafham, and also at Holkham. I examined his flock attentively, and it certainly is a very beautiful one. His wool now averages eight to a tod, equally of hogs and ewes: his Norfolks todded twelve: he is clear that, take the country through, they average half as much again as Norfolks. Before he took the farm there were 50 sheep on it, and a dairy of cows. What an improvement!

In the vicinity of Downham are found all sorts of breeds: towards the river, Lincolns and Leicesters; higher up, Norfolks and South Downs. Mr. SAFFORY likes the South Downs best, but thinks that if as much care and attention had been exerted to improve the breed of Norfolks as the South Downs have experienced, they would by this time have been a very different sheep. Norfolk three-shear wethers sold, in April last, at St. Ives, at 4l. 4s. to 4l. 10s. each.

Mr. PORTER, of Watlington, keeps Leicesters, which he obtained from Mr. FASSET, Mr. CREASY, of Downham, and Mr. WILCOX, of West Walton, near Wisbeach. Letts tups himself, from 7l. 7s. to 10l. 10s.: this year to the amount of 330l.

Mr. MARTIN, of Tottenhill, keeps half-breds: Leicester tups on Norfolk ewes: sold lambs in 1801, at 27l. a score; this year his shepherd sold at 23l. both ewes and wethers. His flock is subject to the rickets. He has much black sand; but unwilling to attribute the malady to soil, as this year he had not 10; but last year 120; and all circumstances of land and food the same. It attacks the lambs at six or eight weeks old.

Mr. ROGERSON, of Narborough, keeps 700 Norfolk ewes on 1200 acres, which he covers with Leicester tups.

Mr. TWIST, of Bretenham, keeps 60 score of breeding Norfolk ewes on 1800 acres of poor land. He had a South Down tup some years ago, from Mr. CROW, but he could not perceive that the breed did better than Norfolks, though they stood the fold to the full as well.

Marshland.—Mr. DENNIS, of Wigenhall, St. Mary, grazes only the best Lincoln wethers: he buys from May Day to Midsummer; keeps them over-year, clipping twice, average price 50s. to 60s. and sells at 65s. to 75s. getting 18lb. in the two fleeces: his good land will carry six per acre, on an average, in summer; in winter, two on three acres, and these will quite preserve their flesh: if the season be favourable, will get something. He thinks that there is no other breed so profitable here; even a stain of the new Leicester is hurtful, as they will not stand the winter so well.

well. Sheep the chief stock, though some Lincoln bullocks. He never gives hay to sheep; nothing but grass; 32 lb. a quarter, his average of fat wethers.

Mr. SWAYNE, of Walpole, prefers the cross between Lincoln and Leicester: he buys them shearling wethers, about Lady-day; last year 3l. to 3l. 10s. each, but has had them at 36s. and 38s. He clips the best twice; three to a tod, which he likes better than heavier fleeces of sheep demanding more food. Some give 17 or 18 lb. of wool. At Michaelmas he culls the worst, or buys cole for them, if reasonable: sells all by Midsummer, making 8s. or 10s. a head, when bought in high, besides the wool. Very few beasts.

I have heard it made a question, who first introduced South Down sheep in Norfolk? When once an improvement has spread so much as to become an object of importance, there are generally many claimants for the merit; and if such claimants are only heard of many years after, but little attention is due to them. With regard to the neighbouring county of Suffolk, I can speak with some accuracy, but should not mention it on this occasion, were not the fact connected with the introduction into Norfolk. In May, 1785, I published an account of an observation* I made in 1784, the year I brought them into Suffolk from Sussex; and being printed at the time, the fact will admit no doubt. I recommended them strong-

---

* It was this: a South Down ram I had got from Sussex, broke, by accident, to a little flock of Norfolk ewes, belonging to a tenant, the effect of which was, his having seven or eight lambs entirely different from all the rest. His lambs were drawn fat by the butcher early in the summer, who, when he came first to make choice, drew every one of South Down breed before he took a single Norfolk, declaring, at the same time, that they were by much the fattest in the flock. The farmer applied to me immediately to save him a ram lamb.

ly

ly to every gentleman and farmer I conversed with on the subject, and, at my persuasion (as many well know), the late Mr. MACRO, of Barrow, purchased that flock which the Earl of ORFORD, after his death, bought and established at Houghton. Mr. MACRO died in 1789.

In a paper printed in the *Annals*, in 1790, I remark: " I have had six and twenty years experience of Norfolk sheep, and once thought so well of them as to carry them into Hertfordshire; but in the advance of my practice, I began gradually to doubt the superior merit of that breed. I thought, that of all the sheep which I had examined particularly, none promised to answer so well for the general purpose of the counties of Norfolk and Suffolk as the South Downs. I began the import in 1784, and in 1790 had 350. I had too much friendship for the late Mr. MACRO, to advise him to try any experiment that I was not clear would answer to him. I repeatedly urged him to try the South Downs; he listened to me with attention for some time, but would not determine, till, having seen the number I kept proportionably to the quantity of land*, and at the same time with some Norfolks, it proved to him that the South Downs were worth attending to; and the journey I persuaded him to take into Sussex, giving him an opportunity to converse with various noted sheep-masters there, he determined to make the experiment: he went over, previous to Lewes fair, and bought a flock of them. The lambs sold well at Ipswich fair. Mr. LE BLANC, at Cavenham, also turned South Down rams to 700 Norfolk ewes: he found no difficulty at Ipswich; and his shepherd, after three years obstinate preference to Norfolks, gave up his old friends, and actually

---

* On 240 acres, 353 sheep, 45 head of cattle, 10 horses, and every year 70 acres of corn. 1790. *Annals*, vol. xv.

set South Downs for his shepherd's stock. Whether the breed should or should not, in the long run, establish itself, I have the satisfaction of feeling that I have done no ill office to my brother farmers by introducing it. From the daily accounts I receive, I have good reason to believe that it will be established."

I may be pardoned, perhaps, if I here remark, that Mr. LE BLANC, of Cavenham, and Mr. MACRO, of Barrow, being at Ipswich fair, and having felt, as well as other farmers, the great advantage that promised to result to the county from the introduction of South Downs, proposed a meeting of the farmers at the White Horse, to present me with a piece of plate, for doing what they thought a public good. A person came to me to tell me what they were about, and I went immediately and requested them to drop the design, which I effected with difficulty; at last they postponed it, as I urged that the result was too little known, and their experience too short to form a satisfactory idea. A farmer afterwards told me that he heard of the intention, and that had it been brought forward he would have voted me an enemy to Suffolk, for endeavouring to change the best breed in England for a race of rats.

Mr. CROW, of Lakenham, informed me, in 1800, that five and twenty years before, he had a few South Downs at Ash Wicken, which he mixed with Norfolk blood, and has since mixed with new Leicester, and he thinks that between the three, he has a breed better than either of them, pure.

FOOD.

FOOD.—WOOL.

Mr. JOHNSON, of Thurning, remarks, that ling keeps sheep very healthy: he has a farm at Holt, where it abounds on the heath; and he loses four sheep at Thurning to one at Holt, which he attributes to the ling.

Captain BEACHER having 700 fatting sheep, and turnips running short, put 200 of them to oats (not ground); he found that the practice would not answer if oats were more than 6s. per coomb, and then not for longer than six weeks: they were fed on a pasture, and the improvement of it very great. He thinks grey pease, or beans would have answered much better.

Mr. MONEY HILL, from the observations he has made, is of opinion that the fleece being very fine, cannot be regarded as a sign of a thriving disposition.

Mr. HILL's prices:

| 1799 | 48s. a tod. | 1801 | 48s. |
| 1800 | 52s. 6d. | 1802 | 51s. 6d. |

Arrangement of Mr. BEVAN's flock of 45 score South Downs: the tups are put to the ewes about the tenth of September, for two months, being fed on the layers and pastures, and are folded on the old layers for wheat: after wheat-sowing they are folded on the pastures and layers till the time of yeaning, during which they lie on the pastures without fold, and have turnips thrown to them, with plenty of good hay. The fattening sheep are on turnips and hay from Michaelmas to the end of March, followed by the hoggits. In April the couples go to cole-seed in hurdles; from cole to rye, from rye to the new layers, if forward enough, otherwise to the water-meadows, till the beginning of May, and from thence to the new layers, being still in hurdles, with a good deal of of room to fall back, and continue so on the layers till about the 10th of June, when the ewes are washed for clipping, and until the lambs are weaned: the ewes then go to fold with the shearlings on the fallows intended for turnips, and the lambs are put to fresh grass reserved for that purpose: all the sheep on turnips and cole having hay, they consume about 25 tons. The general winter provision is 80 acres of turnips, 20 of cole, and 30 of rye, for the spring *: the latter, after feeding, stands for a crop. He values his turnips on the average at 30s. per acre, and cole at 25s. After turnip-sowing the flock is folded on old layers for rye, till the end of August, when the ewes intended for breeding are put to good pasture till the tups are let in.

1802.—The tups now put to the ewes about a week later, and the lambs not weaned till the latter end of June. Provision this year, 100 acres of turnips, 30 cole, 30 rye, for 25 score breeding ewes, 15 score hoggits, 20 tups, 10 score fatting stock; 51 score in all.

Sale of Cull-lambs, Ewes, and Wethers.

	£.	s.	d.
1796, Sold at Kenninghall Fair, July 18	0	16	0
1797,	0	13	9
1798, In September,	0	18	0
1799, In September,	0	16	0
1800, In July,	0	16	6
1801, In September,	1	0	0
1802, In July, many twins,	0	14	6

* Mr. BEVAN ploughs in his rye-stubbles before the shocks are carried to turn in the scattered seed, harrowing in half a peck of cole-seed for sheep-feed in the spring, and finds it of very great service.

PRICE OF WOOL, PER TOD OF 28lb.

		£.	s.	d.
1794, South Downs at	- - - -	1	15	0
1795,	- - - - - - - -	2	2	0
1796,	- - - - - - - -	2	5	0
1797,	- - - - - - - -	1	18	0
1798,	- - - - - - - -	2	0	0
1799,	- - - - - - - -	2	10	0
1800, 1801, }	- - - - - - -	2	8	0
1802, Unsold,	- - - - -	0	0	0

The shearers at Holkham, clip in a day about 23 shearling wethers, or 20 larger sheep.

Mr. CROW, at Lakenham, 643 sheep and lambs, on 32 acres of very good turnips, and 91 acres of grass and 60 acres of stubbles for the winter.

Mr. PURDIS, of Eggmore, has two shepherds' boarded houses on wheels; they contain a bed, and a stove for heating milk, so well contrived that it is heated in ten minutes: he has found the advantage in lambing time so great, that he has no doubt of having saved a great number of their lives, and recommends it strongly to his brother farmers.

Mr. HILL, at Waterden, has a shepherd's house on wheels, for lambing time, and hinted that it was first used at Waterden.

Mr. COKE readily assists not only his tenants, but other neighbouring farmers, in sorting and selecting their South Down ewes, &c. and distributing them in lots to the rams, according to the shapes and qualities of each. He puts on his shepherd's smock, and superintends the pens, to the sure improvement of the flock : his judgment is superior and admitted. I have seen him and the late Duke of BEDford thus accoutred, work all day, and not quit the business till the darkness forced them to dinner.

FOLD.

I found in 1792, at Mr. BEVAN's, what I had often recommended to the public; a yard well fenced in for a standing fold, in sight of the shepherd's windows, for littering and folding in bad weather. 1802, he continues the practice, and is well persuaded of the great advantage: he thinks it is indispensable, and means in future to have his flock in for yeaning, whether the season be good or bad; and has always 15 or 20 load of hay stacked in it for them to help themselves: he finds this not attended with any waste.

Near Brandon there is a practice introduced about ten years ago, said to be from Kent, which is, to fold their flocks for five or six hours in the middle of the day in hot weather.

In laying out the enclosures of the farm of Waterden, from 15 to 50 acres each, much attention was paid in the arrangement to have every field of the farm to open into a lane, that leads through the whole, so that by dividing the flock, for stocking, according to varying circumstances, Mr. HILL can keep at least one-fourth more than when all the breeding ewes and lambs were in one flock, and the food dirtied by driving to fold: by this means there is not a bent on the farm, the stocking being equal. He is not, however, entirely without a fold; when the lambs are weaned (usually about old Midsummer) the ewes are folded for about two months, principally to prevent their breaking pasture, when the lambs are taken from them: and while thus folded, he finds that it takes one-half more land to feed them, than if they were left allotted, as through the rest of the year. That folding lessens the value of the lambs,

lambs, he has not a doubt, and that considerably; they do not bring so high a price as others not folded—this is not opinion, but fact. The ewes are also in doubly better condition from lying still and quiet. That the *teath* will, in certain cases, be unequally given, he does not deny; but it is not difficult to remedy this by the dung-cart; to fold a lot in its own lay, is also a remedy, and is the only sort of folding he can approve. Where there are downs, heaths, or commons, the case is different; there folding may be necessary without question. In regard to the effect on wool, Mr. HILL is clearly of opinion, that folding does not render it finer—it makes the fleece lighter, but never finer.

Folding is generally given up by all who have South Downs; not because they will not bear it, for they bear it better than any sheep in the island, but because the stock is so valuable, that it is worth the farmer's attention to contrive, by every means, to keep as many as possible.

One circumstance, though a small one, deserves mentioning, for the use of those who form separating sheep-pens: Mr. HILL's, at Waterden, have sliding-gates from one to the other, he remarked, that when a pen is full of sheep, the gate cannot be opened with convenience; but by their sliding in the fence, this is avoided.

Mr. ENGLAND, of Binham, does not fold, conceiving it to be merely robbing Peter to pay Paul. When not folded, sheep do with less food, and as to the common objection, of their drawing under hedges for shelter, in storms, &c. so much the better; it is what they ought not to be prevented from doing. The *tathe* is much more than lost in mutton.

Mr. REEVE, of Wighton, never folds: folding from layers, upon fallow, is only robbing one field to enrich another. He is clear in this point; and also in the fact,

that

that if sheep (whatever the breed) are driven by foul weather to a hedge, there is the proper place for them, and not by penning, left to abide the beating of the storm.

Mr. H. BLYTHE, of Burnham, sometimes folds, but never from choice, but solely by reason of the openness of his farm; nor does he approve the practice. And he explained a point, in his manuring for wheat, which comes home to the question:—he never sows tempered land with wheat, without either oil-cake or muck, *except on pieces from which the sheep were not folded while feeding the layers.*

Mr. DURSGATE remarks, that folded sheep certainly demand more food than those which are not folded; a quarter of a ton of rape-cake is equal to the fold; and the flock, without any doubt, suffers more than that value by folding. In short, folding is to gain one shilling in manure, by the loss of two in flesh.

Mr. GODDISON folds Lord CHOLMONDELEY's flock of Norfolk and South Downs: and Mr. BECK, of Massingham, who has 35 score of South Downs, folds.

Mr. BECK, of Riseing, does not fold; and he is very certain that if he did fold, he could not keep any thing like the number of his present flock.

As I rode across a layer of 40 or 50 acres, on Mr. OVERMAN's farm, I observed a great difference in the verdure, to a line across it, the appearance of one side of that line being so much superior to the other; and on my remarking it, I was informed that it was an accidental experiment, which was well worth attention: there was no other difference in management, to make one part of that layer better than another, except the sheep that fed it being from one part of it folded on another arable field during the summer; but from the other part they were not folded at all, but left in the layer night and day. The

difference

difference was very considerable, and might have been discerned half a mile off. This experiment made Mr. OVERMAN give up folding, except when his flock was in a salt-marsh; and Mr. TUTTLE, a neighbour, asserted, he would never fold at all had he no marshes. Nor does Mr. ETHERIDGE, of Stanhow, fold. These facts should be combined with another, that of heaths and sheep-walks, that have been fed with sheep for centuries, but those sheep constantly folded on other lands, are so far from improving, that they are to all appearance as poor as they could have been at any former period.—*Note, some years past.*

Mr. STYLEMAN, at Snettisham, turned his flock loose, and without folding, in 20 acres of *ollond* every night, for the same period that would have folded it in the common manner. The sheep did much better than they would have done had they been folded; the face of the herbage materially improved during the period, and upon ploughing it up for wheat, the crop was equal to what it would have been with folding, and shewed, by a regular verdure, that they had distributed the manure equally in every part.

Mr. STYLEMAN conceives that lambs sell 3s. a head lower on account of folding, than they would do without it; but this is only his opinion. He thinks also that the ewe is much injured.

Mr. PITTS, of Thorpe Abbots, finds that no mucking, on his burning gravels, will do so much good as the fold, and especially on a white clover and trefoil layer for barley.

DISTEMPERS.

When I had formerly the pleasure of being at Houghton, I have often urged Lord ORFORD to break up a heath

heath of black sand, but his Lordship informed me, that the farmers were clear that if he did it, his lambs would be ricketty, by feeding on the turnips or grasses. Mr. GODDISON has, however, laid on 3200 loads of clay on 44 acres of it, in six months, and broke it up: he got great turnips on it, and this year very fine barley, two loads and a half per acre, in the straw. As to the rickets, he cannot assert that there has been none, but quite inconsiderable, not ten affected in a large flock. On this distemper he observes, that the only danger is while the ewes are in lamb; and that after lambing the malady is not acquired.

On a ground noted for causing this distemper, the soil a black sand, heath, but marled and cultivated, a farmer accidentally removed part of his flock during the months of October and November; the flock then moved escaped the rickets, but those left had it. In consequence of this the trial was repeated next year, and the effect the same. It should seem from this case, that the distemper is taken only in the autumn, at whatever time it may appear, and if so, there is very little difficulty in avoiding it.

Mr. COKE's receipt for dressing his flock previous to winter, against lice and ticks:—Two pounds of tobacco, two pounds and a half of soft soap, one pound of white mercury, ground to powder: boil in eight gallons of water one hour. Part the wool down each shoulder, and the breast of the sheep, and twice along each side, into which pour it. This quantity enough for 60 sheep.

Mr. OVERMAN, as soon as his flock is sheared, dresses his lambs to destroy ticks and lice. He boils a pound of arsenick and a pound of soap in about six gallons of water, and then adds 26 gallons more water: in this he dips the lambs, and finds it effective in the destruction of all the vermin: without this precaution they are propagated a-fresh from the lambs to the ewes.

Mr.

Mr. M. HILL has experienced a distemper among lambs which, from the description, should seem to be a species of rickets. In 1799 all were quite healthy; but in 1800, 140 lambs fell lame in the knees and hamstrings, and wasted away much. In 1801 the same. In 1802 no such appearance; all healthy; the only difference in the management, and to which he is inclined to assign the cause, was putting the tups to the ewes the 10th of September in 1798, and 1801 and in the intermediate years, on the 7th of October. I inquired if they had fed on different lands, or if he had broken up any black sandy heath? No such thing. In 1803 quite healthy.

SECT. III.—HOGS.

THE breed of hogs in Norfolk do not demand any particular attention, though a useful pig if well supported: their most usual colour black and white, &c. &c.: carcass long, but wants thickness; legs the same, or at least not short: good breeders.

I found Mr. SALTER, of Winborough, fatting 180 pigs in August, by throwing down pease in a well littered yard, and says the pigs lose none at all: they have the run of a meadow, and he is clear, from long observation, that they fatten much better and quicker than if confined. He assigns 11 score of pease for fatting 200, more or less; and considers it as a profitable application of the crop: he buys them all. He has compared stye-fatting and loose-fatting, and decidedly in favour of the latter.

Mr. HAVERS, at Thelton, has the Suffolk breed, and has also Berkshires; but finds the cross between them better than either separate.

I found

I found a new piggery building by Mr. HAVERS, at Thelton, in which the most singular circumstance is the sties for fatting, being single, for one hog, and so narrow that he cannot turn himself; a range of these on one side and a space for cistern on the other, the whole near the new dairy—perhaps, rather too near: a degree of vicinity is necessary for the milk and whey to flow to the cisterns, but the air around a dairy should be preserved quite uncontaminated.

Mr. WISEMAN, at Happsborough, having occasion to wean some pigs much too young, from the death of a sow, or some other cause, tried boiled pease for them, and the success was so great, that he would never enter largely into breeding or fattening hogs without a furnace and copper for boiling whatever corn might be given.

Mr. JOHNSON, of Thurning, fattens with boiled barley, and by this means made one so fat that he was blind from excess of fat.

Mr. REEVE, of Wighton, every feature of whose husbandry merits attention, condemns a dairy stock as unprofitable to a farm, in respect of manuring: he once had a large dairy, now only 26 cows, and a principal motive for keeping so many, is the right application of it to supporting a large stock of swine.

SECT. IV.—HORSES. OXEN.

1792. MR. OVERMAN's arable, 523 acres: he keeps 21 horses.

In the district of Holkham, 20 to 500 acres in one farm, 16 to 500 acres. The larger the farm, generally the smaller the proportion.

At Snettisham, 30 years ago, 16 horses necessary for 500 acres of arable land.

In Happing hundred, five per 100 acres.

Mr. H. BLYTHE, 30 to 1000 acres of sand, and chalk, drilled.

Mr. REEVE, of Wighton, 700 acres, 22 horses; not four per 100: drilled.

Mr. DURSGATE, at Summerfield, 1050, and 31 horses. At Sedgford, 1240 acres, and 36 horses; very nearly all in tillage, and all drilled.

Mr. STYLEMAN, of Snettisham, four to 100 acres; not the more for drilling.

Mr. POWELL, of Broomsthorp, near Fakenham, with 15 Suffolk horses in his stable, used but one load of hay last winter. He feeds with cut straw and oats, and has no racks. Uses BURRELL's chaff-cutter, paying 2s. for 20 coombs.

Mr. BURTON, of Langley, never lets his horses remain in the stable at night, always turning them into a well-littered warm yard, contiguous to the stable. This is the practice of the farmers in the angle of country formed by Woodbridge, Saxmundham, and the sea.

Two and thirty years ago, Mr. RAMEY, of Yarmouth, was in the regular practice of soiling, which has not since been followed to any thing like the extent to which it ought to be every where carried. The second week in May he began on clover for 20 horses and 7 cows, 5 calves, and his hogs, and found seven acres sufficient for them till the wheat crops were cleared: reckoning the horses and cows at 2s. 6d. a week, the calves at 1s. 6d. and nothing for hogs, it amounted to 9l. 2s. 1d. per acre. A tenant fed stock to the same amount in an adjoining field, and when Mr. RAMEY had eaten 5 acres, this man had eaten and wasted 30.

Mr.

Mr. FERRIER, of Hemsby, is very sensible of the importance of soiling: his team had not been out all summer when I was with him in September.

Mr. FRANCIS, at Martham, in common with every good farmer in the neighbourhood, uses tares for soiling horses, but thinks clover better, and much used every where.

Mr. JOHNSON, of Thurning, soils on clover, in preference to vetches.

Mr. ENGLAND, of Binham, has tares enough for baiting his teams: does not sow turnips the same year after them.

Mr. PRIEST, of Besthorpe, cuts all the hay he uses, for eight horses and fourteen bullocks, and has no doubt of its answering greatly.

Mr. M. HILL, does the work of 1300 acres with forty horses, but he has 200 of pasture.

In 1784, Mr. COKE worked 12 oxen in harness for carting, and found them a very considerable saving, in comparison with horses.

1803. Mr. COKE gave them up some years past, from the difficulty of shoeing them, but more from the inveterate prejudices of the men against them. He was not, however, convinced that the practice might not in many cases be profitable.

Mr. DREW, of Bexwell, tried oxen, and gave them up from being troublesome in a waggon. Mr. CLARK, of Denver, used them many years, but gave them up at last.

Mr. BEVAN had them for a few years, and then gave them up.

Mr. PURDIS, of Eggmore, works 32 Devonshire oxen in yokes and bows, four to a plough. I saw them at work, and was much pleased to see them step out so nimbly, as to be fully equal to the horses ploughing in the same field, in point of movement: they plough an acre and an half in one journey. Mr. PURDIS thinks the saving will be considerable, as he has procured 40 heifers of the same breed for his regular supply.

Mr. M. HILL uses two pair of short-horned oxen, which, walking well, plough two acres a-day; each pair an acre in five hours and an half. He is of opinion, that on farms that employ 22 horses, it would be more profitable to have 16 horses and 8 oxen; but he cannot recommend them for nearly the whole strength; as in hay and harvest it is necessary to be very nimble, the horses in empty waggons trotting fast; this cannot be done with oxen.

Mr. HAVERS, at Thelton, works Devon oxen for carting, and approves of them much.

Mr. THURTELL, near Yarmouth, a few years back, worked bulls; he had two pair, and two of them ploughed as much land as two horses: he has a great opinion of them, and would recommend to any active young farmer going into business to employ them in preference to horses.

CHAP. XIII.

RURAL ECONOMY.

SECT. I.—LABOUR.

THE circumstance in rural economy, which for many years distinguished Norfolk in a remarkable manner, was the cheapness wherewith the farmer carried on his business. This arose not only from a low price of labour, but also from a much greater activity and spirit of exertion amongst servants and labourers, than was to be found in almost any other county of the kingdom. This spirit is still highly commendable here, but by reason of the scarcities throwing the mass of the people on the parish to be supported by rates, it has suffered considerably.

In 1767, I registered the price of labour in West Norfolk at 1s. a day in winter; in spring, 1s. 2d.; for the harvest, 2l. 12s. 6d. to 3l. with meat, drink, and lodging, and lasting from one month to five weeks; hoeing turnips, 3s. and 2s.; ploughing, per acre, 2s. 6d.

Holkham, 1792.—In harvest, 5l. 5s. generally five weeks.

Threshing wheat, 1l. 1s. a last.
————— barley, 9s. ditto.

Thatching stacks, 1s. a yard, running measure, length of stack all widths on an average.

AT BRAMMERTON, 1770.

	£.	s.	d.	
For the harvest, with board,	2	2	0	
— — Hay time,	0	1	6	and beer.
— — Winter,	0	1	0	and ditto.
Mowing grass, per acre,	0	1	6	and ditto.
Hoe turnips,	0	4	0	and 2s.
Clay, per 120 loads,	1	5	0	
First man's wages,	10	10	0	
Second, ditto,	6	6	0	
Lad,	3	0	0	
Dairy-maid,	4	4	0	
Others,	3	0	0	

In the Holkham district, winter and summer, 1s. 6d. Odd men in harvest, 2s. 6d. and 3s. Regular men, 2l. 2s. and board for the harvest.

Snetterton, &c. in winter, 1s. 6d. ⎫ no beer.
——————— —— Summer, 1s. 9d. ⎭

At Hingham, summer and winter, 1s. 8d.

The harvest 42s. to 50s. and board, generally for a month.

A custom is coming in around Waterden, &c. of allowing board-wages to farm servants, instead of the old way of feeding in the house; 8s. a week are given; wages 5l. 5s. This is one material cause of an increased neglect of the Sabbath, and looseness of morals; they are free from the master's eye, sleep where and with whom they please, and are rarely seen at church. A most pernicious practice, which will by-and-bye be felt severely in its consequences by the farmers. Mr. HILL feeds his servants in the old way.

The price is raised at Waterden 6d. a day, in the last two years, and the work worse done. Last winter 2s. a day,

LABOUR. 485

a day, and the same in summer. But Mr. HILL intends next winter to reduce it to 1s. 9d.

At Winborough, in winter, 1s. 8d. ; in summer, 2s. ; for the harvest, 50s. gloves, and 1s. hiring, with board: reaping wheat 10s. 6d. an acre, or 2s. 6d. a day, and board.

Through Loddon hundred, 45s. for the harvest, and board ; and allot twelve acres per man ; some four load an acre, all three, of spring corn, and two of wheat : dibbling wheat, 10s. 6d. In winter, 1s. 6d. ; summer, 2s., which was lately 1s. 6d.

In Fleg hundreds, winter and summer, 1s. 6d. a day, allowing bread-corn at 6s. a bushel : harvest, 50s. and board.

At Martham, dibbling wheat 9s. an acre ; pease, 8s. ; bread-corn at 6s.

In parts of Happing hundred, 2s. winter and summer: harvest, 2l. 12s. 6d. and board.

At Honing, harvest 50s. and board ; winter, 1s. 4d. and 1s. 6d. a day ; summer, 2s. but wheat at 5s. a bushel to the men.

At North Walsham, 10s. a week, winter and summer.

Scotter, 1s. 6d. and beer to harvest ; then 42s. and board.

Reepham, &c. in winter, 1s. 6d. to 1s. 8d. ; in summer, 2s. ; in harvest, all included, 4s. 6d.

Thurning, 2s. to harvest, for which 7l. without board.

Binham, 20d. to 2s. in winter and to harvest, when 52s. 6d. and board, or 6l. 6s. without.

Wighton, harvest 6l. 6s. and sixteen acres a man; six, wheat, and ten spring corn.

Thornham, 1s. 9d. to 2s. the year round, except harvest, then 2l. 12s. 6d. and board ; 6l. 6s. without : filling marle cart, 28s. per 120 loads.

At

At Snettisham, 10s. 6d. a week, winter, and to hay and turnip hoeing; in harvest, 6l. 10s. without board, Hoeing turnips twice, 6s.; mowing grass, 20d. to 2s. 6d.; filling marle, 25s. per 120 loads; 12s. for small low carts half loads.

At Houghton, in winter, 10s. 6d.; in summer, 12s.; in harvest, 50s. to 60s. and board.

At Wigenhall, in Marshland, the average price of reaping wheat, 12s. the statute acre.

Labour, at Lynn, is sunk by the peace:—Sailors wages, from 4l. 10s. a month, to 50s.; and that of corn-porters, from 1½d. a sack to 1d.

At Walpole, in Marshland, 6s. to 7s. a day, general in harvest: some, this year, gave 9s. and 10s. a day: 2s. 6d. after Michaelmas, till seed time is over; 2s. all winter; 2s. 6d. after May-day.

Near Downham, in winter, 20d. and beer; summer, 2s.; hay-time, 3s. 6d.; mowing grass, 2s. to 4s.; mowing barley, 5s.; reaping oats, 10s. to 16s.; wheat, 10s. to 12s.; in harvest, by day, 6s.

At Besthorpe, 1s. 8d. in winter; 2s. at hay, and after harvest till Michealmas; harvest, 45s. and board; 6l. without. No malt.

RUNCTON.

	1770.	1803.
Harvest,	2l. 2s. and board.	2l. 12s. 6d. and board.
Hay, -	1s. 6d. and beer.	2s. and beer.
Winter,	1s. 2d. and beer.	1s. 9d. and beer.
Reaping wheat,	4s. to 6s.	7s. to 8s.
Reaping oats,	4s.	5s.
Mowing barley,	1s. 6d.	2s.
Mowing grass,	2s.	3s.
Hoeing turnips,	4s. and 2s.	4s. 6d. and 2s. 6d.
Threshing wheat,	2s. a quarter.	3s.

	1770.	1803.
Threshing barley,	1s.	1s. 6d.
Threshing oats,	8d.	1s.
Head man,	12l.	12l.
Next ditto,	9l.	9l.
Lad,	5l.	6l.
Dairy-maid,	4l. 10s.	4l. 10
Others,	3l.	3l.

AT SNETTISHAM.

	1770.	1803.
Five weeks harvest and board,	45s. to 50s.	2l. 12s. 6d.
In hay time, a day,	1s. 6d. to 2s.	2s. 6d.
In winter,	1s. 2d.	1s. 9d.
Reaping,	5s. per acre.	12s.
Mowing barley,	1s.	2s.
Mowing grass, artificial,	1s. to 2s.	2s.
———— natural,		4s. 6d.
Hoeing turnips,	4s. and 2s.	7s.
Filling and spreading marle,	25s. per 120 loads,	28s. to 30s.
Threshing wheat, per quarter,	1s. 2d. to 1s. 4d.	1s. 8d. to 2s.
Threshing barley and oats,	8d.	8d. to 10d.
Threshing pease,	1s. 3d.	8d. to 10d.
Head man's wages,	10l. to 12l.	11l. to 15l.
Next ditto,	9l.	7l. to 10l.
Lad,	4l. to 7l.	4l. to 7l.
Dairy-maid,	5l.	5l.
Other ditto,	3l. to 4l.	3l. to 5l.
Women, in harvest,	1s. and board,	1s.
in hay,	9d. and beer,	1s. No beer.
in winter,	6d.	8d. No beer.

Mr. HENRY BLYTHE assures me, that labour, in the vicinity of Burnham, nearly doubled from 1795 to 1801: to satisfy me of this fact, he laid his books before me, in which the fact appeared clearly; and this has taken place on his farm, not at all by any extraordinary works done on it, but merely by the rise of prices: it is not, however, to be accounted for merely by the rise per week, but the men will not perform what they formerly did, and more must, therefore be kept for the same work.

The labour of a sand farm, of 1000 acres, near Holkham, 820l. last year.

Mr. MARSHALL takes every opportunity, in his work, to note the great activity of the farm workmen, in the dispatch of business in this county, much exceeding any other. The observation is pointedly just; insomuch, that there could hardly be a greater improvement than to introduce their system, in this respect, in many other counties, equally adapted to it; but managed, at present, at a far greater expense.

Recapitulation.—General average in harvest, 2l. 8s. 3d. the month, and board; in summer, 1s. 11d. a day; in winter, 1s. 8d. a day.

CHAP. XIV.

POLITICAL ECONOMY.

SECT. I.—ROADS.

DURING the long period in which Norfolk was content with the reputation given to her roads by the observation of CHARLES the SECOND,* without having recourse to the assistance of Parliament, in the establishment of turnpikes, her ways were bad enough, though not so bad as in heavier soils: she has, however, in the last 20 years, made considerable exertions—turnpike gates are erected in all the principal communications; these are kept in good repair, and the roads, in general, must be considered as equal to those of the most improved counties.

In the line from Dereham, 30 miles to Harleston, the direction is diagonally across all the Norwich roads, yet I found this road as good as a turnpike.

SECT. II.—CANALS.

VERY little has been effected in Norfolk by means of canals; merely artificial; but some rivers have been ren-

* That Norfolk should be cut into roads for all the rest of England.

dered

dered navigable, which add considerably to the communications of the county. The Little Ouze is navigable to Thetford; the Yare to Norwich; the Waveny to Bungay, and the Bure to Aylesham; a small branch of the Great Ouze to Narborough. More considerable exertions are rarely found in counties merely agricultural. Without coal-mines, or a great demand for coals, lime, &c. and the establishment of immense manufactures, canals are too precarious a speculation for the ready advancement of great subscriptions.

SECT. III.—FAIRS AND MARKETS.

The principal fairs are those of Harleston, St. Faith's, Hempton, Kenninghall, Harling, Causton, Kipton, Swafham, and Thetford, for wool.

As to markets, there are but three considerable ones in the county, Norwich, Yarmouth, and Lynn: these are plentiful: Norwich approaches to Bath itself, and Yarmouth is most plentifully supplied. The rest are very inferior.

In the district of Holkham, a circumstance has taken place, which I cannot but consider as an extraordinary proof of the spirit of exertion which pervades this county.

Mr. Overman, of Burnham, has a small ship, which he keeps constantly employed in carrying his corn to London, in bringing rape-cake for manure from Holland, London, Hull, or wherever it is to be procured best, and at the cheapest rate. When his farm does not in this manner produce employment, he sends her for coals, or deals,

or

or on any service which times and markets render eligible: and this speculation answers well. He conceives that drilled corn, kept perfectly clean, is a better sample than the common run of broad-cast, and he finds it difficult to get, in the country, a price proportioned to the merit of his productions, and to send the corn by sea to London, does not cost so much as land-carriage to Lynn would do. Entering on a new farm of Mr. Coke's, a year and a half past, and finding many hurdles necessary, he sent his ship to Sussex for a lading of hurdles: there, made much better, of riven oak, and at the same time lighter than others to be had in Norfolk; they cost him 3s. 6d. each, and will last 20 years. Common wattle hurdles cost 1s. 1s. 1d. besides long carriage, and will not last above two years.

Mr. Money Hill, of Waterden, has also a sloop of 50 tons, which goes to sea, with two men and two boys: he built her, and the employment is the same as Mr. Overman's. Mr. Davy has likewise one. They have not yet brought manure from London; though probably tallow-chandlers' graves, woollen rags, and various other articles, would answer as well as rape-cake, at the present price of 7l. to 8l. per ton.

SECT.

SECT. IV.—THE POOR.

An establishment at Snettisham, which has been found of the greatest use to the poor, and has answered every expectation, is a subscription water-mill: it cost 800l. and a miller is employed, at 20s. per week, to grind, at 4d. per bushel, for all persons, whencesoever coming.

PROGRESS OF LABOUR, &c. &c. IN SUNDRY PARISHES, 1793. Communicated by Sir Thomas Beevor, Bart.

Parishes.	Wages in 1752.	Wages in 1772.	Wages in 1792.	Cottages in 1752.	Cottages in 1792.	Poor-Rate in 1752.	Poor-Rate in 1772.	Poor-Rate in 1792.	Payers in 1752.	Payers in 1772.	Payers in 1792.	Labourers.	Observations, &c.
1. Aldborough	1s. 1 pint of beer in W. and 2 pints in S.	1s. the same as in 1752	W. 1s. 2d. 1s. 4d. to 1s. 6d. no allowance of beer	13; mostly double	17; increasing	28l.	90l.	140l.	More payers in 1792 than in any former period, from assessing all cottages except real paupers, and there being a number of little shops in the village			A sufficient number, and when corn has sold low, more than can be employed	1. In 1752, the labouring hand had corn, butter, cheese, &c. at a low price. In 1772, many farmers sold their labourers corn at a reduced price, when more than 20s. per coomb. The present state of living of the poor is much altered in my memory. They live better, except in the article of beer, which the high price of malt precludes their brewing, and drives them to the ale-houses and gin-shops. Most of the labourers kill one or two pigs, and many have cows, where they have commons. The labourers have their corn at 5s. per bushel; and the overplus, of 1s. 3d. per stone, is paid by the farmers to the miller. They have butter at 6d. per pound, and what cheese can be spared, at 2½d. to 3½d. and 4d. per pound; but the latter article is much decreased, and the poor are obliged to go to the shop, and pay 4d. and 5d. per pound.
2. Bradfield	S. and W. 1s.	1s.	1s.	Not known, but more numerous than in 1792	More than are wanted	20l. 3s. 10d.	67l. 17s. 4d.	80l. 17s. 3d.	….	….	14, but chiefly paid by 3	A sufficient number, but not too many	
3. Hethel	W. 1s. S. 1s. 2d.	1s. 2d. 1s. 2d.	1s. 2d. 1s. 4d.	9	21	….	28l. 1s. 7d.	38l. 6s. 9d.	15	15	12	No want of labourers	3. Whenever the price of wheat exceeded 5s. per bushel, it has been sold to the poor at 5s. Wood is frequently given them. They purchase their meal and flour in the proportion of half a stone per head, at 3d. per stone cheaper than the shop price. This is at the parish expense. Where it could be conveniently done, it has been allowed the occupier of a cottage to enclose a piece of land for the growth of potatoes, &c. without any rent for the same; and no labourer, whose rent is under 10l. per ann. is charged to any rates, or is required to pay or perform any statute duty.
4. Hapton	W. 10d. S. 1s.	1s. 1s. 2d.	1s. 2d. 1s. 3d. and 1s. 4d.	10; most were double, and some triple	12	34l. 1s.	43l. 14s. 6d.	84l. 19s. 4d.	9	7	7 besides out-setters	Are rather short, and should be more so, and are obliged to cast tithes	4. The rental of the parish is 550l. In 1752, wheat 3s. 6d. to 4s. and malt 2s. 9d. per bushel. Meat of all kinds, 2½d. to 3d. cheese, from 2½d. to 3½d. per pound. Butter, 3½d. to 4d. per pint. Milk plenty, and cheap. In 1772, wheat 5s. 6d. to 6s. and malt 4s. 4d. per bushel. Meat, 3½d. to 4d. cheese, 3d. to 4d. per pound. Butter, 8d. to 10d. per pound. In 1792, wheat 5s. 6d. and malt 5s. 8d. per bushel. Meat, 4d. to 4½d. cheese, 4d. to 5d. per pound. Butter, 10d. to 1s. per pint. In 1752, the labourer could bake, brew, and hang on the pot, with a piece of meat in it, and ask his master or neighbour to drink. In 1792, the very reverse. The high price of provision, wool, and leather, altogether, affects the labourer so much, that it is impossible for a poor man with a young family to keep them either full or warm, without assistance. Some farmers let them have wheat at 5s. per bushel, and put out a deal of their work, so that the labourer may earn from 18d. to 20d. per day.
5. North Walsham	1s. and 3 pints good strong beer, per diem	1s. and 2 pints of tolerable good beer	W. 1s. S. 1s. 2d. and only 1 pint of miserable small beer	….	Not ascertained, but many unoccupied. In 1792, many families in the workhouse that used to live in cottages	Year 1758, 388l. 9s. 9d.	810l. 7s. 4d.	Year 1791, 889l. 10s. 8d. The books are not made up for 1792, but expect they will be 900l. at least	Have not half the number of farmers that pay to the rates, that we had 40 years ago. The small farmers are swallowed up: I know several instances, where 3, 4, and 5 small farms, are laid into one			Have no want of labourers in the farming line, but rather a want of labour, from the monopolization of farms, and decline of weaving	5. In 1750, the expense of the poor did not exceed 200l. or 250l. per ann. and many years since, it has been near 1000l. From long observation, I perceive the labourer is not less industrious, and spends less money unnecessarily than they did 40 years ago; therefore their present distress is occasioned by the advance of labour bearing no proportion to the advance on the necessaries of life.
6. Swaffham	1s.	1s. 2d.	1s. 3d.	….	120	204l. 2s.	401l. 5s. 6d.	517l. 13s. 2d.	….	….	323	A sufficient number, but not too many	6. Most of the labourers work by the piece, and the common earnings of a good workman is 10s. per week. The poors-rate, in 1736, was 141l. 8s. 11d. per ann. The increase has been much less these last 20 years, owing to the number of box-clubs in this place, which many gentlemen encourage, by being members of them. The firing which the labourer cuts off the common, is carried by their employers gratis.
7. Thurgaton	1s.	1s.	1s. 2d.	….	Sufficient	Year 1762, 67l.	87l.	119l. This year 20l. was expended in law	More payers in 1792 than in any former period: the reason, vide Aldborough, under this column			The same as at Aldborough	7. The observation will apply to this parish, except, as it abounds in small farms, the poor have not the corn, butter, &c. at such reduced prices; but the small farmers pay more than the large ones.
8. Edingthorpe	1s. and 2 pints of good beer	1s. and 2 pints of good beer	1s. 2d. and 2 pints of beer	15	15, just sufficient	24l. 4s.	53l. 2s. 11½d.	73l. 17s. 5d.	18	20	17	If our poor all resided in the parish, they would suffice to do the labour of it	8. Notwithstanding the advance in the overseers'-rate, it is not so well with the poor as in days past: from the scarcity of bark, leather is raised to such a pitch, that it is hardly possible for the poor man to provide himself and children with shoes.
9. Trunch	1s.	1s.	1s. 2d.	16	36, more than sufficient	48l. 13s. 9½d.	212l. 9s. 9d.	155l. 9s. 4½d.	32	32	28	A sufficient number	
10. Witchingham	1s.	….	1s. 3d.	….	Encreased, but not sufficient for all	Cannot get exact information; but think the poor-rate is doubled in 40 years in the sum collected, though not by the pound, owing to a rise in rentals			….	….	Number of payers diminished one-third, by the consolidation of neighbourhood of farms	Not a sufficient number resident in the parish; but the overflow of the neighbourhood seldom leaves any want of labourers	10. Labour per diem is rarely adopted; almost every thing is put as task-work, at which an active hard-working man may earn from 1s. 6d. to 2s. per day; and the prices of this work have risen in 40 years full one-fifth.
11. Wymondham	W. 1s. S. 1s.	1s. 1s. 2d.	Average, 1s. 3d.	….	Considerably less now than in 1752	….	….	….	….	….	….	No want of labourers	11. It has been an invariable custom, for the labourer, whenever he does hedging or ditching-work, to be allowed a bunch of wood per day.
12. Cawston	W. 1s. S. 1s. 2d.	W. 1s. S. 1s. 2d.	W. 1s. 2d. S. 1s. 3d. to 1s. 4d. The greatest number of employers allow no beer. In none of these 3 periods the labourer had very few advantages, with respect to public charity or contribution	Not known	There are at least 170 dwelling-houses, but how many come under the name of cottages, I know not: there are a sufficient quantity, but not too many	126l.	317l.	454l.	65	67	62	A sufficient number, but not too many	12. The prices of threshing grain have remained the same for 40 years. Many new cottages have been built within 20 years, and the population certainly increases.
13. South Town*	W. 1s. 4d. S. 1s. 4d.	W. 1s. 2d. S. 1s. 4d.	W. 1s. 4d. S. 1s. 6d.	Being a large parish, have plenty of small tenements	….	….	….	In general, moderate	….	….	Uncertain	A sufficient number	Harvest wages go for house-rent, and gleaned corn is sold, to maintain wife and children in harvest.

Weekly Expenses of a Man, Wife, and Three Children, when Wheat is at 24s. per Coomb.

	s.	d.
Bread corn,	6	0
Shop goods,	2	0
Clothing,	1	0
Utensils, firing, &c.	1	0
	10	0

* Is an incorporated parish, within the hundreds of Mutford and Lothingland.

THE POOR. 493

The necessary and unavoidable expenses of a labourer and his wife, without any family, for one year, calculated at the price of flour 4s. and meal 3s. per stone, and the other articles at present price, 1799*.

	£.	s.	d.	£.	s.	d.
Cottage rent - - -				2	10	0
One peck of coals per day, 20 weeks, or 140 days, at 3½d. per peck	2	0	10			
Half a peck per day, 8 weeks, or 56 days - -	0	8	2			
A quarter of a peck per day, 24 weeks, or rather 169 days	0	12	4			
				3	1	4
Soap, 4 ounces per week, at 2½d. which is per annum -	0	10	10			
Oil or candles, at 6d. per week, for 20 weeks - -	0	10	0			
Do. do. at 3d. per do. for 32 weeks	0	8	0			
				1	8	10
Shoemaker, one pair of shoes, one pair of highloos, and mending	0	16	0			
Stockings, two pair - -	0	5	0			
Hat - -	0	2	0			
Slops, jacket, &c. - -	0	9	0			
Breeches - - - -	0	5	0			
Two shirts - -	0	10	0			
				2	7	0
Amount				£.9	7	2

* Communicated by an active Magistrate.

THE POOR.

	£.	s.	d.
Amount brought forward	9	7	2
Woman's apparel	1	15	0
Sixpence per day, for food for each person	18	5	0
Sixpence per do. additional food in the harvest month, per man	0	14	0
Expense of tools	0	8	0
	£.30	9	2

In the above account no allowance is made for the wear and tear of the furniture, and sundry small articles used in the house.

EARNINGS.

	£.	s.	d
Harvest, 4 weeks	4	14	6
48 weeks, at 8s. per week	18	4	0
Woman's gleanings	0	14	0
48 weeks, at 1s. per week	2	8	0
	£.26	0	6

HOUSES OF INDUSTRY.

MITFORD AND LAUNDITCH.—MINUTE, 1792—INCORPORATED 1776.

Paupers	404
House built for	600
Parishes	50
Annual revenue	£.3960
Borrowed, at 4 per cent.	£.15,000 including 1500l. laid out by the treasurer.
Have paid off	£.5000
Earnings	£.500
Land	60 Acres.
Cows	10

A wind-mill.

MIT·

MITFORD AND LAUNDITCH HOUSE OF INDUSTRY.

Years.	Average weekly number Paupers, ending Christmas.	Number of Births.	Deaths under one month.	Total Deaths	Weekly and annual maintenance Paupers in the House. £. s. d.	Payments, annual, out of House. £. s. d.	Salaries and Law Expenses £. s. d.	Debt at Midsummer. £. s. d.
1795	548	23	2	89	0 1 4¾ / 3 12 7	1342 15 9	489 4 4	11,399 4 7
1796	523	20	1	59	0 1 11 / 4 19 8	2923 4 2	481 8 10	13,253 7 1
1797	476	8	0	37	0 1 0 / 4 0 2 6½	1253 19 10	462 10 0	10,457 14 4
1798	442	15	1	31	0 1 7 / 4 2 4	1249 5 5	467 12 4	10,411 11 7
1799	459	19	2	42	0 1 4 / 5 0 2½	2675 15 3	473 3 0	10,254 8 1
1800	589	19	1	38	0 1 1¼ / 2 9¼	12,499 6	467 10 4	16,661 6 2
1801	577	22	2	58	0 1 2½ / 3 11 8½	12,025 4	491 18 6	12,534 16 5
1802*	324	3	0	......	0 2 7½ / 7 13 10	...... †	112 5 3	10,856 17 6, at Lady-day, 1802.

* Christmas to Lady-day.
† Only three quarters of the year to Michaelmas.
‡ In the parishes.

MITFORD AND LAUNDITCH.—1802.

Paupers in the house, 2d of Aug. 1802: 244 old and young.

Earnings—Upon the 21st of June last there were in the house 52 paupers: 86 women; 4 boys and girls above 16, and under 18; 25 above 10, and under 16; 48 above 5, and under 10; and 38 under 5.—Earnings then about 8l. or 9l. a week.

Poor-rates lowered.

Income uncertain.

Salaries	£	£ s. d.
Surgeon, House,	55	
Out-Surgeons	45	
Ditto,	40	225 0 0
Ditto,	45	
Ditto,	40	
Chaplain,		40 0 0
Clerk,		42 0 0
Governor and Matron, exclusive of tea and sugar,		84 0 0
Miller and Baker,		15 12 0
Brewer and Shoemaker (paupers)		15 12 0

Debt remaining 9000l. and 1502l. to the Treasurer.

Poor-rates higher or lower.—Each parish, by a late act of parliament, pays according to the number sent into the house—have been much lower than those not incorporated.

The circumstances which occasioned an application to parliament, to enable the parishes to keep their own poor, paying to the standing charges of the house—a proceeding quite new, were these: several gentlemen thought that the parish of East Dereham
were

were a burthen to the corporation. It was agreed between the gentlemen who took the active part for a new act, and the gentlemen of East Dereham, that that parish should by the new act maintain their own poor, and accordingly a clause was inserted for that purpose, and Dereham has built a workhouse, and inclosed ten acres of the common waste for their own poor.

HOUSE OF INDUSTRY—HACKINGHAM—MINUTE, 1792.

	£.	s.	d.
Paupers, - - - 215			
Will hold - - - 300			
Expense of building, 25 years ago,	7000	0	0
All paid off, and a balance in hand of	515	0	0
Parishes, - - - 41			
Poor-rates lowered one-eighth.			
Annual income, - -	1866	0	0
Spin wool; some hemp.			
Annual earnings, - -	250	0	0
Salaries, Surgeon, -	105	0	0
Chaplain, - - -	30	0	0
Clerk, - - -	30	0	0
Governor, - -	40	0	0

and ½d. per lb. on all wool spun, which is 12l. more. Baker 6s. a week and board.

FOREHOE HOUSE OF INDUSTRY, BUILT 1776.

	Average number of Paupers.	Earnings.		
		£.	s.	d.
From Midsummer, 1788 to 1789,	306	532	19	2¼
1789 to 1790,	256	471	4	8½
1790 to 1791,	239	390	12	0
1791 to 1792,	220	436	15	5¾
1792 to 1793,	206	357	14	4½
1793 to 1794,	243	366	0	9
	1470	2555	6	6

Average

THE POOR.

	£.	s.	d.
Average number for 6 years, from 1788 to 1794, } 245	425	17	9
Average earnings for each pauper per year,	1	14	9
Ditto for ditto per week,	0	0	8

In the above earnings are included all the servants' labour, for which they are not paid.

Sixty acres of land farmed.

Income, till 1796, fixed at 2888l. per annum, from twenty-three parishes.

	Income.		No. in the House.
1796. Lady-day,	£.1084	- - -	309
Midsummer,	- 1445	- - -	244
Michaelmas,	- 1807	- - -	224
Christmas, -	- 1807	- - -	269
	£.6143	Average,	261
1797. Lady-day,	£.1084	- - -	242
Midsummer,	- 722	- - -	234
Michaelmas,	- 722	- - -	237
Christmas, -	- 722	- - -	270
	£.3250	Average,	245
1798. Lady-day,	£.903	- - -	253
Midsummer,	- 903	- - -	117
Michaelmas,	- 722	- - -	224
Christmas, -	- 722	- - -	247
	£.3250	Average,	235

1799.

THE POOR. 499

	Income.		No. in the House.
1799. Lady-day,	£.903	-	242
Midsummer,	903	-	233
Michaelmas,	722	-	240
Christmas,	1084	-	284
	£.3612	Average,	249
1800. Lady-day,	£.1807	-	315
Midsummer,	2530	-	334
Michaelmas,	2530	Suppose*	334
Suppose	2530	Suppose	334
	£.9397	Average,	329

A different arrangement must be made, in order to add the earnings, and deduct the allowances paid out of the house, as in these respects the year's accounts are made up at Midsummer.

			Average Persons.
Parochial income for the year, ending Midsummer, 1797,	£.5420		
Earnings,	378		
	£.5798		
Allowances,	1131		
	£.4667	-	242
1798. Parishes,	£.3250		
Earnings,	347		
	£.3597		
Allowances,	1012		
	£.2585	-	244

* These three articles, of course, not made up; there is very little doubt but the sum is exactly, and the number very nearly stated.

1799.

						Average Persons.
1799.	Parishes,	-	-	£.3250		
	Earnings,	-	-	355		
				£.3605		
	Allowances,	-	-	913		
				£.2692	-	236
1800.	Parishes,	-	-	6143		
	Earnings,	-	-	461		
				£.6604		
	Allowances,		-	3210		
				£.3394	-	293

In order to compare the dear and cheap years, we should take the average of 1798 and 1799 as cheap, and of 1797 and 1800 as dear.

1798 and 1799.	Parishes,	-	£.3250		
	Earnings,	-	351		
			£.3601		
	Allowances,	-	962		
			£.2639	-	240
1797 and 1800.	Parishes,	-	£.5781		
	Earnings,	-	419		
			£.6200		
	Allowances,	-	2170		
			£.4030	-	267

ANNUAL

ANNUAL EXPENSE, PER HEAD.

	£.	s.	d.
1797.—242 persons cost 4667l. which is per head, per annum,	19	5	8
1798.—244 persons 2585l. which is per head, per annum,	10	11	10
1799.—236 persons 2692l. which is per head, per annum	11	8	1
1800.—293 persons 3394l. which is per head, per annum,	11	11	8
	£.52	17	3
Average,	£.13	4	3

HOUSEKEEPING.

1798.	The expense,	£.1256
1799.	—— ——,	1235
1800.	—— ——,	2229

EARNINGS.

	£.	s.	d.
1797.—242 persons earned 378l. which is per head, per annum,	1	11	2
1798.—244 persons earned 347l. which is per head, per annum,	1	8	5
1799.—236 persons earned 355l. which is per head, per annum,	1	10	1
1800.—293 persons earned 461l. which is per head, per annum,	1	11	5

Many are old and decrepid, and many are children.

THE POOR.

MORTALITY.

Parishes.	1797	1798	1799	1800
Barnham Broom,	0	1	0	0
Barford,	1	0	0	1
Brandon Parva,	0	1	0	1
Banberg,	0	0	0	1
Bowthorpe,	0	0	0	1
Carleton Forehoe,	0	0	0	0
Cossey,	1	1	0	1
Coulton,	0	0	1	1
Crownthorpe,	0	0	0	2
Causton,	0	1	0	0
Deepham,	0	2	0	0
Easton,	0	1	0	1
Hingham,	2	2	1	6
Hackford,	0	0	0	0
Kimberley,	0	1	1	1
Marlingford,	0	0	0	0
Morley, St. Peter,	1	1	0	1
———, St. Botolph,	0	0	1	1
Runhall,	0	0	1	0
Wymondham,	4	16	6	16
Wicklewood,	0	0	0	2
Wramplingham,	0	0	0	0
Welborne,	1	1	0	0
	10	28	11	35

Proportion annually.

In 1797, of 242 persons, 10 died — 1 in 24
— 1798, of 244 persons, 28 died — 1 in 9
— 1799, of 236 persons, 11 died — 1 in 20
— 1800, of 293 persons, 35 died — 1 in 8

A HOUSE

A HOUSE OF CORRECTION.

That at Wymondham is one of the best managed in the kingdom. I viewed it with much pleasure, for the extreme cleanliness throughout; and in the persons of the prisoners, as well as in every other circumstance, is highly worthy of praise. The earnings shew how well it is conducted.

		£.	s.	d.
1797.—April to July, earnings of 19 prisoners, exceed their maintenance by	-	7	2	2
July to September, of 15,	- - -	5	14	8
September to December of 15,	- - -	10	6	7
1798.—December to April, of 21,	- - -	11	18	2
April to June, of 14,	- - - -	8	9	2
June to September, of 11,	- - - -	2	7	8
September to December, of 12,	- -	4	13	2
1799.—December to March, of 15,	- -	10	1	3
March to July, of 14,	- - - -	9	18	5
July to October, of 12,	- - - -	2	5	6
October to January, of 24,	- - - -	3	4	1

And the proportion of their earnings which the prisoners receive for themselves, has, in some cases, amounted to sums which have established them in industrious callings, such as from 4l. to 9l. One BROWN had the latter sum, with which he set up as a basket-maker, and is now in good business, and with a good character. The main hinge upon which this turns, is the employment being the hemp manufacture; bunching, heckling, and spinning hemp, are, by far the most profitable occupations they can be put to.

SECT. V.—COMPARISON OF TIMES.

THE Board of Agriculture having, in consequence of a requisition from the Corn Committee of the House of Commons (1804), procured returns from the several counties, of the expenses on arable land in 1790 and 1803, I am permitted to insert here the result of their inquiries for the county of Norfolk, which will be found in the following Tables.

DAY-WORK.

Communication.	Price in Winter, 1790. Per Week.		Price in Winter, 1804. Per Week.		Price in Summer, 1790. Per Week.		Price in Summer, 1804. Per Week.		Price in Harvest, 1790. Per Week.		Price in Harvest, 1804. Per Week.	
	s.	d.	s.	d.	s.	d.	s.	d.	s.	d.	s.	d.
No. 1. Samuel Tayler	7	....	10	....	8	....	10	....	21	....	35	....
2. M. Gooch	7	....	10	....	9	....	12	....	18	....	24	....
3. Rich. Fowell......	7	....	9	....	8	....	10	....	16	6	24	....
4. Henry Styleman ..	7	6	10	6	9	....	12	....	15	....	24	....
5. St. John Priest	9	....	10	....	10	....	11	....	21	....	24	....
6. Anonymous	7	....	9	....	8	....	10	....	14	6	18	....
7. William Curtis	8	....	10	6	12	....	15	....	12	....	16	....
8. Thomas Thurtell ..	7	6	10	6	10	6	13	6	10	6	13	4
9. William Bircham ..	7	6	10	....	9	....	11	....	24	....	27	....
10. Baker Rackham....	7	....	10	....	8	....	12	....	10	....	13	....
11. John Repton	6	....	9	....	9	....	12	....	10	6*	10	6
12. Thomas Nelson....	8	....	12	....	10	....	15	....	21	....	30	....
13. James Crowe	7	....	13	....	8	....	15	....	21	....	39	....
14. Stephen Reeve	7	6	9	6	8	6	11	....	24	....	28	....
15. Henry Burton	7	6	9	....	9	....	12	....	18	....	30	....
16. Edward Howman..	7	....	9	....	8	....	10	6	10*	....	12	6
17. Daniel Sewell	7	....	10	....	9	....	12	....	21	....	30	....
18. W. M. Hill	7	....	10	....	9	....	12	....	27	....	37	6
19. John Mosely......	6	....	10	....	7	....	12	....	13	6	24	....
20. John Shearing	7	....	10	6	9	....	15	....	23	....	34	....
21. William Foster, jun.	7	....	10	....	8	....	12	....	21	....	33	....
22. Richard Ferrier....	7	....	9	....	8	....	10	6	10	6	12	6
23. John Wagstaffe ...	7	....	9	....	8	....	10	6	10	....	12	6
24. ⎫	..7	....	10	....	8	....	12	....	10*	....	12*	6
25. ⎬ William Palgrave	..7	....	10	....	8	....	12	....	20	....	28	....
26. ⎪	..8	....	10	....	9	....	12	....	20	....	25	....
27. ⎭	..7	....	9	....	8	....	12	....	24	....	39	....
28. John Reeve	7	....	10	....	9	....	12	....	15	....	18	....
29. John Keddle......	7	....	10	....	8	....	12	....	18	....	24	....
Average	7	$2\frac{1}{4}$	10	$0\frac{3}{4}$	8	$8\frac{1}{4}$	12	....	17	$2\frac{3}{4}$	24	
Per Cent.	.... $43\frac{39}{67}$		....		.... $38\frac{54}{417}$		....		$41\frac{1}{4}$			

* Denotes with board.

NORFOLK.]

WAGES.

ommunication.	Head Man's Wages, 1790. Per Ann.			Head Man's Wages, 1804. Per Ann.			Second Man's Wages, 1790. Per Ann.			Second Man's Wages, 1804. Per Ann.		
	£.	s.	d.	£.	s.	d.	£.	s.	d.	£.	s.	d.
1.	0	9*	0 per wk.	0	12	0 per wk.	0	8	0 per wk.	0	11	0 per wk.
2.	0	12	0 ————	0	13	6 ————	0	9	0	0	10	0 ————
3.	10	0	0	12	10	0	8	0	0	10	0	0
4.	7	0	0	11	0	0	4	0	0	7	0	0
5.	11	0	0	13	0	0	8	0	0	9	0	0
6.	9	9	0	10	10	0	7	7	0	8	0	0
7.	10	0	0	10	0	0	4	0	0	5	0	0
8.	8	8	0	10	10	0	4	4	0	5	5	0
9.	9	0	0	10	0	0	7	0	0	8	0	0
0.	0	7	6 per w.	0	11	0 per w.	0	7	0 per w.	0	10	6 per w.
1.	7	7	0	9	9	0	5	5	0	6	6	0
2.	10	10	0	12	12	0	8	0	0	10	0	0
3.	0	10	6 per w.	0	15	0 per w.	0	9	0 per w.	0	12	0 per w.
4.	9	10	0	11	0	0	6	10	0	7	10	0
5.	0	10	6 per w.	0	15	0 per w.	0	7	6 per w.	0	12	0 per w.
6.	8	0	0	10	0	0	5	0	0	7	0	0
7.	0	9	0 per w.	0	13	0 per w.	0	8	0 per w.	0	12	0 per w.
8.	9	9	0	10	10	0	5	0	0	7	0	0
9.	0	7	6 per w.	0	12	0 per w.	0	6	6 per w.	0	10	0 per w.
0.	10	0	0	14	0	0	7	0	0	10	0	0
1.	0	9	0 per w.	0	12	0 per w.	0	8	0 per w.	0	11	0 per w.
2.	9	9	0	12	12	0	5	5	0	7	7	0
3.	9	0	0	12	0	0	6	0	0	8	0	0
4.	0	7	6 per w.	0	10	6 per w.	0	7	0 per w.	0	9	6 per w.
5.	9	0	0	12	12	0	6	0	0	8	8	0
6.	9	0	0	12	0	0	6	0	0	8	0	0
7.	9	0	0	14	0	0	6	0	0	9	0	0
8.	0	8	0 per w.	0	12	0 per w.	0	7	0 per w.	0	12	0 per w.
9.	0	8	0 ————	0	12	0 ————	0	7	0 ————	0	10	0 ————
erage	9	2	0	14	8	0	6	8	0	8	6	0
Cent.				.... $58\frac{22}{91}$						.... $29\frac{1}{16}$		

The shillings per week are reckoned as pounds, to form the average: thus, 10s. 6d. per week stands for 10l. 10s. being in me proportion.

RFOLK.]

PIECE-WO

Communication.	Reap Wheat, 1790. Per Acre.		Reap Wheat, 1804. Per Acre.		Mow Barley, 1790. Per Acre.		Mow Barley, 1804. Per Acre.		Thresh Wheat, 1790. Per Qr.		Thresh Wheat 1804 Per Q
	s.	d.	s.	d.	s.	d.	s.	d.	s.	d.	s.
No. 1.	5	..	9	..	1	..	1	6	2	..	4
2.	5	..	8	..	2	..	2	6	2	..	4
3.	6	..	8	..	1	6	2	3	2	..	3
4.	6	..	9	..	2	..	3	..	1	8	4
5.	8	3	10	..	1	9	2	3	2	4	4
6.	6	..	7	..	3	..	3	6	2	..	5
7.	7	..	8	..	..	..	..	..	1	10	2
8.	6	..	8	6	1	6	2	..	2	3	3
9.	6	..	8	6	2	..	2	6	2	..	3
10.	5	..	10	..	1	6	2	6	2	..	3
11.	5	6	8	..	1	..	1	6	2	..	3
12.	7	..	9	..	1	9	3	..	2	4	4
13.	5	..	9	..	1	6	3	6	2	..	3
14.	6	..	9	..	1	10	2	9	2	..	3
15.	7	..	9	6	2	..	3	..	2	..	3
16.	6	..	8	..	1	..	1	6	2	..	3
17.	6	..	10	..	1	3	2	..	2	..	4
18.	5	6	8	..	2	..	2	6	2	..	2
19.	3	..	5	..	..	10	1	6	1	4	3
20.	5	..	8	..	1	3	2	..	1	8	3
21.	5	6	9	..	1	6	2	3	2	..	3
22.	5	..	8	..	1	6	2	..	2	..	2
23.	6	..	9	..	1	6	2	6	2	..	3
24.	5	..	9	..	1	6	2	6	2	..	3
25.	6	..	10	6	1	..	2	..	2	..	3
26.	7	..	10	..	2	..	3	..	2	..	3
27.	5	6	10	6	2	..	3	..	2	..	3
28.	5	..	7	..	1	6	2	..	1	8	4
29.	5	..	7	6	1	6	2	6	2	..	3
Average	5	8¾	8	7¼	1	6½	2	4¾	1	11½	3
Per Cent.	....	50 2/11	....		....	55 15/37	....		....	72 16/47	....

NORFOLK.]

…resh …rley, …790.	Thresh Barley, 1804.		Filling Earth, 1790.		Filling Earth, 1804.		Filling Dung, 1790.		Filling Dung, 1804.	
…r Qr.	Per Qr.		Per Yard.		Per Yard.		Per Load.		Per Load.	
d.	s.	d.	s.	d.	s.	d.	s.	d	s.	d.
..	1	6	..	2¼	..	3	..	1¾	..	2¼
..	1	4	..	2½	..	3	..	2	..	3
..	1	4	..	2½	..	3	..	2	..	2½
10	1	8	..	1½	..	3	..	1½	..	3
2	1	6	..	2	..	2½	..	3	..	4
2	1	6	..	2¼	..	3	..	2¼	..	2¼
11	1	4	..	3	..	3	..	2½	..	2½
2	1	6	..	2	..	3	..	2½	..	4
..	1	6	..	1½	..	1½	..	2½	..	2½
..	1	6	..	2	..	3	..	2	..	3
..	1	6	..	2½	..	3	..	2½	..	3
..	1	6	..	2	..	3	..	2½	..	3
..	1	8	..	2	..	3	..	2	..	3
..	1	6	..	2	..	2½	..	2	..	2½
..	1	6	..	2	..	3	..	2	..	3
..	1	6	..	2¾	..	3½	..	1½	..	2½
..	1	6	..	2¼	..	3	..	2½	..	3
10	1	3½	..	2½	..	3	..	2	..	2½
10	2	..	..	1½	..	3	..	1½	..	3
10	1	6	..	2	..	3	..	1¾	..	2½
..	1	6	..	1½	..	2	..	2	..	3
..	1	4	..	..	..	..	..	..	..	..
..	1	4	..	2½	..	3¼	..	2	..	3
..	1	6	..	2	..	3	..	2	..	3
..	1	6	..	2	..	3	..	2	..	3
..	1	6	..	2	..	2½	..	2	..	2½
..	1	6	..	2	..	3½	..	2	..	3
..	1	2	..	2	..	3	..	2½	..	3
..	1	6	..	1½	..	2	..	1½	..	2
11¾	1	5¾	..	2	..	2¾	..	2	..	3
... 51$\frac{3}{47}$			.... 37½				 50			

Communication.	Tire, 1790. Per lb.		Tire, 1804. Per lb.		Plough-Irons, 1790. Per lb.		Plough-Irons, 1804. Per lb.	
	s.	d.	s.	d.	s.	d.	s.	d.
No. 1.		4		5		4		5
2.		3		4		4		6
3.		3		$4\frac{1}{2}$		4		6
4.		4		$4\frac{1}{2}$		4		7
5.		$3\frac{1}{2}$		$4\frac{1}{2}$				
6.		$3\frac{1}{2}$		$5\frac{3}{4}$	1 /	1 / 4	1 /	6 / 6
7.								
8.						4		6
9.		$3\frac{1}{2}$		$4\frac{1}{2}$		$4\frac{1}{2}$		6
10.		$3\frac{1}{2}$		$5\frac{1}{2}$		4		7
11.		4		6		4		6
12.		$3\frac{1}{2}$		$5\frac{1}{4}$				
13.		$4\frac{1}{2}$		7		incr. $\frac{1}{3}$ / 4		.. / 5
14.		$3\frac{1}{2}$		$4\frac{3}{4}$		4		6
15.		$3\frac{3}{4}$		$4\frac{1}{2}$		$4\frac{1}{2}$		6
16.		$3\frac{1}{2}$		$4\frac{1}{2}$		4		7
17.		$3\frac{1}{2}$		5	24*		40*	
18.		$4\frac{1}{2}$		5		5		6
19.		$2\frac{1}{2}$		5		3		6
20.		$3\frac{1}{2}$		$5\frac{1}{2}$		4		6
21.		$3\frac{1}{2}$		5		5		7
22.		4		6		4		6
23.		$3\frac{1}{2}$		5		4		7
24.		$3\frac{1}{2}$		4		4		7
25.		$3\frac{1}{2}$		$4\frac{1}{2}$		4		6
26.		$3\frac{1}{2}$		$4\frac{3}{4}$		4		6
27.		$3\frac{1}{2}$		5		4		6
28.		3		$4\frac{1}{4}$		5		6
29.		$3\frac{1}{2}$		5		4		6
Average		$3\frac{1}{2}$		$4\frac{3}{4}$		$3\frac{1}{2}$		6
Per Cent.			$35\frac{5}{7}$				$71\frac{3}{7}$	

* Per plough per year.   † The return is

Chains, 1790. Per lb.		Chains, 1804. Per lb.		Shoeing, 1790. Per Shoe		Shoeing, 1804. Per Shoe.	
s.	d.	s.	d.	s.	d.	s.	d.
	5		6		4		6
	5		7		4		6
	5		$7\frac{1}{2}$		$4\frac{1}{2}$		6
	$5\frac{1}{2}$		$9\frac{1}{4}$		4		7
	4		6		4		7
	$4\frac{1}{2}$		$7\frac{1}{2}$		4		7
					6		$7\frac{1}{2}$
	$5\frac{1}{2}$		7		6		7
	4		8		5		8
	4				4		7
	4		6		4		7
	$\{$incr. $\frac{1}{3}$ $\}$ 4		$5\frac{1}{2}$		$4\frac{1}{2}$		$7\frac{1}{2}$
	5		8		$5\frac{1}{2}$		$7\frac{1}{2}$
	5		7		$4\frac{1}{2}$		6
	5		8		4		7
	5		7	10*		15*	
	$5\frac{1}{2}$		6		$5\frac{1}{2}$		7
	$2\frac{1}{2}$		7		$3\frac{1}{2}$		7
	4		7		5		$7\frac{1}{2}$
	6		8		6		8
	4		6		4		6
	5		7		5		7
	4		8		5		8
	6		8		6		8
	5		8		6		8
	5		$7\frac{1}{2}$		$4\frac{1}{2}$		6
	6		7		6		8
	4		6		4		6
	$4\frac{3}{4}$		7		$4\frac{1}{2}$		7
		$47\frac{7}{19}$				$64\frac{12}{17}$	

6d. which is as 4d. to 6d.

ARTISANS.

Communi-cation.	Carpenter. By Day, 1790.		Carpenter. By Day, 1804.		Mason. By Day, 1790.		Mason. By Day, 1804.		Thatcher. By Day, 1790.		Thatcher. By Day, 1804.		Collar-Makers' Work. 1790.			Collar-Makers' Work. 1804.		
	s.	d.	s.	d.	s.	d.	s.	d.	s.	d.	s.	d.	£.	s.	d.	£.	s.	d.
No. 1.....	1	10	2	6	1	10	2	6	2	....	3	....	..	..	..	..	..	..
2.....	1	8	2	6	2	8	3	9	2	8	3	9	..	2	....	..	2	8ª
3.....	1	8	2	4	2	8	3	9	2	8	3	9	..	2	....	..	2	8ᵇ
4.....	1	8	2	4	2	....	2	4	2	....	2	4	..	..	..	..	..	..
5.....	1	8	2	4	2	....	2	6	2	....	2	6	..	3	6	..	5	....
6.....	1	9	2	2	1	9	2	2	3	....	4	....	12	..	..	18	..	..
7.....	2	3	2	8	2	3	2	8	2	6	2	6	..	2	....	..	3	....
8.....	2	3	3	6	2	....	3	....	2	....	3	....	..	2	....	..	3ᶜ	....
9.....	1	8	2	9	1	8	2	9	1	8	2	9	..	2	....	..	2	8ᵈ
10.....	1	8	2	2	1	8	2	2	2	6	3	2	..	2	....	..	4	4¼ᵉ
11.....	1	8	2	....	3	6	4	2	2	6	3	....	..	..	..	..	..	..
12.....	1	8	2	4	2	....	2	6	2	....	2	6	..	3	6	..	5	....
13.....	2	....	3	....	2	....	3	....	2	....	3	6	..	2	....	..	2	8ᶠ
14.....	1	8	2	6	1	8	2	6	1	8	2	6	..	2	....	..	2	8ᵍ
15.....	1	6	2	6	1	6	2	6	1	6	2	6	..	..	..	..	..	..
16.....	1	6	2	3	1	8	2	3	1	9	2	....	..	..	..	..	..	..
17.....	1	10	2	2	2	....	3	....	2	6	3	6	1	1	....	1	10ʰ	....
18.....	1	8	2	4	1	8	2	4	..	..	..	..	..	..	..	..	..	..
19.....	1	2	2	6	1	2	2	6	1	2	2	6	..	1	2	..	2	6
20.....	1	6	2	6	1	6	2	6	2	6	3	6	1	1	....	1	10ⁱ	....
21.....	1	8	2	3	1	8	2	3	2	0	2	0	..	..	..	..	..	..
22.....	2	....	3	....	2	....	3	....	1	9	2	6	..	..	..	..	..	..
23.....	1	9	2	3	1	9	2	3	1	9	2	3	..	2	....	..	2	8ᵏ
24.....	1	8	2	3	1	8	2	3	2	6	3	2	4	..	..	8ˡ	..	..
25.....	1	8	2	2	1	8	2	2	2	....	3	....	..	2	....	..	4	....
26.....	1	8	2	6	1	8	2	6	1	8	2	6	..	2	....	..	2	8ᵐ
27.....	2	....	3	4	2	....	3	4	2	6	4	6	..	2	....	..	3	4
28.....	2	....	2	6	2	....	2	6	..	..	..	..	..	2	....	..	2	8ⁿ
29.....	2	....	2	6	2	....	2	6	1	6	2	6	..	2	....	..	2	8ᵒ
Average	1	9¾₁	2	5¼	1	11	2	8	2	....	2	11	..	1	11¼	..	2	8
Per Cent.	..... 36⁶⁸⁄₈₇				..... 39³⁄₂₃				..... 45⅚				.... 38⁵⁶⁄₉₃					

ª Increase one-third.   ᵇ Increase one-third.   ᶜ Increase one-half.   ᵈ Increase one-third.   ᵉ Increase as 5 to 11.   ᶠ Increa one-third.   ᵍ Increase one-third.   ʰ Per diem, with materials.   ⁱ Harness per horse.   ᵏ Increase one-third.   ˡ Per an ᵐ Increase 33 per cent.   ⁿ Increase one-third.   ᵒ Increase one-third.

NORFOLK.]

RENT, TITHE, AND PARISH TAXES.

Communication.	Rise of Rent, from 1790 to 1804.	Tithe. Per Acre, 1790.		Tithe. Per Acre, 1803.		Parish Taxes. In the Pound, 1790.		Parish Taxes. In the Pound, 1803.	
		s.	d.	s.	d.	s.	d.	s.	d.
No. 1.	50 per cent.	3	6	4	6	1	8	5	4
2.		3	....	5	....	2	6	4	....
3.		3	....	5	....	3	....	4	....
4.	28$\frac{4}{7}$ per cent.*	2	6	4	6	1	6	3	....
5.		3	....	4	....	3	....	6	....
6.		2	....	3	4	3	....	5	9
7.		....	....	....	....	....	....	....	....
8.	50 per cent.†	4	....	5	6	2	....	4**	....
9.	25	2	5	4	....	6	....	8	....
10.	50 ‡	3	....	4	....	2	....	2	....
11.	20	3	6	4	6	4	6	8	....
12.	42$\frac{6}{7}$	2	3	4	6	3	....	6	....
13.		....	....	....	....	3	....	4††	....
14.	25 per cent.	2	9	4	9	2	6	3‡‡	....
15.	40	3	6	6	....	4	....	7	....
16.	20	2	6	4	6	4	2	10	8
17.	20	1	....	1	6	1	....	2	....
18.	16$\frac{2}{3}$ ‖	2	3	3	6	1	8	2	6
19.		....	....	....	....	....	6	2	6
20.	50 per cent. ¶	2	3	4	3	1	3	4	....
21.	33	....	....	....	....	4	....	6	....
22.	33	4	....	6	....	....	....	....	....
23.	33	3	6	4	6	4	6	9	....
24.	60	3	....	4	6	3	6‖‖	3	6‖‖
25.	33	3	....	5	6	2	2	5	6
26.	25	3	....	5	....	2	6¶¶	3¶¶	....
27.	40	4	6	7	....	2	6	2	6
28.		2	2	3	....	2	7	4	2
29.		....	....	....	....	3	....	7	3
		2	10$\frac{1}{4}$	4	6$\frac{1}{4}$	3	9	4	10$\frac{3}{4}$
Aver. per Ct.	35 per cent.	..... 58$\frac{54}{137}$				.... 30$\frac{5}{9}$			

* 14s. to 18s.   † 16s. to 24s.   ‡ 14s. to 21s.   ‖ 12s. to 14s.   ¶ 11s. to 15s. 6d.
** Increase one-half.   †† Increase one-third.   ‡‡ Increase one-fifth.   ‖‖ House of Industry.
¶¶ Increase 20 per cent.
NORFOLK.]

CULTIVATION.

Communication.	Expense of an Acre of Turnips.						Expense of an Acre of Barley					
	1790.			1804.			1790.			1804.		
	£	s.	d.	£	s.	d.	£	s.	d.	£	s.	d.
No. 1.												
2.	1	13	6	2	10	6	1	13		2	1	4
3.	1	13	6	2	10	6	1	13		2	1	4
4.		10		1				6	6		13	
5.	4	8	6	5	6	10	2	19		3	16	6
6.												
7.												
8.												
9.	4			4	10			18	6		18*	
10.	1	19	6	3	15	8	1	4	2	1	6	8
11.	2	5		3								
12.	4	8	1	6	10	8	3	1	3	4	3	3
13.	4			6			1	10		2		
14.												
15.	2	10		3	10		1	5		1	15	
16.	2	11	6	3	11	9		18		1	6	3
17.	3	1		4	8		1	14		2	3	6
18.	1	1		1	7			19	6	1	4	
19.		19	6	1	3	6		13	8		19	6
20.	3	5		6	5		3	15		5		
21.												
22.												
23.	3	15		5	5		2			2	15	
24.	1	19	6	3	15	8	1	4	2	1	6	8
25.	3			4	5	6	1			1	10	
26.	4	10		5			1*	4	6		18*	
27.	4	10		7		6	1	3	6	1	18	
28.	1	1		1	8			19	8	1	4	3
29.	2			3	15		1	7		1	10	
Average	2	13	8½	3	18	1½	1	9	11¼	1	18	7
Per Cent.				45 $\frac{595}{1289}$						28 $\frac{568}{719}$		

Turnips 45
Barley 28
Wheat 16
                           ——
Average 29

* Drilling has lowered the expense.

COMPARISON OF TIMES.

Expense of an Acre of Wheat.

	1790.			1804.	
£.	s.	d.	£.	s.	d.
1	17	6	2	10	4
1	17	6	2	10	4
..	3	6	..	7	..
3	8	6	4	12	..
..	..	..	..	..	..
..	..	..	..	..	..
3	..	..	2	13*	..
1	11	..	2	6	..
..	..	..	..	..	..
3	7	1	4	17	..
4	..	..	6	..	..
..	..	..	..	..	..
1	15	..	2	10	..
..	14	9	1	..	..
2	2	6	2	12	..
1	12	4	1	18	5
1	1	..	1	8	5
3	..	..	3	19	..
..	..	..	..	..	..
4	..	..	5	10	..
1	11	..	2	6	..
2	10	..	3	19	6
3	10	..	3	10	..
1	8	..	2	12	..
1	12	4	1	18	5
1	16	6	3	15	..
2	3	$2\frac{1}{4}$	2	10	$2\frac{1}{2}$
		$16\frac{532}{2075}$			

MANURE.

	Expense of Manure.			
	1790.		1804.	
Communication.	s.	d.	s.	d.
No. 1.	1	6	2	..
2.	2	6	4	..
3.	2	6	4	..
4.	2	6	6	..
5.	..	..	..	..
6.	..	..	..	..
7.	2	6	2	9
8.	2	6	4	..
9.	2	6	3	4
10.	2	..	4	..
11.	2	..	4	6
12.	..	..	..	..
13.	1	4	3	..
14.	2	6	3	..
15.	3	..	6	..
16.	2	6	3	7
17.	2	6	2	9
18.	2	6	3	$4\frac{1}{2}$
19.	..	..	..	..
20.	2	6	4	4
21.	..	..	..	..
22.	..	..	..	..
23.	..	..	..	..
24.	1	6	6	..
25.	2	..	4	6
26.	..	..	..	..
27.	1	6	4	6
28.	..	..	..	..
29.	2	6	3	$7\frac{1}{2}$
Average	2	$4\frac{1}{4}$	4	2
Per Cent.			$76\frac{112}{113}$	

EXTRACTS FROM THE LETTERS ACCOMPANYING THE PRECEDING RETURNS.

COMMUNICATION, No. 8.

PARISH OF GORLSTON.

I could grow wheat in 1790 at 40s. a quarter, with a greater profit to myself than I can now at 60s. a quarter, and other grain in proportion. I am confident, if something is not done by the Legislature, to enable the farmer to carry on his business with spirit, it must ultimately be a very great injury to society. I assure you, a great relaxation in improvement has taken place in this neighbourhood, in consequence of the low price of grain.

COMMUNICATION, No. 11.

PARISH OF HEVINGHAM.

	£.	s.	d.
Annual disbursements on the average of three years preceding the building of the Buxton House of Industry, amounted to	1345	17	8¾

Disbursements from October 10, 1802, *to October* 10, 1803.

	£.	s.	d.
Out-door relief to aged and infirm	124	5	11
———— to sick	19	18	9
———— to clothing boys and girls put out to service	24	4	8½
Extra miscellaneous parish charges	71	17	2
In-door expenses for provisions	58	2	7
———— for clothing	21	12	0¼
———— for payments of debt on house, &c.	115	14	7½
Interest on 1337*l*.	66	17	0
	£. 502	12	9¼
Receipts by rate, at 6s. 6d. in the pound	£. 502	12	9¼

	£.	s.	d.
Debt on house, &c.	1337	0	0
Paid off	115	14	7½
Total unpaid	£. 1221	5	4½

JOSEPH ALDERSON, Visitor.
THOMAS RACKHAM, Guardian.

To Mr. John Repton, Oxnead.

PARISH OF BUXTON.

	£.	s.	d.
Annual disbursements on the average of three years preceding the building of the Buxton House of Industry, amounted to	1370	3	7¾

Disbursements from October 10, 1802, to October 10, 1803.

	£	s	d
Out-door relief to aged and infirm	117	5	5
———— to sick	19	15	3
———— to clothing boys and girls put out to service	13	18	8½
Extra miscellaneous parish charges	68	9	9
In-door expenses for provisions	239	8	5
———— for clothing	72	1	0¼
———— for payments of debt on house, &c.	116	13	2¼
Interest on 1518l. 1s. 6d.	75	18	0
	£.723	9	9
Receipts by rate, at 17s. in the pound	£.723	9	9

Debt on house, &c.	£.1518	1	6
Paid off	116	13	2¼
Total unpaid	£.1401	8	3¾

JOSEPH ALDERSON, Visitor.
WM. JEWELL, Guardian.

To Mr. John Repton, Oxnead.

COMMUNICATION, No. 17.

COMPARISON OF TAXES.

1790.	Taxes.	£.	s.	d.
Rent, 250l. 13s.	Windows, &c.	2	16	0
	Chaise duty	3	3	0
	Balance in favour of 1790	37	8	0
		43	7	0

1804.	Taxes.	£.	s.	d.
Rent, 300l. 13s.	Windows	4	15	0
	Male servant	2	0	0
	Chaise	5	5	0
	Horse tax riding	6	12	0
	Cart-horses	12	10	0
	Dog tax	1	0	0
	Property tax	11	5	0
		43	7	0

commu-

COMMUNICATION, No. 28.

PARISH OF WIGHTON.

One material circumstance is not inquired after, viz. the increase of assessed taxes. By the establishment in 1790, I paid 26l. 7s.; I now pay 47l. 4s. 6d. exclusive of the property tax, and the increase of taxation upon *every* article of consumption.

COMMUNICATION, No. 29.

PARISH OF MARSHAM.

Remarks to be annexed to the Statement respecting the Rates of Labour, and various Charges on Arable Farms, in the Year 1790, and in 1804.

The difference in the price of labour, between the year 1790 and 1804 (which I have taken from my farm expenditure books in those years), cannot be attributed to the price of bread corn; as I find, by reference to the receipt-book in the former of those years, that my wheat was sold at an average of 3l. per quarter, and barley at 1l. 4s. per quarter; but must be traced to other causes; one of which I conceive to be from the high price of bread-corn in the years 1800 and 1801, which obliged the employers of workmen in all the different trades connected with the farming interest, to give a considerable increase of wages, to correspond with the parish allowance granted by the magistrates to those who had no employment during that scarcity; and although the price of bread-corn is reduced considerably more than half since those years, the other necessaries of the poor are, since the year 1790, increased
nearly

nearly double in their price, as grocery of all kinds, shoes, the little malt they are enabled to purchase, &c. that the labourers enjoy but little benefit from their increase of wages.

There is but little work done by the day in the Norfolk harvests. The usual method has been, engaging a proper quantity of labourers, in proportion to the number of acres of corn, allowing five acres of winter corn, and ten acres of summer corn, to a man's share, and giving them such sum, with board and lodging, to finish the harvest; which, if the weather be favourable, is ended in four weeks; frequently, when the harvest is early, and the weather good, in three, having the assistance of the carters and ploughmen kept on the farm. The difference of wages I find, by my books, to be one-third increased since the year 1790; in which year I gave 1l. 13s. per man; the last harvest 2l. 10s. per man. The increase of wages I attribute to the recent practice of the large occupiers of lands in this county giving 7l. per man, the labourers finding their own victuals, drink, and lodging; their wives not liking to have the trouble and fatigue of providing provision in their houses, which has compelled those who adhere to the old custom to give the increased price. The day-labour I have stated.

It is usual in this county, where labourers are employed as carters or ploughmen, and have the care of a team of horses during the year, to give them the harvest-wages equal to those retained for that purpose, with their board. This is prudent, as, being so engaged, they have an interest with the rest in ending the harvest as soon as possible; which, if continued at their common wages, in all probability the business would be retarded.

The advance in the price of threshing, between the two years in question, may be attributed to the same causes

causes as the advance of other labour connected with the farm, with the general objection labourers now have to that employment, whose capacities are adapted to any other method of obtaining a livelihood; and many will go miles for employment, sooner than thresh corn.

There is no doubt but the great increase in the price of blacksmiths' work, since the year 1790, may be accounted for from the dearness of foreign iron, which was then sold at 1l. 3s. per cwt.; is now at 1l. 10s. per cwt. Coals enhanced 1s. per chaldron; and the advance of journeymen wages, which has been in proportion with other labour.

A very material charge on arable farms (not noted in the Statement), since the year 1790, is the advance on wheelwrights' work, which has been greater in proportion than any other, owing to the increase in the price of timber, ash-timber, which is materially useful in their employment, having had a rise from 2l. 5s. per load, which was then the general price, to 5l. per load, the present one, and other timber in the like proportion; the advance in the price of their working tools, and the additional wages given to their men.

In this county, thatchers are seldom employed by the day, only on small breaches that cannot be conveniently measured. The price, per day, for man and labourer, is advanced, since the year 1790, 1s. per day. Measured work, in that year, was charged at 4s. 2d. per square of 100 feet, and 1s. per square for materials: it is now 5s., and 2s. for materials. Hay and corn ricks are usually done by the square yard: in 1790 it was $0\frac{3}{4}$d. per yard; in 1803 $1\frac{1}{4}$d. Here seems to be a greater advance than in any other labour connected with the farmer; which may be supposed to have its cause from few learn-

ing

ing the business; as it does not give constant employment throughout the year, there is not an inducement.

Collar-makers' work is, since the year 1790, increased at least one-third from the advance of leather, hemp, iron, and the addition to the journeymen's wages. The usual method of repairing harness is, for the farmers to have them done at their own houses, the collar-maker charging 1s. per day, for labour, per man, the employer boarding them, and finding food for their horses; the collar-maker charging for the materials used, which they always take with them.

The difference in the expense of preparing and raising an acre of turnips since the year 1790, is chiefly owing to the advance on manure. Flag and turf ashes, which, in the neighbourhood of large heaths and commons, is eagerly sought for for that purpose, and carried to a considerable distance, is enhanced in its price, from 1s. per cart-load to 4s.; 10 loads of which is the usual quantity used per acre. Stable-dung, and street-muck, as it is usually termed, collected in large towns, has had nearly the same advance. Malt-dust, or cooms, as they are provincially called, have had an equal rise, from 1s. 6d. per sack to 3s. 6d. This, in different parts of the county, is frequently used as a manure for turnips. The carriage is certainly a considerable saving; and where no other can be procured, it is necessary to use it, as the turnip requires always to be manured for; but I must confess it never met my approbation, when recourse could be had to any other.

The little difference in the raising the barley-crop between the years in question, as described in the Statement, is merely the alteration in the expense of tillage; as barley in this county is rarely manured for, being sowed
after

after wheat or turnips that has the preceding year had that operation.

Wheat—The expenses of raising of which (as it is usual to manure for it in this county), has had the same additional charges as the turnip crop, from the great rise of manures. Soot, a manure frequently used for that purpose, is sold at double the price it was in the year 1790, being then 6d. per bushel, now 1s. Rape, or oil-cake, an excellent manure for cold wet lands, has since that time sold at the same proportional advance.

The reader will perceive, on consulting the preceding table of the expense of cultivation, that the object is incompletely ascertained. Some correspondents returned only the amount of labour; others excluded rent and rates; others omitted seed, &c. The returns from some other counties were still more deficient. To remedy the omissions, the Board ordered a second letter to be written, requesting an answer to the following question: *What are the charges upon* 100 *acres of arable land, under the following distinct heads?*

	1790.	1803.
Rent,	-	-
Tithe,	-	-
Rates, &c.	-	-
Wear and tear,	-	-
Labour,	-	-
Seed,	-	-
Manure purchased,	-	-
Team,	-	-
Interest of capital,	-	-

The following tables contain the result of these inquiries for the county of Norfolk

COMPARISON OF TIMES.

Communication. No.	Rent 1790. £. s. d.	Rent 1803. £. s. d.	Tithe 1740. £. s. d.	Tithe 1803. £. s. d.	Rates and Taxes 1790. £. s. d.	Rates and Taxes 1803. £. s. d.
1	80 0 0	100 0 0	20 0 0	25 0 0	20 0 0	30 0 0
2	75 0 0	100 0 0	15 0 0	20 0 0	25 0 0	40 0 0
3	65 0 0	78 0 0	8 7 0	13 7 0	9 14 0	22 5 0
4	80 0 0	125 0 0	15 0 0	20 0 0	15 0 0	30 0 0
5	80 0 0	100 0 0	20 0 0	25 0 0	28 0 6	48 0 0
6	100 0 0	150 0 0	17 10 0	25 0 0	8 0 8	37 0 0
Average	80 0 0	108 16 8	15 19 6	21 7 10	17 13 5¼	34 12 2¼
Per cent.	....	$36\frac{8}{92}$	....	29	....	96

Wear

COMPARISON OF TIMES.

Communication No.	Wear and Tear 1790.			Wear and Tear 1803.			Labour 1790.			Labour 1803.			Seed 1790.			Seed 1803.		
	£.	s.	d.	£.	s.	d.	£.	s.	d.	£.	s.	d.	£.	s.	d.	£.	s.	d.
1	13	5	0	18	10	0	90	13	0	116	13	0	45	0	0	45	0	0
2	20	0	0	30	0	0	90	0	0	120	0	0	26	0	0	34	0	0
3	13	0	0	19	10	0	54	0	0	78	10	0	4	0	0	4	0	0
4	11	0	0	15	10	0	94	0	0	110	19	0	44	10	0	48	5	0
5	16	0	0	20	10	0	96	0	0	126	0	0	32	0	0	37	0	0
6	35	0	0	42	0	0	145	0	0	204	0	0	0	0	0	0	0	0
Average	18	0	10	24	5	0	94	18	10	126	0	4	30	6	0	33	13	0
Per cent.	$34\frac{7}{13}$						33									$11\frac{17}{363}$		

COMPARISON OF TIMES.

No.	Manure 1790 £. s. d.	Manure 1803 £. s. d.	Team 1790 £. s. d.	Team 1803 £. s. d.	Interest 1790 £. s. d.	Interest 1803 £. s. d.	Total 1790 £. s. d.	Total 1803 £. s. d.
1	15 0 0	20 0 0	20 0 0*	20 0 0	34 0 0	48 0 0	423 3 0	339 0 0
2	20 0 0	30 0 0	40 0 0	60 0 0	50 0 0	50 0 0	464 0 0	361 0 0
3	10 0 0	7 0 0	6*	6 0 0	25 0 0	30 0 0	258 12 0	195 1 0
4	0 0 0	0 0 0	60 0 0	75 0 0	25 0 0	30 0 0	454 14 0	344 10 0
5	7 0 0	10 0 0	75 0 0	90 0 0	30 0 0	40 0 0	496 0 0	384 0 0
6	2 0 0	2 10 0	78 0 0	78 0 0	40 0 0	40 0 0	578 18 2	425 16 8
Average	10 16 0	13 18 0	46 10 0	54 16 8	39 13 4	341 11 3¼	445 17 10¼	
Per cent.	28 19/27		18		16 272/463		'30½	

Communication.

* It is sufficiently clear from these sums, that the question was misunderstood. Suppose four horses per 100 acres arable—how are they to be kept for 6l. or for 20l.?

RECA-

RECAPITULATION.

LABOUR.

	Per Cent.	Average.
Rise in the price in winter, from 1790 to 1803	43	
——————— summer	38	
		40
——————— harvest	36	
——————— reaping wheat	50	
——————— mowing barley	55	
		47
——————— head man's wages	58	
——————— second man's do.	29	
		43
——————— threshing wheat	72	
——————— threshing barley	51	
		61
——————— filling earth	37	
——————— filling dung	50	
		43
		234
Fractions make it -		237

Divide by 5 - - 47*, general rise in husbandry labour.

* This may not militate so much with the result of the second series of labour as some may at first sight imagine. This table resulting from very different data; the five different divisions of labour being considered as equal in amount. To vary the calculation, without authority in the papers for so doing, would be taking too great a liberty.

ARTISANS.

ARTISANS.

	Per Cent.
Blacksmith—Rise in the price of tire, from 1790 to 1803	35
——————————————— plough-irons	71
——————————————— chains	47
——————————————— shoeing	64
Average	54
Carpenter	36
Mason	39
Thatcher	45
Collar-maker	38
Average of artisans	42

RENT AND TAXES, &c.

Rise of rent, from 1790 to 1803	35
———— tithe	58
———— rates	30

CULTIVATION OF ARABLE LAND.

Average rise on an acre of turnips, barley, and wheat	29
Average rise by the tables for 100 acres	30

	Per Cent.
Labour	47
Artisans	42
Rent	35
Tithe	

				Per Cent.
Tithe	-	-	-	58
Rates	-	-	-	30
Manure	-	-	-	76
Cultivation	-	-	-	29
Average	-	-	-	45

In remarking on the preceding particulars, I am, in the first place, to note, that the Board is not in the least committed in drawing any of these averages. That Body simply ordered circular letters to be written; and every reply stands distinctly on the personal authority of the writer. There ends the authority of the papers as I received them. The calculations, to draw them into one view, I have made, for the satisfaction of such readers as might wish to know what such a general result would be.

It does not, however, follow, that, supposing the authority of the letters correct, the averages would be the same, when a certain rise per cent. is deduced from them.

In this sketch, for instance, which gives 45 per cent. every one of the seven articles should be of equal importance, which is far from being the case. Manure, which, if bought, stands so high in the list, may in fact be the lowest, and, in bad times, perhaps is so *. Artisans count for one, as well as labour and rent; but that article is of far inferior importance to either. The same may be said of tithe and rates.

The same observation is applicable to the particulars from which these sums are drawn. In that of Labour, servants' wages count with summer and winter labour;

---

* If purchased manure be left out, the average of the other articles would be 40 per cent.

and

and filling earth and dung the same; but it is sufficiently obvious, that, in fact, no such proportion holds. In the article of Artisans, if the particulars are examined, there will be found a still greater disproportion in their importance to the farmer. If these circumstances be not kept in the reader's mind, he must necessarily be deceived.

In order to attain a nearer approximation to the real fact, or to ascertain that the view of the subject now given be indeed accurate, it may contribute to the reader's satisfaction to combine the results of the two series of replies; to take the proportions of the expenses to each other from the second letter, and the rise per cent. from the first; these proportions not appearing in the first correspondence.

The average expense of 100 acres in 1790, returned as above, is:

	£.	s.	d.
Rent - - - -	80	0	0
Tithe - - - -	15	19	6
Rates - - - -	17	13	5¼
Wear and tear - - - -	18	0	0
Labour - - -	94	18	10
Seed - - - - - -	30	6	0
Manure - - -	10	16	0
Team - - - -	46	10	0
Interest of capital - - -	34	0	0
Total -	£.341	11	3¼

Now, if the rise upon these be estimated from the first series of letters, viz. Rent 35, Tithe 58, Rates 30, Wear and Tear 42, Labour 47, Manure 76, and taking the advance in the articles, Seed, Team, and Interest, from the answers to the second letter (not having place in the first), the result would stand thus:

	1790.			Rise per Cent.	1803.		
	£.	s.	d.		£.	s.	d.
Rent	80	0	0	35	108	0	0
Tithe	15	19	6	58	25	0	0
Rates	17	13	5¼	30	22	10	0
Wear and tear	18	0	0	42	25	0	0
Labour	94	18	10	47	138	0	0
Seed	30	6	0	11	33	13	0
Manure	10	16	0	76	19	0	0
Team	46	10	0	18	54	16	8
Interest of capital	34	0	0	16	39	13	4
Total	£.341	11	3¼		£.465	13	0

Which is a rise of $36\frac{12}{34}$ per cent.

And this I take to be as near the truth as these data will permit an estimate to arrive.

POPULATION OF THE COUNTY OF NORFOLK.

SECT. VI.—POPULATION OF THE COUNTY OF NORFOLK, ASCERTAINED IN CONSEQUENCE OF THE ACT 41 GEO. III. 1800.

Hundreds.	Houses.			Persons.	Occupations.		
	Inhabited.	By how many Families occupied.	Uninhabited.		Chiefly employed in Agriculture.	In Trade, Manufactures, or Handicraft.	In all other Occupations.
Blofield	557	700	22	3351	834	183	2334
Brotherscross	595	655	15	2932	740	237	1955
Clackclose	1910	2403	32	11,720	4155	1158	6024
Clavering	758	919	6	5116	1495	438	2302
Depwade	1070	1444	7	7780	2346	784	3599
Diss	972	1424	11	7072	1270	999	4803
Earsham	1046	1384	20	6955	1456	653	4846
Erpinghams	3283	3984	50	18,530	4706	1666	10,143
Eynesford	1422	1452	39	8175	1794	636	4969
Fleggs	738	941	9	4820	1034	337	3110
Forehoe	1509	1903	27	9508	3385	1923	4200
Freebridge	2308	3030	40	14,585	5755	847	4388
Gallow	1066	1335	15	6305	1539	605	3530
Greenhoes	2628	3121	86	14,793	6003	1877	5845
Grimshoe	756	963	8	4616	1442	425	2504
Guiltcross	743	1017	14	5317	1101	1331	2887
Happing	878	1093	10	5095	1043	396	3656
Henstead	627	791	13	4041	1642	461	1808
Holt	1439	1537	48	7486	1609	856	3896
Humbleyard	537	769	7	4055	967	236	2852
Launditch	1385	1792	21	9484	2215	766	6513
Loddon	879	968	12	5389	1590	477	2790
Mitford	1312	1639	13	7960	2345	1003	4812
Shropham	1002	1439	4	6487	1792	683	3982
Smithdon	1008	1255	25	5963	1727	360	3876
Taversham	721	1003	12	5111	1073	486	3550
Tunstead	1617	1750	55	8393	1935	839	5344
Walsham	567	691	15	3501	1101	197	2203
Wayland	739	944	13	4790	3028	1186	378
Yarmouth Town	3081	3541	78	14,845	15	1399	13,431
Thetford	483	513	9	2246	149	367	1730
Lynn	1965	2437	47	10,096	97	2103	7896
Norwich	8016	9093	747	36,832	408	12,267	24,157
—— Gaol	......	......	......	......	......	......	22
	47,617	57,930	1523	273,371	61,701	38,181	160,313

CHAP. XV.

OBSTACLES.

ROOKS.

" SELDOM attempted to be shot in East Norfolk, where a notion prevails, and is, perhaps, well founded, that rooks are essentially useful to the farmer, in picking up worms and grubs, especially the grub of the cockchafer, injurious to meadows and marshes."—*Mr. Marshall.* Confirmed in the following note of Mr. JOHNSON, of Thurning.

I cannot but notice two growing evils with us, of which but little notice is taken:—1st, the number of insects in the lands, owing to the loss of rooks, by felling so many rookeries, and not taking care of what are left; 2d, the increase of mice, and, were I to give my opinion as to quantity and damage done, but few would give credit to it: I have, at different times, had five mice killed to every coomb of corn moved off the stacks in the summer season, and sometimes double that quantity; besides being on every other part of the premises, corn and grass pieces not excepted. Some are driven into the barns and stacks in wet seasons; but when wheat stands long on the shock, we are sure to have most mice in our barns and stacks, except where they are driven away by some other vermin:—in my memory there were 20 grey owls, where there are now one, and though the country was in a rougher state, we had not so many mice; the owls prey very

very much on them, and in wet weather they are more exposed to the owl than to any other vermin. The grey owl is destroyed by the game-keepers, and by felling the pollards. I have seen a young hare in their nests, but never saw a young pheasant or partridge :—the white, or church owl, are not so destructive to game; and were there places made within side the top of one end of every barn, like a box, for them to pass through as they come into the barn, they would there make their nests, and become more numerous, and be of great service.

S. JOHNSON."

THE END.

Printed by B. M'Millan,
Bow-Street, Covent-Garden.